Green Energy and Technology

More information about this series at http://www.springer.com/series/8059

Ana Silva · Jorge de Brito
Pedro Lima Gaspar

Methodologies for Service Life Prediction of Buildings

With a Focus on Façade Claddings

Springer

Ana Silva
Instituto Superior Técnico
Universidade de Lisboa
Lisbon
Portugal

Pedro Lima Gaspar
Faculty of Architecture
Universidade de Lisboa
Lisbon
Portugal

Jorge de Brito
Instituto Superior Técnico
Universidade de Lisboa
Lisbon
Portugal

ISSN 1865-3529 ISSN 1865-3537 (electronic)
Green Energy and Technology
ISBN 978-3-319-33288-8 ISBN 978-3-319-33290-1 (eBook)
DOI 10.1007/978-3-319-33290-1

Library of Congress Control Number: 2016937506

Printed on acid-free paper

This Springer imprint is published by Springer Nature
The registered company is Springer International Publishing AG Switzerland

Contents

Chapter 1
Introduction

1.1 Introduction

Buildings are complex systems composed of several elements, which are assembled to respond to a number of needs—functional and symbolic—according to set of legal and environmental requirements and to potentially accommodate users with different levels of demand (Kyle 2001). In fact, many expectations are created around constructions; according to Allen (1995), buildings have to fulfil safety requirements, protecting users from the exterior environment, providing privacy and security to people, connecting them to the outside world through doors and windows. Furthermore, construction elements are designed to be used for a certain period of time (their service life), performing all the functions for which they were designed without excessive maintenance costs. However, buildings and their components initiate an inevitable degradation process as soon as they are put into use and progressively lose their performance capacity over time.

For a long period of time, the construction industry was only concerned with minimizing the initial costs and there were no practical and reliable methods for service life prediction or for management of the maintenance of buildings throughout their life cycle (Lounis et al. 1998). Currently, due to the progressive degradation of the built heritage and the high costs associated with their construction, maintenance and repair, the study of the durability of buildings and their elements has been assuming an increasingly important role (Beer et al. 2011). This growing interest is due to several factors, including: (a) the interest shown by the owners to establish requirements for the service life of structures during the design stage; (b) awareness of the stakeholders in the construction sector that the quality and global costs of buildings comprise both the construction costs and the maintenance and repair costs; (c) the knowledge that durability is a key part of the quality and performance of the construction elements; (d) the awareness that the

A. Silva et al., *Methodologies for Service Life Prediction of Buildings*,
Green Energy and Technology, DOI 10.1007/978-3-319-33290-1_1

visual appearance and the ageing of the structures contribute to their loss of performance (DuraCrete 2000); (e) the awareness of the owners and insurers that the failure of construction elements has implications not only on the building's performance but also may cause serious damage to users (in this context, it becomes very important to evaluate the risk of failure of construction elements, their main mechanisms of failure and the impact/severity of rupture).

Besides the reasons mentioned above, service life prediction also assumes an important role in the economic and environmental sustainability of the construction sector. In fact, the knowledge of the service life of building elements is necessary in order to apply economic analysis tools, such as LCC (life-cycle cost), a methodology that accounts for the total cost of building elements during their life cycle (Lounis and Daigle 2008). Likewise, in environmental terms, there are tools such as LCA—life cycle assessment that quantify the environmental impacts of materials and construction processes throughout the building's life cycle (Optis and Wild 2010). However, currently, various studies in the literature that apply LCC and LCA tools do not adequately contemplate the service life of the construction elements, using instead standard values (Adalberth 1997; Keoleian et al. 2001; Thormark 2002); generally, these studies apply an average service life value for a given building element, without considering the specific conditions that it will be subjected to, thus introducing additional uncertainty—or errors—in the analysis (Aktas and Bilec 2012). Time and again, it has been suggested that LCA and LCC analyses need to apply realistic service life data, adjusted to the construction element under analysis (Grant et al. 2014). However, the service life of any given element may vary significantly from building to building, since each building is a prototype that responds to unrepeatable conditions. Therefore, service life prediction models intend to express degradation patterns not so much directed to a specific building, but identifying the common points between different buildings, according to a number of key characteristics, degradation mechanisms and established performance requirements.

A survey conducted by Brisch and Englund (2005) to a group of researchers, standardization committees, universities, manufacturers, associations and consultants related to the construction area revealed that 63 % of the experts surveyed believe that service life prediction methods are extremely important and only 6 % consider that these models are not relevant. However, when considering the use of these methods, only 40 % use this type of methodologies and 13 % assume that they are totally ignorant on this subject. To make matters worse, data related with the durability and service life of buildings and materials are not even included in the majority of the architecture and construction projects. Therefore, even though the usefulness of the service life prediction methodologies is widely accepted in the scientific community, their application is still incipient. Furthermore, despite the number of studies regarding the service life and maintenance of buildings and components that have been published in recent decades, the application of such methodologies still presents some limitations, mainly due to the complexity of degradation phenomena and the lack of reliable tools for their modelling. Overcoming this situation, involves the adoption of a number of assumptions within service life prediction methodologies that must be properly grounded. As mentioned by Field (2009), to model a real world phenomenon, whatever it may be, it is necessary to build models that cover information gathered in this actual situation. Martin (1985) refers that a scientific model can be seen as an abstraction of a real system and must include two conflicting attributes: realism and simplicity. The model should function as a reasonably accurate approximation of the real system, containing a number of important parameters for their correct description; however, it should not be so complex that makes it impossible to understand and manipulate. The definition of efficient and accurate tools to evaluate the degradation of construction elements over time is becoming increasingly important, allowing estimating the time after which it is necessary to intervene. A study performed by Moser (2004) discussed the work done by different authors in this area and concluded that more studies were needed in order to identify the parameters that influence the service life of the construction elements, being necessary to create viable mathematical relationships that allow applying these methods.

1.2 Motivation

The deterioration of buildings does not occur uniformly: since buildings are composed of various subsystems, which degrade at different rates, it can be assumed that they are composed of several layers of durability, with different service lives, which are distinct from the structure's service life. Façade claddings have a fundamental role in the performance of buildings, functioning as the first and foremost protective layer from environmental degradation agents of the wall and the structure. As a result, the buildings' envelope—sometimes also referred to as the skin of the building—is very prone to defects, with direct consequences on the quality of the urban space, on the comfort of users and in repair and maintenance costs (Kirkham and Boussabaine 2005). The degradation of the envelope can lead, in some situations, to structural problems to such an extent that the rehabilitation of the built heritage usually always implies the evaluation and monitoring of the external surface of the buildings (ASCE/SEI 30-00: 2000). Finally, in recent years, due to the use of innovative materials and complex constructive technologies, external surfaces have gained an increased importance, being currently considered and designed as part of the building (Schittich 2002).

Currently, stakeholders in the construction sector select a given cladding system considering various parameters such as: the visual appearance; the thermal and acoustic performance of the claddings; the type of support; the cost of the materials applied; among others. Generally, designers base their decision on commercial documents, being relatively easy to fulfil the performance requirements at "instant zero", i.e. at the instant that the building is put into use. However, designers rarely considered other properties, whose analysis is more complex, such as: (i) the ageing of materials in situ, i.e. the interaction between time and the elements that constitute the cladding system; (ii) the interaction between the materials applied in the cladding and the environmental exposure conditions; (iii) the potential effects of changes in material's performance in the overall performance of the assembly (Fagerlund 1985). In fact, in most cases, the information available on a given material is not directly operable when its performance over time is unknown for a given set of environmental exposure conditions. Therefore, for an effective selection, use and maintenance of a façade's cladding system it is essential to know with some accuracy the following aspects: (i) the façade's expected service life; (ii) the forecasted degradation mechanisms—chemical, physical and mechanical; (iii) the properties of materials applied on the façade and their execution conditions; (iv) ensure the compatibility between the cladding system and the structure; and (v) the effects of environmental exposure conditions and of the use of the building.

The study of the durability of façade claddings during their service life is a fundamental dimension for most decisions related to the built environment. Data on the their durability allow assessing their overall cost, considering both the initial cost (in the construction phase) and the maintenance costs combined with the service life, enabling the selection of the most economically attractive proposal. This book intends to contribute to the study of durability and expected service life of buildings, based on the working assumption that the service life of façade claddings (and other non-structural elements) can be modelled by different mathematical approaches with different levels of accuracy and complexity, leading to results with various degrees of richness of information.

1.3 Scope, Objectives, Methodology and Background of the Proposed Research Work

1.3.1 Scope of the Study

Service life prediction of buildings depends on the characterization of the materials, associated with the development of mathematical and computational models able to describe the physical degradation phenomena (NMAB 1996). As mentioned by Yatim et al. (2005), a model to predict the service life of buildings can result from the evaluation of existing buildings as well as from the statistical analysis of the variables that influence their service life. In this sense, this book intends to apply and develop different methodologies for service life prediction of façade claddings, by performing through the book and especially in Chap. 7, a comparative analysis of the applicability and validity of the proposed models. For this, advanced statistical tools are applied, thus combining the study of the degradation of façade claddings under real in-use conditions (based on an extensive fieldwork) with the development of mathematical models. The proposed models are divided into different innovation levels:

- As a starting point, models previously applied to the service life prediction are used, such as: a simple regression analysis or degradation curves (Shohet et al. 1999; Gaspar and de Brito 2008b; Silva et al. 2011a); a classic approach to the factor method (Bourke and Davis 1999; Gaspar and de Brito 2008a; van Nunen 2010; Silva et al. 2012); Markov chains—less common in the service life prediction of non-structural elements, but which have been widely applied in the monitoring of bridges' degradation state—models based on non-linear regression analysis, as is the particular case of Gompertz and Weibull curves (Garrido et al. 2012);
- These models are then expanded into new approaches. The simple regression model evolves into a multiple linear regression model, thus introducing more than one explanatory variable in the model. These models are further expanded to multiple non-linear regression models which, in addition to incorporating several variables, adopt non-linear relationships between them;
- Other models, which have been previously applied in the construction industry and whose applicability to service life prediction seems appropriate are analysed, such as artificial neuronal networks;
- In a different investigation approach, models that usually are not applied to the service life prediction are used in this book. Logistic regression has been applied in the literature to predict the remaining life of patients and their survival within the medical sciences; it is assumed that a model able to accurately predict the life of a patient can also be capable of predicting the service life of façade claddings (with the appropriate adjustments naturally). Likewise, fuzzy logic models are applied to service life prediction, leading to results similar to those obtained with artificial neural networks, but using linguistic concepts, more subjective, to model the degradation phenomena;
- Finally, new prediction models are defined in the book, such as a stochastic approach to the factor method, whose modelling depends on an algorithm developed by the authors.

It should be noted that other models could be analysed. However, the aim of this study is not to describe and implement all the possible models for service life prediction but instead provide the decision-maker with a set of reliable models with different characteristics, which can be applied according to user-established criteria. In each chapter, two methods are analysed (which produce similar information), providing alternative ways to solve the problem under analysis.

1.3.2 Background of the Proposed Research Work

This book follows a series of studies concerning the service life prediction of non-structural components performed at Instituto Superior Técnico. Data regarding the evaluation of the degradation condition of façade claddings are collected during an extensive fieldwork that is used to benchmark the results obtained by each model proposed. The research has been published in various international journals, co-authored by the authors of this book. From those, the following studies are highlighted:

- Gaspar P., de Brito J. Service life estimation of cement-rendered facades. Building Research & Information 2008a; 36(1): 44–55;
- Gaspar P.L., de Brito J. Quantifying environmental effects on cement-rendered facades: A comparison between different degradation indicators. Building and Environment 2008b; 43(11): 1818–1828;
- Gaspar P. Service life of constructions: Development of a method to estimate the durability of construction elements. Application to renderings of current buildings (in Portuguese). Ph.D. Thesis in Engineering Sciences, 2009, Instituto Superior Técnico, Technical University of Lisbon, Portugal;
- Bordalo R., de Brito J., Gaspar P., Silva A. Service life prediction modelling of adhesive ceramic tiling systems. Building Research and Information 2011; 39 (1): 66–78;
- Silva A., de Brito J., Gaspar P. Service life prediction model applied to natural stone wall claddings (directly adhered to the substrate). Construction and Building Materials 2011a 25(9): 3674–3684;
- Silva A., Dias J.L.R., Gaspar P.L., de Brito J. Service life prediction models for exterior stone cladding. Building Research and Information 2011b; 39(6): 637–653;
- Silva A., de Brito J., Gaspar P.L. Application of the Factor Method to maintenance decision support for stone cladding. Automation in Construction 2012a; 22(3): 165–174;
- Silva A., de Brito J., Gaspar P.L. Probabilistic analysis of the degradation evolution of stone wall cladding (directly adhered to the substrate). Journal of Materials in Civil Engineering 2013a; 25(2): 227–235;
- Silva A., Dias J.L.R., Gaspar P.L., de Brito J. Statistical models applied to service life prediction of rendered façades. Automation in Construction 2013b; 30: 151–160;
- Silva A., Gaspar P.L., de Brito J. Durability of current renderings: a probabilistic analysis. Automation in Construction 2014; 44: 92–102;
- Chai C., de Brito J., Gaspar P., Silva A. Predicting the service life of exterior wall painting: techno-economic analysis of alternative maintenance strategies. Journal of Construction Engineering and Management 2014; 140(3): 04013057;
- Emídio F., de Brito J., Gaspar P., Silva A. Application of the factor method to the estimation of the service life of natural stone cladding. Construction and Building Materials 2014; 66: 484–493;

- Galbusera M.M., de Brito J., Silva A. Application of the factor method to the prediction of the service life of ceramic external wall claddings. Journal of Building Performance of Constructed Facilities 2014; 10.1016/j.conbuildmat.2014.05.045, 19–29;
- Silva A. Statistical modelling of service life prediction of façade's claddings (in Portuguese). Ph.D. Thesis in Engineering Sciences, 2015, Instituto Superior Técnico, University of Lisbon, Portugal;
- Silva A., Neves L.C., Gaspar P.L., de Brito J. Probabilistic transition of condition: render facades. Building Research and Information 2015a, DOI:10.1080/09613218.2015.1023645;
- Silva A., de Brito J., Gaspar P.L. Stochastic approach to the factor method: durability of rendered façades. Journal of Materials in Civil Engineering 2015b; 10.1061/(ASCE)MT.1943-5533.0001409, 04015130;
- Vieira S.M., Silva A., Sousa J.M.C., de Brito J., Gaspar P.L Modelling the service life of rendered façades using fuzzy systems. Automation in Construction 2015; 51: 1–7;
- Silva A., de Brito J., Vieira S.M., Gaspar P.L. Fuzzy systems in service life prediction of natural stone claddings. Journal of Performance of Constructed Facilities 2016a, 10.1061/(ASCE)CF.1943-5509.0000860, 04016005;
- Silva A., Gaspar P.L., de Brito J. Comparative analysis of service life prediction methods applied to rendered façades. Materials and Structures 2016b; DOI 10.1617/s11527-016-0832-6, pp. 1–18.

1.3.3 Objectives and Methodology

This book intends to accomplish the following objectives:

I. Develop a set of effective methodologies for service life prediction of façade claddings to be used by different stakeholders in the construction sector;
II. Understand the evolution of the degradation of various types of façade claddings according to their age;
III. Analyse the influence of the claddings' characteristics in their degradation process (identifying degradation patterns associated with specific degradation mechanisms) such as: the environmental exposure conditions; the quality of materials; the design and execution level; the use and maintenance conditions;
IV. Rank the most relevant characteristics to the explanation of the degradation process, analysing the causal relationships between deterioration agents;
V. Predict the service life of façade claddings with a well-known precision, achieving an average value associated with a set of dispersion measures;
VI. Perform a comparative analysis of the proposed methodologies, providing some recommendations for the implementation of each model to a specific application.

The proposed models produce different types of information, from absolute values to probabilistic distributions that characterize the estimated service life of each case study. The results achieved in this book can be applied to aid decision-making in the planning of maintenance and rehabilitation actions. Additionally to the applications provided in this book, these methodologies can be applied to other construction elements and to buildings in other environmental contexts with the appropriate modifications. This book is comprised of eight chapters:

- **This chapter** describes the motivation and main goals of the book;
- **Chapter** 2—**Service life and durability of assemblies**: presents a survey of the existing literature on this subject. This chapter also provides the description of the methodology applied during the fieldwork and the criteria adopted in the definition of the degradation condition of the façade claddings analysed, as well the estimation of the overall condition of the claddings analysed in the book;
- **Chapter** 3—**Deterministic models**: presents the models based on regression analysis, divided in three parts:
  - **Section** 3.1—**Simple regression analysis (linear and non-linear)**: these models are based on degradation curves, which express the loss of performance of the constructive elements over time;
  - **Section** 3.2—**Multiple linear regression analysis**: where the theoretical concepts associated with this tool are applied to the service life prediction of façade claddings, evaluating the most significant variables to the description of the degradation of the claddings analysed. Based on this methodology, the causal relationships between variables and their implication on the estimated service lives of the claddings are analysed;
  - **Section** 3.3—**Multiple non-linear regression analysis**: addresses the application of various non-linear models to the service life prediction of claddings.
- **Chapter** 4—**Stochastic models**: establishes the theoretical concepts and the practical application of these models to the service life prediction of the claddings. Two methodologies are proposed in this chapter:
  - **Section** 4.1—**Logistic regression**: evaluates the probabilistic transition between degradation conditions over time and according to the façades' characteristics, also providing the probability of the claddings to reach the end of their service life;
  - **Section** 4.2—**Markov chains**: proposes the application of a continuous-time Markov chain model to the service life prediction of the claddings analysed. The application of this model produces similar results to those obtained by logistic regression, still allowing evaluating the mean time of permanence in each degradation condition.

- **Chapter** 5—**Computational models**: applies computational techniques, allowing simulating automated intelligent behaviours. This chapter describes the fundamental concepts for the application of the methods proposed, followed by the definition of the mathematical formulation that allows estimating the severity of degradation of façade claddings according to the explanatory variables. This chapter is divided in two subsections: **Sect.** 5.1—**Artificial neural networks**; **Sect.** 5.2—**Fuzzy logic systems**.
- **Chapter** 6—**Factor models**: addresses the methodology proposed in the international standard for durability (ISO 15686: 2011) for service life prediction of buildings and components. This chapter is divided in two subchapters:
  - **Section** 6.1—**Classic approach**: estimates the service life of claddings in a purely deterministic way. In this approach, the durability sub-factors are quantified as absolute values;
  - **Section** 6.2—**Stochastic approach**: proposes a probabilistic approach to the factor method, assigning a probability distribution to each of the durability factors. In this method, the estimated service life of each case study is given by a probability distribution function. To illustrate the application of this methodology, an example of the determination of the estimated service life of one case study is provided for each of the claddings analysed.
- **Chapter** 7—**Comparative analysis of service life prediction methods**: presents a comprehensive analysis of the models proposed in the book, discussing their advantages, limitations and the wealth of information produced. In this chapter, the predictive and classification capabilities of the proposed models are analysed. A multi-criteria analysis is also performed in order to model the preferences of different decision-makers, providing several suggestions and recommendations for the adequate selection and application of the proposed service life prediction models;
- **Chapter** 8—**Conclusions**: summarizes the main findings and contributions of the methodologies proposed in the book.

References

Adalberth K (1997) Energy use during the life cycle of buildings: a method. Build Environ 32 (4):317–320

Aktas CB, Bilec MM (2012) Service life prediction of residential interior finishes for life cycle assessment. Int J Life Cycle Assess 17(3):362–371

Allen E (1995) How buildings work, 2nd edn. Oxford University Press, New York, USA

ASCE/SEI 30-00: 2000 (2000) Guideline for condition assessment of the building envelope. Consulted in 12 July 2013. Available online on http://ascelibrary.org/doi/book/10.1061/9780784404836

Beer HR, Kerber C, Schläpfer J-L, Rieche G, Unold H, Wehrle S (2011) Longterm field studies and residual service life prediction of FPO roofing membranes. In: 12th international conference on durability of building materials and components, Porto, Portugal

Bourke K, Davis H (1999) Estimating service lives using the factor method for use in whole life costing. In: 8th international conference on durability of building materials and components, Vancouver, Canada, pp 1518–1526

Brisch C, Englund F (2005) Service life prediction methods—outcome of TFPC questionnaire. Cost E37 workshop, 19–21 June 2005, Oslo

DuraCrete Probabilistic performance based durability design concrete structures—final technical report. General guidelines for durability design and redesign. The European Union—Brite EuRAm III, May, European Union, 2000

Fagerlund G (1985) Essential data for service life prediction. Problems in service life prediction of building and construction materials. NATO ASI series 95:113–138

Field AP (2009) Discovering statistics using SPSS, 3rd edn. Sage Publications, California, USA

Garrido MA, Paulo PV, Branco FA (2012) Service life prediction of façade paint coatings in old buildings. Constr Build Mater 29:394–402

Gaspar P, de Brito J (2008a) Service life estimation of cement-rendered facades. Build Res Inf 36 (1):44–55

Gaspar PL, de Brito J (2008b) Quantifying environmental effects on cement-rendered facades: a comparison between different degradation indicators. Build Environ 43(11):1818–1828

Grant A, Ries R, Kibert C (2014) Life cycle assessment and service life prediction. A case study of building envelope materials. J Ind Ecol 18(2):187–200

Keoleian GA, Blanchard S, Reppe P (2001) Life-cycle energy, costs, and strategies for improving a single-family house. J Ind Ecol 4(2):135–156

Kirkham RJ, Boussabaine H (2005) Forecasting the residual service life of NHS hospital buildings: a stochastic approach. Constr Manag Econ 23(5):521–529

Kyle BR (2001) Toward effective decision making for building management. In: APWA international public works congress, NRCC/CPWA/IPWEA seminar series "innovations in urban infrastructure", Philadelphia, USA, pp 51–69

Lounis Z, Daigle L (2008) Reliability-based decision support tool for life cycle design and management of highway bridge decks. In: Annual conference of the transportation association of Canada, Toronto, Canada

Lounis Z, Lacasse MA, Vanier DJ, Kyle BR (1998) Towards standardization of service life prediction of roofing membranes. In: Wallace TJ, Rossiter Jr WJ (eds) Roofing research and standards development. ASTM STP 1349, vol 4. American Society for Testing and Materials

Martin JW (1985) Service life prediction from accelerated aging test results using reliability theory and life testing analysis. In: Problems in service life prediction of building and construction materials, NATO ASI series, vol 95, pp 191–211

Moser K (2004) Engineering design methods for service life prediction. CIB W080/RILEM 175 SLM: service life methodologies prediction of service life for buildings and components, task group: performance based methods of service life prediction, Trondheim, Norway, pp 52–95

NMAB (1996) Accelerated aging of materials and structures: the effects of long-term elevated temperature exposure, NMAB-479. National Materials Advisory Board, National Research Council, National Academy Press, Washington, DC

Optis M, Wild P (2010) Inadequate documentation in published life cycle energy reports on buildings. Int J Life Cycle Assess 15(7):644–651

Schittich C (2002) In detail: building skins: concepts, layers, materials, 1st edn. Princeton Architectural Press, New York, USA

Shohet I, Rosenfeld Y, Puterman M, Gilboa E (1999) Deterioration patterns for maintenance management—a methodological approach. In: 8th international conference on durability of building materials and components, Vancouver, Canada, pp 1666–1678

Silva A, de Brito J, Gaspar P (2011a) Service life prediction model applied to natural stone wall claddings (directly adhered to the substrate). Construction and Building Materials 25(9):3674–3684

Silva A, de Brito J, Gaspar PL (2012) Application of the factor method to maintenance decision support for stone cladding. Autom Constr 22(3):165–174

Thormark C (2002) A low energy building in a life cycle—its embodied energy, energy need for operation and recycling potential. Build Environ 37(4):429–435

van Nunen H (2010) Assessment of the sustainability of flexible building. The improved factor method: service life prediction of buildings in the Netherlands, applied to life cycle assessment, 1st edn, Aeneas, The Netherlands

Yatim JM, Tapir SH, Usman F (2005) Evaluation of building performance using neural network: study on service life planning in achieving sustainability. In: 8th international conference of the application of artificial intelligence to civil, structural and environment engineering, Rome, Italy, paper 46

Chapter 2
Service Life and Durability of Assemblies

2.1 Theoretical Foundations

The quality and the durability of buildings are an essential dimension for the quality of life, and are critical for the social and economic stability of contemporary societies (Wekesa et al. 2010). In fact, the building stock represents approximately 50 % of the wealth of most European countries (Long et al. 2001). The increased degradation of the built heritage is due to the natural ageing process of materials and components, which begins as soon as buildings are put into use, associated with the scarcity of resources and—in some countries—a yet incipient culture of building's maintenance. According to Haagenrud (2004), when adopted, an attitude of "build and let it stay" leads to an accelerated loss of building's performance. This loss of performance, which manifests itself in ever higher levels over time, expressed by the presence of defects, leads to the inability of the buildings to fulfil the requirements for which they were designed (Chai et al. 2014). Therefore, the knowledge concerning the service life of buildings and their components assumes a very important role, allowing understanding how to manage a vast—and often aged—built park.

The service life prediction is a field that leads to more rational and sustainable solutions. In the last decades, the possibility of using data on the service life of materials and components during the design and construction phase has been thoroughly analysed (Frohnsdorff and Martin 1996; Soronis 1996). This interest arose due to two main factors: (i) environmental factors, since the energy crisis led to a scarcity of resources and an increased awareness of the notion of sustainability (witnessing a growing concern in terms of the environmental impact of the construction sector); (ii) economic factors, since the management and maintenance of the built stock is a very significant dimension of the economy of societies (Hovde 2002). Moreover, the changes in construction methods, the onset of durability

A. Silva et al., *Methodologies for Service Life Prediction of Buildings*,
Green Energy and Technology, DOI 10.1007/978-3-319-33290-1_2

problems and the need to understand the performance of new building materials also contribute to the pursuit of knowledge regarding the durability of buildings and their components.

This chapter aims to establish the first approach to the service life prediction methods applied to the façades claddings, with three main objectives: (i) perform an extensive survey regarding the concepts related with the service life and durability of building components, the methodologies and standardization related to this matter; (ii) define the information collected during the fieldwork, which allows the physical and visual characterization and further quantification of the degradation condition of the claddings analysed; iii) the characterization of the samples collected during the fieldwork.

2.2 Service Life of Buildings

Making an analogy with living beings, it can be said that all buildings have a life cycle during which they "are born", "get older" and "die" (Gaspar 2009). Often does this analysis appear in the literature on the subject, namely regarding life cycle assessment analysis, referred to as "cradle to grave". The building's life cycle comprises all stages of its life, i.e. the time elapsing since it is placed in use, after construction (birth—corresponding to the stages of design and construction) and ending at the instant from which it is unable to meet the acceptable minimum performance requirements (ISO 15686: 2011). The concept of service life is relatively simple, although different definitions can be found in the literature, sharing common ideas

- According to ISO 15686: 2011 (Service life planning), service life can be defined as the period of time after installation in which the buildings or their parts meet or exceed the minimum performance requirements;
- ASTM (1990) shares the same definition of ISO 15686: 2011 and also mentions that during this period of time the building and its elements must be subjected to periodic maintenance;
- The Canadian standard CSA S478-95 (Guideline on Durability in Buildings) refers that the service life can be described as the period of time during which the building or its components fulfil the requirements for which they were designed, without unexpected costs or maintenance and repair actions;
- EOTA, in the document Assumption of Working Life of Construction Products in Guideline for European Technical Approvals and Harmonized Standards (EOTA 1999), characterizes service life as the period of time during which the performance of the product is maintained at a level compatible with the fulfilment of the essential requirements.

The concept of service life is often confused with that of durability, leading to the misuse of the terms. Contrary to the concept of service life, durability is not related to a period of time but instead corresponds to the building's ability and that of their components to show an adequate performance during their life cycle. According to the Canadian standard (CSA S478-95: 2001), the concept of durability refers to the ability of a building or its components to achieve the best performance in a given environment or location, without having to be subjected to significant corrective measures or to the repair or replacement of its elements. In turn, the New Zealand Building Code includes a functional requirement, referring that the elements and construction materials as well as construction methods should be sufficiently durable, thereby ensuring that the building meets all the performance requirements during the building's service life, without having to be subjected to significant rehabilitation works (Lee et al. 2008). In reality, durability cannot be seen only as an intrinsic quality of a material; simple changes in the construction details may promote a higher protection of a building element against the degradation agents, contributing to the increase of its service life (John et al. 2002). The service life prediction methodologies gain therefore an increasingly important role in the implementation of durable buildings, which should be designed based on: (i) the knowledge of the performance of the materials over time; (ii) the knowledge regarding the capacity of the material withstanding the degradation mechanisms it will be subjected to, under a given set of environment exposure conditions; (iii) the characteristics of the construction and its context (Wyatt 2005). In the end, although being different concepts, the service life and durability of constructions are closely related and necessary to understand the life of buildings; they are therefore relevant concepts in the construction process, either at the design stage or in the use phase, thus allowing reducing the maintenance costs, increasing the comfort of users and the sustainability of the solutions adopted (Moreno 2012).

A study performed by Bordeau (1999) for the CIB W82 commission reveals that in some languages (Dutch, Finnish, Romanian, among others), the word "sustainable" can be translated by "durable". This "overlapping" between the two concepts seems logical, since the increase in the durability and service life of the constructions is an acting vector in the search for sustainability, in order to decrease the environmental impact of buildings, because it directly contributes to a more rational management of resources, reducing the production and disposal of waste (DeSimone and Poppof 1998). In recent decades, the concept of sustainable development has even led to a reassessment of the contribution of the construction industry in the quality of life (Pearce 2003). The sustainable development of the construction sector can be analysed in three dimensions: (i) environmental; (ii) social; (iii) and economic (Mateus et al. 2008). A sustainable development can only be achieved in these three levels if measures to reach sustainability are adopted during all the phases of the buildings' life cycle. In particular, during the use phase, it is extremely important to carry out detailed planning of the maintenance needs (Daniotti et al. 2008). Inversely, various authors (Takata et al. 2004; Donca et al. 2007) refer that maintenance actions, if properly implemented and planned in a rational and technically informed manner, allow increasing the buildings' service life (thereby contributing to improved durability and sustainability of buildings). In practice, one of the most effective ways of optimizing maintenance actions in buildings is through the knowledge how the building and their components deteriorate over time, estimating the instant after which it is necessary to intervene (Talon et al. 2005). Generally, without this information, maintenance actions may generate high costs associated with unnecessary interventions or urgent repairs, which could be avoided. According to Vanier (1999), before implementing any maintenance policy, a set of questions should be answered, in which it is necessary to evaluate the condition of the element in question and its remaining service life, so as to define the extent of the intervention that should be performed.

2.2.1 Criteria that Influence the End of the Service Life of Buildings

During its life cycle, a building should meet a set of performance requirements, e.g. safety, watertightness, compatibility to the substrate, visual comfort, durability, among others. However, in some situations, the building's components fail to fulfil these requirements after some time. Obsolescence is also a concern for all owners and users, for even if the building may fulfil the initial requirements, it may fail to respond to new expectations or requirements especially when compared to (often newer) available alternatives. Sarja (2005) defines obsolescence as the inability of the building or its parts to adapt over time to the functional, economic and cultural requirements. The usefulness of the buildings may also be compromised by their inability to accommodate changes over time (Slaughter 2001); in fact, throughout their life cycle, all the buildings experience changes, e.g. changes in its occupants or their needs and expectations, renovations and/or extensions, the ageing and replacement of components and systems (Cowan 1963; Brand 1997; Augenbroe and Park 2002).

Construction assets suffer various types of depreciation throughout their life cycle, in a process that begins as soon as the building is put into use (Rikey and Cotgrave 2005), eventually leads to the end of the building service life. This process is due to several factors: (i) physical deterioration; (ii) economic obsolescence; (iii) functional obsolescence; (iv) technological obsolescence; (v) changes in the social context; (vi) obsolescence due to the building envelope; (vii) legal obsolescence; (viii) aesthetic obsolescence; and (ix) environmental obsolescence (Flanagan et al. 1989; Flores-Colen and de Brito 2010). Hovde (2002) refers as the main reasons to establish the end of the service life of a building: (i) technical aspects; (ii) economic aspects; (iii) environmental reasons; (iv) planning reasons—e.g. when the demolition of a building or a structure leads to the construction of a railway or other construction for public service; (iv) society requirements or technological development. On the other hand, Moser (2004) proposes a more strict range of reasons, suggesting three main reasons to establish the end of the service life of a building: (i) safety; (ii) functionality; and (iii) aesthetics. Gaspar (2009) refers that despite the relative simplicity of the service life concept, the service life is extremely difficult to predict or simulate through models, since it "depends on the definition of acceptance criteria, which varies to the time, place and the stakeholder and even with the social, economic, political, aesthetic, environmental context of the building under analysis". Generally, new performance requirements (usually more demanding) are enforced all the time, leading to a constant investment to keep up with them and to delay the degradation trend affecting the building elements. Aikivuori (1999) states that, even if the building maintains its original properties (not showing visible degradation), the end of its service life is inevitable due to the evolution of the acceptance criteria, for example by comparing the existing situation and a new available constructive solutions. In the definition of the service life prediction methods it is therefore necessary to understand and incorporate the maximum acceptable degradation level for the element under analysis—i.e. the minimum acceptable level of performance—after which the element has reached the end of its service life. This theoretical limit is not generally easy to specify (Moser 2004); in fact, as mentioned by Iselin and Lerner (1993), there are no rational criteria to guide the decision to intervene, but rather subjective and programmatic criteria. A study performed by Aikivuori (1999) shows that in only 17 % of the situations, the decision to intervene is taken based on the building's deterioration and in 44 % of cases maintenance actions are performed based on subjective criteria. This study also concludes that, when the decision to intervene is based on technical criteria only, depending on the building's degradation condition, the rehabilitation action takes place later than it would if the criteria were subjective (e.g. due to the aesthetic criteria). Sarja (2005) drew similar conclusions, referring that the most prevailing reasons for heavy refurbishments or demolitions of buildings are (i) obsolescence under changing use, (ii) demands of users and (iii) general requirements of the society; and indeed not the buildings' degradation as we are led to believe.

Regardless of the variability of the acceptance criteria, the service life of buildings can be distinguished into three main categories: (i) physical service life; (ii) functional service life; and (iii) economic service life (Marteinsson 2003). The physical or technical service life is related with the deterioration of the materials and building elements. The deterioration of the construction elements occurs systematically, which implies that its failure rate increases over time (Zhang and Gao 2011). The physical deterioration of buildings is mainly due to the action of the degradation agents (whether physical, chemical or mechanical) and the natural ageing process. In several situations, design and construction errors, or the application of inadequate materials, can contribute to the reduction of the physical service life of buildings. The functional service life is directly related with the expectations and demands of the users; Davies and Szigeti (1999) refer that, as is the case of computer applications or the motor industry, the functional obsolescence of buildings occurs when they are unable to follow the users' requirements. In the real word, the service life of buildings is often conditioned by economic reasons; Brand (1997) states that the economic service life can be defined as the time elapsing since the construction is placed into use until the instant that it is replaced by a more profitable solution. The author also refers that buildings do not reach the end of their economic service life while the cost/benefits ratio is more attractive than the alternatives. In this case, the end of the economic service life is reached when the cost of replacing an element by another is lower than that of maintaining the existing one. Naturally, over time, and with the emergence of new constructive solutions (more economical, more durable and requiring less maintenance), buildings became economically obsolete.

This book addresses the physical service life of façade claddings, evaluating the physical and visual degradation of the claddings analysed, as described ahead in Sect. 2.5. In this study, the end of the service life is reached when the façade shows a degradation condition considered inadmissible, i.e. when the maximum acceptable degradation level is reached. This level is a theoretical limit that seems realistic considering the local context of the study (the Portuguese context) but, as referred before, this limit is subjective and difficult to define, since it varies according to the users' demands, changing over time.

2.3 Methodologies for Service Life Prediction

In the last decades, different methodologies for service life prediction and different tools to support the decision-making process regarding the maintenance of construction have been put forward. Nevertheless, these models have several shortcomings, essentially due to: the complexity of the degradation phenomena; the lack of understanding of the degradation mechanisms and how these affect the construction elements; the scarcity of reliable methods to quantify the durability and service life of buildings (Lounis et al. 1998). In fact, the service life prediction of buildings and their components can be a complex and time-consuming process that is associated with many factors (Hovde 2004), such as the quality of materials, the design and execution level, the indoor and outdoor environment conditions, as well the use and maintenance conditions (ISO 15686-1: 2011). The researcher must choose the best approach to the problem that he/she wants to model, considering the advantages and limitations of each methodology (Freitas et al. 2013). Generally, in the literature, the service life prediction methods have been grouped into deterministic, stochastic and engineering models (symbiosis between the other two methodologies) (Moser 2004; Lacasse and Sjöström 2004). This classification is also used in the next sections to describe the existing methodologies for service life prediction.

2.3.1 Deterministic Models

The deterministic models are based on the study of the degradation factors affecting the elements under analysis, on the understanding of the degradation mechanisms and, finally, on their quantification translated into degradation functions. These degradation factors are then translated into formulas that express their action over time until the maximum acceptable degradation level of the element analysed is reached. These methods have significant advantages: are easy to understand and apply; can be easily implemented; and maintain their operability even when all the variables related with the modelled phenomena are unknown. However, these approaches have been subjected to several criticisms, essentially due to the simplistic way with which they deal with complex phenomenon such as the service life of building's components. According to Paulo et al. (2014), the deterministic models are unable to capture the random nature of the degradation phenomena. Various authors (Hovde 2000; Mc Duling et al. 2008) state that in the deterministic models, the service life is given by an absolute value that does not provide any information regarding the degradation process or related to the probability of transition from one degradation condition to another. Nevertheless, these methods have produced practical solutions to the problem and are widely used; they have also provided the basis for the international standard for the durability of buildings (ISO 15686: 2011).

Within the deterministic methods, empirical methods have been developed in order to evaluate the durability (or loss of performance) of a building or its components under real conditions of service, at different stages of its service life, based on data collected during field work (Shohet and Paciuk 2004; Gaspar and de Brito 2008). In these methods, the characterization of the buildings' degradation condition is performed through the definition of a classification system for defects and degradation ratings in order to express the physical and functional degradation of the elements under analysis, which can be converted into quantitative information. Based on this information, the loss of the material's capacity to answer the requisites demanded can be expressed by degradation functions (Bordalo et al. 2011). In practice, after collecting the fieldwork information regarding the degradation condition of the façades—which needs to be converted to quantitative, numerical data—service life can be estimated using a graphical procedure and a statistical analysis of the evolution of degradation over time. The evolution of degradation is thus represented graphically by degradation curves that can be associated with specific degradation mechanisms (Shohet et al. 1999; Florentzou et al. 1999). This methodology is implemented and described in Part 3A of this book.

2.3.2 Stochastic Models

Due to the uncertainty associated with the building's performance, it is often necessary to use probabilistic models to predict the service life of the construction elements (Ross 1996). In the probabilistic methods, the degradation of construction is regarded as a stochastic process, which is described by a set of random variables that define probabilistic parameters affecting the average degradation curve (Moser 2003).

These methods are usually quite complex and require an extensive data collection to obtain sufficiently representative samples, which is not always possible, due to time and cost constraints (Re Cecconi 2002). According to this author, the translation of these models into real situations can be complex, which implies that these methods are only profitable on large-scale projects. In fact, these methods are often beyond the reach of a common designer, due to the impossibility of this acquiring all the necessary knowledge in the time required for the application of the methodology (Moser and Edvardsen 2002). Stochastic models are usually associated with three main drawbacks, which can restrict their application: (i) the high complexity of the mathematical expressions used; (ii) the large amount of data required necessary to validate the model, which should preferably be collected over a long period of time; (iii) the great dependence of the fieldwork.

In the literature, there are several studies addressing the application of probabilistic methods for service life prediction

- Markov chains are a probabilistic method widely used in the study of the deterioration of buildings. This method is based on the assumption that a deterioration model can be defined based on a limited number of conditions, evaluating the probability of transition between a degradation level to the next one (Frangopol et al. 2004);
- Using a development of the Markovian transition matrix, Mc Duling et al. (2008) suggest a hybrid model between artificial intelligence and fuzzy logic (neuro-fuzzy artificial intelligence) that allows translating the rate of changes between degradation states, incorporating the effects of maintenance actions in the buildings service life prediction;
- Liang et al. (2001) suggest a model based on fuzzy logic to evaluate the service life of bridges, in which the end of service life is defined based on a minimum security index, by assessing the state of degradation of the existing elements;
- Leira et al. (1999) propose the Trend plots method, which is a simple statistical tool for forecasting rehabilitation needs, requiring a large amount of information regarding the durability of a given element to be applied;
- The European project Energy Performance Indoor Environment Quality Retrofit (EPIQR) proposes a tool for diagnosis and support of decision-making regarding the maintenance actions, based on the classification of the buildings' degradation condition in four states, ranging from "a" (best condition) and "d" (worse state condition, requiring an immediate intervention) (Balaras et al. 2000);

- The MEDIC method (*Méthode d'Évaluation de scénarios de Dégradation probables d'Investissements Correspondants*), proposed by Flourentzou et al. (1999), is based on the EPIQR tool and uses the theory of conditional probabilities (Bayes' theorem) to evaluate the transition between degradation states over time, allowing evaluating the residual service life of the buildings, thus predicting the investments necessary to the buildings' rehabilitation;
- Some probabilistic methods are based on the concept of fault and decision trees, assessing the durability of structures based on the definition of deterioration levels and the conditional probability of transition until their failure (Faber and Gehlen 2002).

Despite these approaches, the majority of the probabilistic models proposed in the literature focus one type of material only—usually reinforced concrete—subjected only to a degradation agent (usually chloride attack) (Lounis et al. 1998; Abraham 2002; Siemes and Edvardsen 1999; Edvardsen and Mohr 2000).

2.3.3 Engineering Models

The engineering "design" methods are a symbiosis between the two previous methods. These models are as easy to learn, understand and implement as the deterministic methods, but allow describing the degradation process in a stochastic way (Re Cecconi 2002). Usually, engineering models possess an acceptable level of complexity and are implemented using probabilistic data, leading to simple deterministic equations. According to Daniotti and Spagnolo (2008), engineering methods can be used to identify in a more analytical way the degradation phenomena, thus allowing the implementation of this information in the design stage or establishing a methodical planning of maintenance actions. Some of the best known engineering methods are

- The Failure Modes and Effects Analysis (FMEA) method, which was initially developed by the aerospace industry and later used in mechanical and electrical engineering; it has rarely been used in the construction sector (Lair 2003). Nevertheless, when applied to the construction sector, this method can be used for the certification of construction products, allowing obtaining an as complete as possible list of the degradation agents and failure mechanisms that can act in building elements during their use phase (Talon et al. 2005);
- The Performance Limits method, whose aims are to evaluate the durability of building components by simulating their performance over time, until their physical or performance limit has been reached. In this method, the degradation phenomena is evaluated by a chain of events: agents → actions → effects → deterioration (Daniotti and Spangolo 2008);
- Besides these methods, the probabilistic approach of the factorial method can also be seen as an engineering model. In this approach, the quantification of the

durability factors is performed based on probability density functions instead of absolute values, (Moser 1999; Aarseth and Hovde 1999; Moser 2004). Therefore, this model combines the simplicity of the factorial method and the probabilistic analysis of the degradation factors.

2.4 Normative Framework for the Service Life Prediction of Buildings

Currently there are numerous standards and guidelines that intend to establish standardized methodologies to evaluate the durability and service life of buildings and their components. In 1979, the Architectural Institute of Japan decided to organize a technical commission in order to systematize the concept of durability of constructions. This study led to the creation of the first normative document addressing the durability and service of life of buildings and their components: the Japanese guide developed in 1989, later translated to English under the title "(Japanese principal) guide for service life planning of buildings" (AIJ 1993). The Japanese guide establishes a set of recommendations for the service life prediction of buildings, their components or equipments, assuming that the end of their service life is determined by physical deterioration or obsolescence (Rudbeck 2002). This methodology is the basis of the factorial models, leading to a number of studies and standards currently published. Also, in Japan, with the approval of the Housing Quality Assurance Law (HQAL 2000), a series of measures were imposed in order to monitor the degradation of buildings and their components (Nireki 1996). Other guiding documents for the rehabilitation of buildings were put forward, such as the Guide to Condition Assessment for Refurbishment (1993) and Design Guide Refurbishment (1999).

In 1992, the British Standards Institute published standard 7543 for durability "British guide to durability of building elements, products and components" (BS 7543: 1992) that lists various methods to estimate the service life of construction products: (a) through past experience, using similar buildings, subjected to similar use and climatic conditions; (b) by evaluating the degradation level of the elements in a short period of use or exposure, estimating the value for which the durability limit is reached; (c) through accelerated ageing tests, which is a complex approach, due to the need to simulate real situations that have many variables to be considered. BS 7543 proposes defining the service life of buildings as a function of the type of use; buildings are classified into five categories: temporary buildings, with a service life of less than 10 years; short-lived buildings, such as storehouses, with a service life of at least 10 years; average buildings, such as industrial buildings, with a service life of at least 30 years; current buildings, such as new housing, hospitals and schools, with a service life of at least 60 years; long-living buildings, such as public buildings, with a service life of at least 120 years. The standard also prescribes that façade claddings must guarantee a service life similar to that of the building, with proper periodic maintenance.

Inspired by the Japanese guide, the International Organization for Standardization (ISO) proposed an international standard for the service life prediction of buildings, based on a recommendation of RILEM (International Union of Testing and Research Laboratories for Materials and Structures) (Frohnsdorff et al. 1996). Currently, the ISO 15686 "Building Service Life Planning" standard has eleven parts that define the general principles, framework and procedures of the proposed service life prediction methodology. Moreover, this standard defines the functional performance criteria to be fulfilled in the design phase and throughout the buildings' life cycle, which ultimately contribute to the definition of the end of the service life of the elements under analysis (Hed 1999). The ISO 15686 standard is one of the most relevant information sources regarding service life prediction and is composed of the following parts (Sjöström and Davies 2005; Haapio 2008): (i) ISO 15686-1: 2011, which defines the general principles to be adopted in the design phase, in order to ensure the buildings' durability; (ii) ISO 15686-2: 2012 (Service life prediction procedures); (iii) ISO 15686-3: 2002 (Performance audits and reviews); (iv) ISO 15686-4 (Data requirements/data formats); (v) ISO 15686-5: 2008 (Life cycle costing); (vi) ISO 15686-6: 2004 (Procedure for considering environmental impacts); (vii) ISO 15686-7:2006 (Performance evaluation for feedback of service life data from practice); (viii) ISO 15686-8: 2008 (Reference service life and service life estimation); (ix) ISO 15686-9: 2008 (Service life declarations); (x) ISO 15686-10: 2010 (Using requirements for functionality and ratings of serviceability during the service life); (xi) ISO 15686-11 (Terminology).

Other documents and guidelines have been developed within the service life prediction of buildings (Rudbeck 2002; Lacasse and Sjöström 2004; Kooymans and Abbott 2006), in countries like: (i) **Nordic countries**, which established a joint committee in 1976 to develop a Nordic model for the creation of standards based on the performance of buildings; (ii) **Norway**, which established a standard that specifies the performance criteria, as recommendations for maintenance and rehabilitation of buildings (NS 3422: 1994); (iii) **Denmark**, through the Danish Building Defects Fund committee that, since 1986, promotes projects (such as Quality-Assurance Danish and Liability Reform) that intend to reduce the defects in buildings, improving their performance (e.g. through the knowledge of the behaviour of building materials in service); (iv) the **Netherlands**, where the first standards based on performance requirements were defined, which subsequently formed the basis for the development of the European Directive on Construction Products; (v) **New Zealand Building Code**, which establishes the service life of buildings, depending on the easiness of the access to the component under analysis, the easiness of repair and the defects detection; (vi) **USA**, through the Partnership for Advancing Technology in Housing (PATH) and the American Society for Testing and Materials (ASTM); (vii) **Canada**, with the Standard S478: Guideline on durability in buildings, which describes the main methodologies for service life prediction.

In addition to the normative documents, a set of reference documents has been published that propose estimated service lives (average, standard values) for construction elements: (i) the HAPM (Housing Association Performance Management), which establishes a range of estimated service life values for a wide range of building components, depending on the materials' characteristics and their exposure conditions; (ii) the NAHB (National Association of Home Builders) that provides an estimated service life for the construction elements based on the opinion of experts and manufacturers.

Internationally, some I&D companies developed studies in the service life prediction area, namely: (i) TNO (Netherlands Organisation for applied scientific research), which applies the knowledge on the service life of current buildings in order to reduce the maintenance costs of these elements; (ii) Building Research Establishment (BRE), which provides expertise services in the detection and correction of defects and the context of life cycle analysis (economic and environmental) of the construction elements. Other organizations have developed studies concerning the durability and service life of structures, such as the International Association for Building Materials and Structures (RILEM) and the International Council for Research and Innovation in Building and Construction (CIB). Within CIB, some commissions address the durability and maintenance of buildings, such as the committees: (i) W60—Performance concept in building; (ii) W70—Management maintenance and modernization of buildings facilities; (iii) W086—Building Pathology; and (iv) W094—Design for durability. On this subject, the W080 commission—Prediction of service life of building materials and components—with the cooperation of the Committee TC59 Technical of RILEM, has been actively working within the service life prediction of constructions.

2.5 Degradation Phenomena of Façade Claddings

Buildings are composed by different components that reach the end of their service life in different stages of the buildings' life cycle. Various authors (Brand 1997; Slaughter 2001; Gaspar 2009) subdivide the building in durability layers, i.e. in construction subsystems whose degradation occurs at different rates, among which are the "structure", the "skin", the "systems" and the "interior lay-out and finishes". The façade can be seen as the "skin" of the building, contributing to increase the durability of the structure, protecting it from the environmental agents. Since the cladding is the most exterior layer of the building, and therefore more exposed to agents causing degradation, it is also more prone to defects. In fact, a research carried out by the BRE concluded that façades are the building component most affected by pathological manifestations, representing 20 % of the defects detected in current buildings (Watt 1999). Façades also present a very important aesthetic function since they represent the public image of the building. Their visual degradation not only affects the quality of urban space and impact on the perception of buildings and—indirectly—of their owners, but is also a major concern for the latter, since in the majority of the cases, maintenance and rehabilitation actions are performed based on the appearance of the building only (Balaras et al. 2005).

2.5.1 Data Acquisition Methods to Establish Service Life Prediction Models

Service life prediction is a multidisciplinary research field, which should be based on the knowledge acquired from materials' science, laboratory testing and from the behaviour of building elements in service conditions (Sjöström 1985). An efficient evaluation of the service life of construction elements must take into account various random factors, such as the natural ageing of materials and the environmental exposure conditions.

The data related with the durability of building components can be obtained through various methods. Clifton (1993) states that the methodologies for the service life prediction of structures can be based on: (i) previous experience; (ii) the performance of the materials analysed under similar conditions; (iii) laboratory tests, e.g. accelerate ageing tests; (iv) mathematical models to describe the physical and mechanical degradation processes; and (v) through the application of reliable stochastic models. Gaspar and de Brito (2008) refer that in the definition of models to simulate the degradation of façades claddings over time two approaches can be used: (i) laboratory testing; and (ii) the evaluation of the condition of the elements under analysis, in real situations, based on in situ surveying work. Laboratory testing presents an important role in assessing and understanding the performance of materials and construction elements, for which they employ a single arbitrary set of stringent conditions (say, the combined effects of temperature and moisture variations). Norvaišienė et al. (2003) refer that this approach has as main advantage, the speed with which they can obtain results, regardless of the relevance of the experimental conditions, compared with the study of degradation of materials in real service conditions. Rimestad (1998) and Augenbroe and Park (2002) state that the field data on the performance of buildings should be used whenever possible, which can also be applied to guide the definition of the accelerated ageing tests. However, according to Frohnsdorff and Martin (1996), such artificial reduction of the degradation effects fails to simulate the holistic effect of the environmental actions that together contribute to the degradation of the building elements. A number of authors further indicate that the relationship between the exposure conditions in the laboratory and real life in-use situations is not generally known or easy to establish, even in the case of natural ageing tests in monitored conditions (Kus and Carlsson 2003). In fact, after installation, a building is exposed to a number of environmental agents whose combined actions are not easily modelled and reproduced.

To assess the durability of buildings through the analysis of their behaviour in service, various techniques can be used ranging from destructive tests carried out in situ to visual inspections. Destructive testing provides accurate information on the performance and characteristics of materials and building components, but they are often expensive and require repair work afterwards. On the contrary, non-destructive assessment methods—most notably, visual inspections of building elements—are generally less expensive, faster to carry out and may be very useful in providing relevant information regarding the degradation process of materials (Meola et al. 2005). Often do they also provide relevant enough early warnings of problems that may occur that can be avoided by repair work or further investigated by on site destructive testing.

In this study, the description of the degradation condition of the façades analysed is based on visual inspections. Visual inspections are an easily grasped method but present some limitations since their accuracy depends significantly on the experience/background and classification criteria of the surveyor. This method also depends on the atmospheric conditions at the time of the inspection (e.g. the difficulty of detecting defects in smooth and dark claddings when under direct solar exposure). On the other hand, this method does not usually require costly equipment, and it is often perfectly adequate to determine the degradation condition of the elements under analysis. Usually, a straightforward visual inspection is considered sufficient to evaluate the degradation state of the façades claddings, and it is sufficient for the surveyor to collect in situ the data regarding the defect type, its intensity and extension (Straub 2003).

2.5.2 Systemic Analysis of the Façades Degradation Condition

In this study, four most current types of claddings are analysed, based on their use in Portugal

- Natural stone claddings—this type of claddings has a large variability concerning its construction technology; it can comprise various types of stone, different thicknesses and different types of finishing. Additionally, there are several fastening technologies, e.g. it can be directly adhered to the substrate or it can be applied indirectly with the application of metallic elements. Stone claddings correspond to 14.6 % of the claddings existing in Portugal (Census 2001). The sample analysed comprises 203 stone claddings; the oldest building was built in 1891 and later rehabilitated in 1948, and the newest building dates from 2008. In this study, a comprehensive sample is analysed, representative of the type of natural stone found in Portuguese claddings;
- Rendered façades—these claddings are traditionally composed of a binder (cement, lime or both), sand, water and other minor constituents; the composition of the mortar (content of the various constituents) depends on their function. Generally, traditional renderings are applied in three distinct layers (spatterdash, base coat and finishing layer), with a small thickness to facilitate

their drying and a gradual decrease of the binder's content from the inside to the outside layer, thus minimizing occurrence of cracking in the cladding (Kus 2002). In Portugal and in most European countries, rendered façades are the most common type of cladding (Flores-Colen et al. 2010). In this study, 100 rendered façades are analysed with different typologies and ages under 60 years;

- Painted surfaces—Portuguese standard NP 41 (1982) defines paint as a liquid, in paste or solid material, with a pigmented composition, which is converted after a period of time into a solid film, coloured and opaque, when applied in a thin layer on a suitable surface in the state in which it is provided or after fusion, dilution or dispersion in volatile products. Generally speaking, paints are constituted by pigments, fillers, binder or vehicle (fixed and volatile) and additives. Painted surfaces, like rendered façades, are a very common cladding solution adopted in Portugal. In this study, 220 painted surfaces are analysed; these paintings were applied on rendered façades or over existing paintings belonging to the building stock of Lisbon, with a range of ages of 18 years;
- Ceramic tiling systems—this cladding system comprises the ceramic tile, the substrate, the adhesive material needed to ensure the bond between the tile and the substrate, and the filling material for the joints (Chew 1999). Ceramic claddings represent a small percentage of façade claddings in Portugal (5.5 % in buildings built between 1946 and 2001) (Flores-Colen et al. 2008) and their use is often associated with fashion trends (Bordalo et al. 2011). For this reason, the sample analysed (195 case studies) comprises a wide range of construction dates, with an historical peak for the use of this cladding in the period 1920–1949, due to socio-economic reasons.

This study is intrinsically related to various studies performed at Instituto Superior Técnico of the University of Lisbon (as mentioned in Part 1), already published in a number of ISI-indexed journals that address the service life of construction elements. Therefore, this study applies fieldwork data gathered and analysed in previous studies, namely: (a) the 203 stone claddings are composed by a comprehensive sample collected by Silva et al. (2011) and Emídio et al. (2014); (b) the 100 rendered façades analysed are initially analysed by Gaspar and de Brito (2008); (c) the 220 painted surfaces are inspected by Chai et al. (2014); and (d) the ceramic tiling systems are collected by Bordalo et al. (2011) and Galbusera et al. (2014).

All these studies apply the same methodology in the systemic analysis of the degradation condition of the façades analysed (Fig. 2.1). The first step for the application of this methodology is the characterization of each case study; for that, complementary information (location, drawings, documentation from the Municipalities and other relevant data) was collected for each case study. Furthermore, an inspection and diagnosis file is created, where the gathered data can be grouped into two categories: façade condition and durability related data. Concerning the data related with the façade's condition, the following information is collected (Silva et al. 2013b): (i) a detailed survey of the façade's dimensions; (ii) the list and description of the defects detected, their extent within the façade and their location; (iii) an indication of the probable causes; (iv) a list of the diagnosis means and inspection techniques adopted in data collection; (v) the definition of the severity of the defects detected to be used later on the definition of the degradation models. The durability related data include the following information: (i) the identification number of the case study; (ii) the building's location; (iii) date and nature of the last intervention on the façade; (iv) construction size and geometry; (v) material's characteristics (type of material, colour, texture, among other parameters); (vi) design factors; (vii) level of detail and execution; (viii) potential critical points in the façade (e.g. balconies or protruding elements); (ix) environmental exposure conditions (temperature, exposure to damp, wind-rain action); (x) type of use.

The survey of the façades during fieldwork is complemented with photographs, direct measurements and diagrammatic sketches. Subsequently, this information is processed by computer, using image processing applications, computer-aided design and calculus spreadsheets. Additionally, the users and owners of the buildings inspected are also inquired. During fieldwork, the inspector should carefully characterize the building's pathological situation, since defects with accidental causes or due to vandalism cannot be used to feed the degradation models, since they are unpredictable and do not represent the natural degradation process of the elements under analysis. In this methodology, the variable 'age' is given by the period of time elapsing between the last significant intervention in the cladding–e.g. generalized repair, with partial or total replacement of the cladding—and the inspection date. The information concerning the last intervention is obtained through the documentation from the Municipalities or provided by the owners of the buildings surveyed.

2.5.2.1 Definition of Degradation Conditions

Currently, there are various methods to assess the degradation condition—often referred to as degradation level—of building's components. Usually, these methods are established based on the importance rating of the construction elements, the rating of the defects and the definition of the condition parameters associated with the defects (Straub 2003). Several authors have established classification systems for defects and degradation ratings in order to express the visual and functional deterioration of the elements analysed (Balaras et al. 2005; Shohet et al. 1999; Marteinsson and Jónsson 1999; Freitas et al. 1999; Brandt and Rasmussen 2002; Chew 2005). Generally, these classification systems consist of rating the defects according to a scale of discrete variables, which varies from the most favourable condition (no visible degradation) to the most unfavourable condition (generalized degradation or loss of functionality). In more simplified approaches, each degradation condition is established based on a set of reference characteristics analysed in real situations during the fieldwork. Even though they are easy to use and interpret, these classification systems present some limitations since the scales adopted provide qualitative degradation parameters only (instead of quantitative ones). To

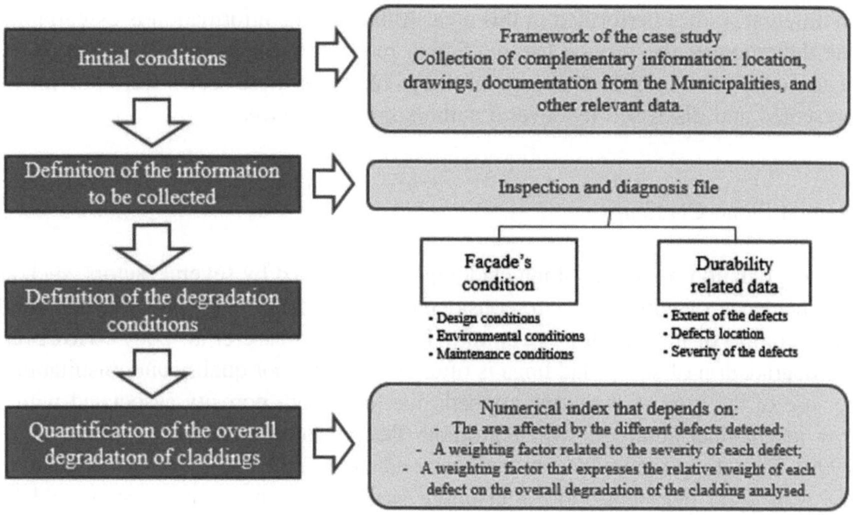

Fig. 2.1 Methodology used in the systemic analysis of the façades degradation condition

overcome these limitations, Shohet and Paciuk (2004) suggest a moderately complex system in which each degradation level is associated with the extent of the defect detected. In these models, the degradation severity is rated based on the defects detected at the inspection only. This method shows some limitations due to the difficulty of knowing with certainty the future consequences of the defects detected; therefore, Florentzou et al. (1999) suggest the implementation of a risk concept in the definition of degradation levels, in which the degradation conditions are defined as a function of the consequences of the problems detected, leading to the definition of risk in probabilistic terms that generally varies from 0 (no probability of occurrence) to 1 (maximum probability of occurrence). Van Winden and Dekker (1998) propose a degradation scale with six levels, where the weight of the proposed degradation levels is not linear; the authors relate each degradation condition to a numerical weight, thus providing an exponential relationship between the degradation levels.

Based on the different degradation scales proposed by various authors, in this study, the degradation conditions of the façade claddings analysed are defined according to the extent of the façade affected by each group of defects and taking into account the importance of the different defects within each group in terms of gravity. The degradation conditions considered vary between A (no visible degradation) and E (generalized degradation); condition D corresponds to the end of the claddings service life, assuming that a cladding with a degradation condition above this level is no longer capable of performing the function for which it was designed and requires comprehensive rehabilitation actions in order to restore its essential properties. In the next sections, the various degradation scales adopted for the different claddings analysed are proposed. These scales were previously defined in different studies performed in this area. Although this information is essential for the definition of the service life prediction models, the detailed discussion of the scales adopted is outside the scope of this book, since these scales were previously presented and discussed by several authors duly referenced.

Natural Stone Claddings

The degradation of natural stone claddings is influenced by several factors, such as the incidence of wind and rain, considered as one of the main physical agents of deterioration of natural stone (Camuffo 1995; Barberousse et al. 2007). However, the degradation of stone claddings is often due to the poor quality and unsuitability for use of the type of the stone applied, due to its high porosity associated with a low mechanical strength, which promote the presence of defects (Shohet et al. 1999). This study focuses on stone directly adhered to the substrate—often glued to the wall—and does not address cladding systems in which the stone is held by fasteners. The defects in stone claddings studied are arranged into four groups (Silva et al. 2011):

- **Visual or surface degradation** (Fig. 2.2)—these defects generally affect the cladding, not contributing to the failure of the cladding system, and do not represent a risk to its integrity. These are the most common defects, appearing prematurely in some cases due to design or execution errors;
- **Joint defects** (Fig. 2.3)—which include situations in which the cladding's joints do not function adequately and may even lead to the occurrence of new defects;
- **Loss of bond to the substrate** (Fig. 2.4)—which characterize the situations when stone material is lost either in small proportions (scaling) or significant proportions (loss of adhesion); these defects usually compromise the integrity of the cladding, jeopardizing the safety demands of the cladding and sometimes constitute a risk for people and property;
- **Loss of integrity** (Fig. 2.5)—which affects irremediably the stone, due to the alteration of the physical and chemical properties of the natural stone, also leading to the visual deterioration of the cladding.

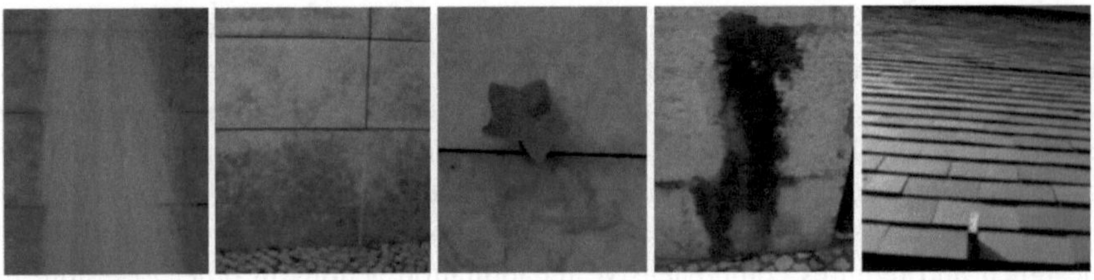

Fig. 2.2 Visual or surface degradation defects in natural stone claddings (from *left* to the *right*): efflorescence, localized stains (rising dump), parasitic vegetation, biological growth and flatness deficiencies

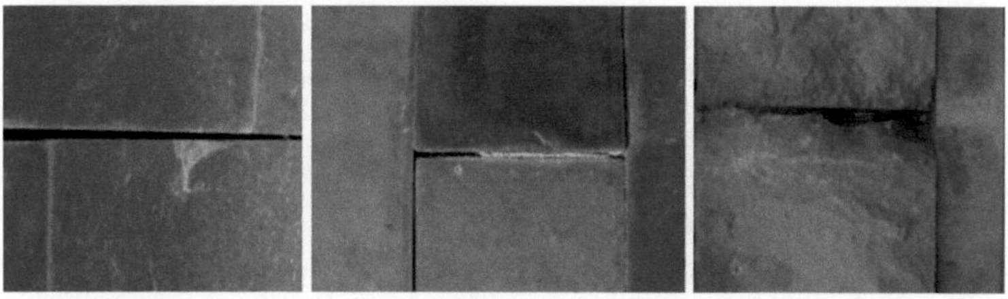

Fig. 2.3 Joint defects (from *left* to the *right*): lack of linearity between the joints of stone plates, material loss—open joint and scaling of stone near the joints

Fig. 2.4 Bond-to-substrate defects in natural stone claddings (from *left* to the *right*): loss of an element in the stone cladding and partial loss of stone material

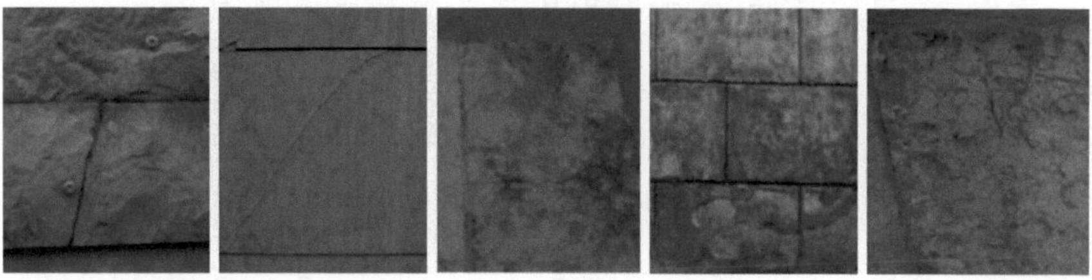

Fig. 2.5 Loss-of-integrity defects in natural stone claddings (from *left* to the *right*): fracture, cracking, erosion, exfoliation and alveolarization of stone plate

Table 2.1 shows the degradation conditions for natural stone claddings.

The degradation levels for the defects in joints and those related to loss of bond to the substrate leap directly from condition A to condition C, because it was considered that conditions below condition B did not represent the gravity of these defects from the moment they are detected.

Table 2.1 Proposed degradation levels for natural stone claddings (data sourced from Silva et al. 2011)

Degradation level	Defects		Area of cladding affected (%)	Illustrative example
Level A ($S_{w,rp} \leq 1$ %)	No visible degradation		–	–
Level B *Good* (1 % < $S_{w,rp} \leq 8$ %)	Visual or surface degradation defects	Surface dirt	>10	
		Moisture stains	≤15	
		Localized stains		
		Colour change		
		Flatness deficiencies	≤10	
	Loss-of-integrity defects	Material degradation[a] ≤ 1 % plate thickness	–	
		Material degradation[a] ≤ 10 % plate thickness	≤20	
		Cracking width ≤ 1 mm		
Level C *Slight degradation* (8 % < $S_{w,rp} \leq 20$ %)	Visual or surface degradation defects	Moisture stains	>15	
		Localized stains		
		Colour change		
		Moss, lichen, algae growth	≤30	
		Parasitic vegetation		
		Efflorescence		
		Flatness deficiencies	>10 and ≤50	
	Joint anomalies	Joint material degradation	≤30	
		Material loss—open joint	≤10	
	Bond-to-substrate defects	Scaling of stone near the edges Partial loss of stone material	≤20	

(continued)

Table 2.1 (continued)

Degradation level	Defects		Area of cladding affected (%)	Illustrative example
	Loss-of-integrity defects	Material degradation[a] $\leq$ 10 % plate thickness	$\leq$20	
		Material degradation[a] > 10 % and $\leq$ 30 % plate thickness	>20	
		Cracking width $\leq$1 mm	$\leq$20	
		Cracking width >1 mm and $\leq$5 mm	$\leq$20	
		Fracture	$\leq$5	
Level D *Moderate degradation* (20 % < $S_{w,rp}$ $\leq$ 45 %)	Visual or surface degradation defects	Moss, lichen algae growth	>30	
		Parasitic vegetation		
		Efflorescence		
		Flatness deficiencies	>50	
	Joint defects	Joint material degradation	>30	
		Material loss—open joint	>10	
	Bond-to-substrate defects	Scaling of stone near the edges Partial loss of stone material	>20	
		Loss of adherence	$\leq$10	
	Loss-of-integrity defects	Material degradation[a] > 10 % e $\leq$ 30 % plate thickness	>20	
		Material degradation[a] > 30 % plate thickness	$\leq$20	
		Cracking width > 1 mm and $\leq$ 5 mm	>20	
		Cracking width $\geq$ 5 mm	$\leq$20	
		Fracture	>5 and $\leq$ 10	

(continued)

Table 2.1 (continued)

Degradation level	Defects		Area of cladding affected (%)	Illustrative example
Level E *Generalized degradation* ($S_{w,rp} \geq 45$ %)	Bond-to-substrate defects	Loss of adherence	>10	
	Loss-of-integrity defects	Material degradation[a] > 30 % plate thickness	>20	
		Cracking width > 5 mm		
		Fracture	>10	

[a]Material degradation is meant to be every anomaly that involves loss of volume of the stone material

Rendered Façades

According to Gaspar and de Brito (2005), the extensive application of renderings as exterior claddings is essentially due to their low cost and the low technology required to their execution when compared to the other types of cladding systems. However, the low investment in this type of cladding often leads to high levels of degradation in the façades (Freitas et al. 1999). The diagnosis of the degradation in rendered façades is not easy, given the variability of causes that can contribute to a given defect. Nevertheless, during their life cycle, rendered façades present a relatively clear degradation pattern, which starts with the presence of surface dirt, soot, stains, and in some situations small cracks, which evolve over time, until the loss of adhesion of the renderings (Gaspar and de Brito 2008). The defects present in rendered façades can be grouped in three main categories

- **Visual or surface degradation defects** (Fig. 2.6)—essentially characterized by the presence of stains, often associated with the presence of damp (Chew and Ping 2003); these defects are the less severe, not compromising the claddings' service life;
- **Cracking** (Fig. 2.7)—these defects are more severe than staining, however, they do not usually jeopardize the claddings' safety and are a natural consequence of the claddings behaviour (Bone et al. 1989; Bonshor and Bonshor 2001);

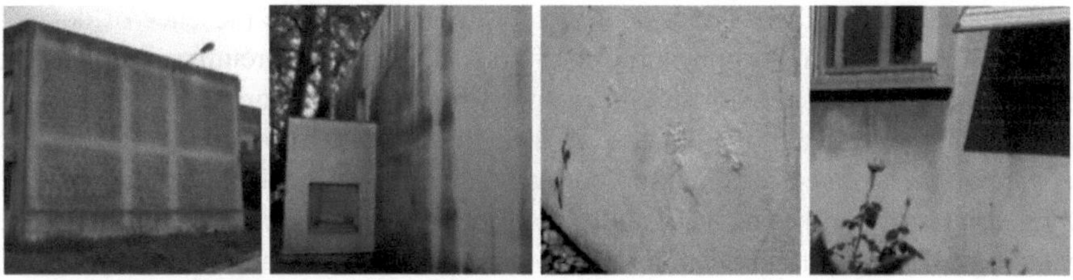

Fig. 2.6 Visual or surface degradation defects in rendered façades (from *left* to the *right*): thermophoresis, biological growth, efflorescence, colour change (data sourced from Gaspar 2009)

Fig. 2.7 Cracking defects: mapped cracking in rendered façades (*left*) and oriented cracking (*right*) (data sourced from Gaspar 2009)

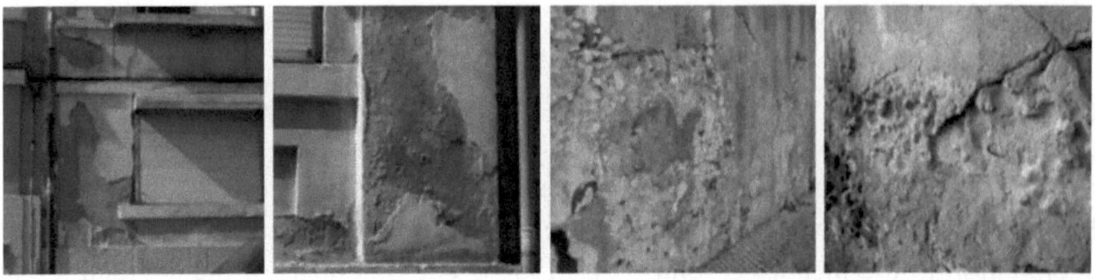

Fig. 2.8 Loss of adhesion or cohesion in rendered façades (from the *left* to the *right*): detachment, pulverulence, arenization, erosion (data sourced from Gaspar 2009)

- **Loss of adhesion or cohesion** (Fig. 2.8)—this group of defects corresponds to the most serious situations and are caused by different factors, such as the combination of cracking with the presence of water (Hansen et al. 1999), often leading to the end of the service life of the renderings.

Based on this classification of the defects, five degradation conditions are established for rendered façades, where condition A corresponds to the most favourable situation and condition E to the most serious one (generalized degradation). Table 2.2 shows the proposed degradation conditions for rendered façades.

Painted Surfaces

In painted surfaces, there are several factors that promote the presence of defects, such as the design and application conditions, the drying circumstances and the environmental exposure conditions, among others. The degradation of painted surfaces not only leads to the loss of their aesthetic performance but may also jeopardize their protective function. For painted surfaces, the degradation levels are defined in terms of the intensity and type of defect, regardless of the extent of the pathological manifestation. The extent is taken into account at a later stage, when the overall level of degradation of the façade coating is defined. In this case, this assumption is adopted since there are several Portuguese standards currently enforced (NP EN ISO 4628-1: 2005; NP EN ISO 462———8-2: 2005; NP EN ISO 4628-4: 2005; NP EN ISO 4628-5: 2005; NP EN ISO 4628-7: 2005) that allow establishing the criteria to evaluate the degradation conditions of painted surfaces. Based on these standards, four main defects that affect paint coatings are considered (Chai et al. 2014)

- **Staining/colour change** (Table 2.3 and Fig. 2.9)—which mainly affect the aesthetic appearance of paintings, usually occurring at the early stages of the paintings service life;
- **Chalking** (Table 2.4)—it generally occurs after loss of gloss and causes wear, detachment and loss of material. Given the difference (in terms of the coating's durability) between small and moderate/high scale chalking, it was decided to adopt a nonlinear degradation scale, considering very low degradation levels

Table 2.2 Proposed degradation levels for rendered façades (data sourced from Silva et al. 2014a)

Condition levels	Severity of degradation (%)	Physical and visual assessment	Illustrative example
Level A	<1	Complete mortar surface with no deterioration. Surface even and uniform. No visible cracking or cracking $\leq$ 0.1 mm. Uniform colour and no dirt. No detachment of elements	
Level B	1–5	Non-uniform mortar surface with likelihood of hollow localized areas determined by percussion, but no signs of detachment. Small cracking (0.25–1.0 mm) in localized areas. Changes in the general colour of the surface. Eventual presence of microorganisms	
Level C	5–15	Localized detachments or perforations of the mortar. Hollow sound when tapped. Detachments only in the socle. Easily visible cracking (1.0–2.0 mm). Dark patches of damp and dirt, often with microorganisms and algae	
Level D	15–30	Incomplete mortar surface due to detachments and falling of mortar patches. Wide or extensive cracking ($\geq$2 mm). Very dark patches with probable presence of algae	
Level E	>30	Incomplete mortar surface due to detachments and falling of mortar patches. Wide or extensive cracking ($\geq$2 mm). Very dark patches with probable presence of algae	

(conditions A and B) and moderate or high levels (levels D and E) of the defect; finally, in advanced degradation situations, very pronounced chalking may cause total or local obliteration of the paint pellicle, leaving the rendering bare; this is considered to signal the end of the cladding's service life (level E);

- **Cracking** (Table 2.5 and Fig. 2.10)—the ranking by degradation level of this defect is based on visual patterns, in terms of cracking frequency. Contrary to

Table 2.3 Definition of degradation levels for stains/colour change defects in painted surfaces (data sourced from Chai et al. 2014)

		Level A	
Intensity of the change		No degradation visible	
Defects characterization		Without visible defects	
Examples			
Description		No perceptible changes	
		Level B	
Intensity of the change		Slight or little perceptible changes	
Defects characterization		Uniform surface dirt; Change in colour or brightness	
Examples			
Description	Little perceptible uniform surface dirt		Little perceptible change in colour
		Level C	
Intensity of the change		Moderate or quite perceptible changes	
Defects characterization		Uniform surface dirt; Change in colour or brightness	
Examples			
Description	Quite perceptible uniform surface dirt		Quite perceptible change in colour
Intensity of the change		Slight or little perceptible changes	
Defects characterization		Localized surface dirt; Humidity stains; Efflorescence	
Examples			
Description		Light humidity stains	
		Level D	
Intensity of the change		Moderate or quite perceptible changes	
Defects characterization		Humidity stains; Efflorescence; Biological growth; Localized surface dirt	
Examples			
Description	Quite perceptible humidity stains		Quite perceptible biological growth
Examples			
Description		Quite perceptible localized surface dirt	Quite perceptible efflorescence
Intensity of the change		High or very perceptible changes	
Defects characterization		Uniform surface dirt; Change in colour or brightness	
Examples			
Description	Very perceptible uniform surface dirt and change in colour	Very perceptible Localized surface dirt	Very perceptible change in colour (discoloration)
		Level E	
Intensity of the change		High or very perceptible changes	
Defects characterization		Biological growth	
Examples	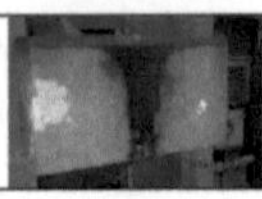		
Description		Very perceptible biological growth	

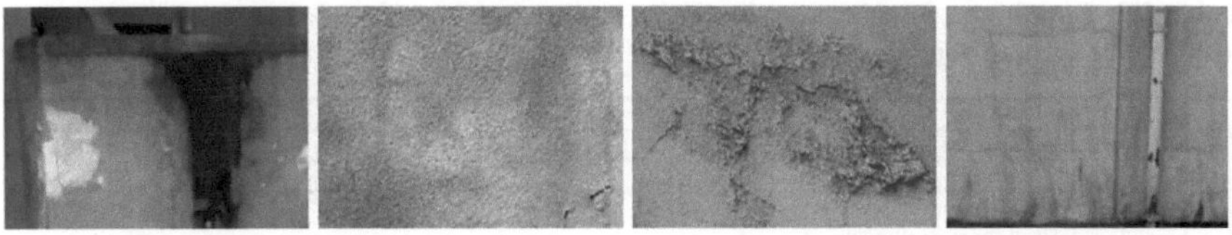

Fig. 2.9 Visual or surface degradation defects in painted surfaces (from *left* to the *right*): biological growth, colour change, efflorescence and damp stains (data sourced from Chai 2011)

Table 2.4 Definition of degradation levels for chalking defects (data sourced from Chai et al. 2014)

Degradation level	Quantity
Level A	No degradation visible
Level B—Good	Clearly perceptible
Level D—Moderate degradation	Quite perceptible
Level E—Generalized degradation	Very perceptible

other materials, in paintings, cracking is usually less than 1 mm wide, which makes it difficult to distinguish the size without magnifying equipment (Table 2.6);

- **Loss of adherence** (Table 2.7 and Fig. 2.11)—these defects usually arise from the combination of various other defects and degradation mechanisms. Therefore, two situations are considered: (i) blistering and (ii) peeling. Due to the severity of these defects, every blistering occurrence has a severity degree of C or higher and every peeling occurrence is condition D;

Ceramic Tiling Systems

Ceramic tiles present a huge variety of sizes, colours, textures, among other characteristics, thus being considered in Mediterranean countries as one of the noblest materials for cladding (Silvestre and de Brito 2009; Bovea et al. 2010). Although there are very old buildings that testify the durability of this type of cladding (e.g. buildings more than a hundred years old with their ceramic cladding intact), in recent decades there has been a significant decrease in the use of this type of cladding. Various authors (Mansur et al. 2008; Wetzel et al. 2010) refer as the main reason for the decline in the use of ceramic tiling system the high incidence of defects throughout their life cycle. Shohet and Laufer (1996) refer that adherent ceramic tiles are extremely susceptible to design and execution errors, as well as to the choice of materials applied. Bordalo et al. (2011) also mention other causes for the premature failure of these claddings, such as the growing demands of the construction market, which implies the reduction of the construction delivery times and the poor education of technicians involved at design and execution phases.

Table 2.5 Definition of degradation levels for peeling and blistering defects in painted surfaces (data sourced from Chai et al. 2014)

Level A			
Defect characterization	No degradation visible		
Level C			
Defect characterization	Blistering		
Quantity and size of the areas affected (greatest size)	Small amount and size up to 3 cm		
Visual scale [NP EN ISO 4628-2, 2005 e NP EN ISO 4628-5, 2005]			
Level D			
Defect characterization	Blistering		
Quantity and size of the areas affected (greatest size)	Small amount and size between 3 and 5 cm	Moderate amount and smaller than 3 cm	
Visual scale [NP EN ISO 4628-2, 2005 e NP EN ISO 4628-5, 2005]			
Defect characterization	Peeling		
Quantity and size of the areas affected (greatest size)	Small amount (area affected up to 1%) e size up to 3 cm		
Visual scale [NP EN ISO 4628-2, 2005 e NP EN ISO 4628-5, 2005]			
Level E			
Defect characterization	Blistering		
Quantity and size of the areas affected (greatest size)	Larger than 5 cm, regardless of the amount	Dense pattern regardless of the size	Moderate amount e and dimension between 3 and 5 cm
Visual scale [NP EN ISO 4628-2, 2005 e NP EN ISO 4628-5, 2005]			
Defect characterization	Peeling		
Quantity and size of the areas affected (greatest size)	Dense and moderate pattern (affected area higher than 1%) regardless of the size	Small amount and larger than 5 cm	
Visual scale [NP EN ISO 4628-2, 2005 e NP EN ISO 4628-5, 2005]			

Fig. 2.10 Cracking defects in painted surfaces (data sourced from Chai 2011)

Table 2.6 Definition of degradation levels for cracking defects in painted surfaces (data sourced from Chai et al. 2014)

Degradation level	Level A	Level B Good condition	Level C Slight degradation	Level D Moderate degradation	Level E Generalized degradation
Quantity	No visible degradation	Small number of cracks	Moderate width or number of cracks	Considerable number of cracks	High number or density of cracks
Visual scale [NP EN ISSO 4628-4, 2005]		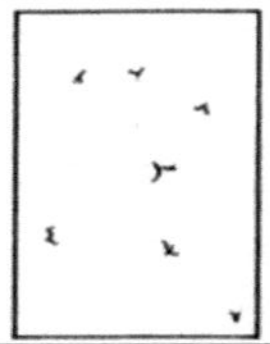	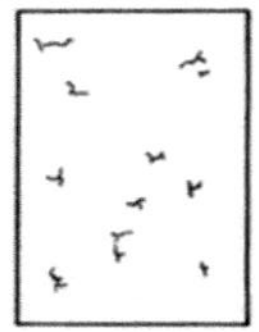	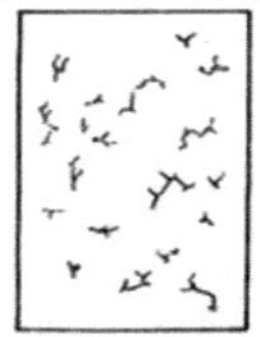	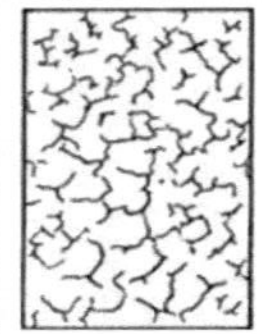

Table 2.7 Proposed degradation levels for ceramic tiling systems (data sourced from Bordalo et al. 2011)

Condition level	Defects		Area of cladding affected (%)	Illustrative example
Level A ($S_{w,rp} \leq 1$ %)	No visible degradation		–	–
Level B *Good* (1 % < $S_{w,rp} \leq 6$ %)	Visual or surface degradation defects	Surface dirt	–	
		Small surface craters	≤10	
		Wear or scratches		
		Crushing or scaling of the borders		
		Change of shine and/or colour		
		Damp stains		
	Cracking	Cracked glazing[a]	–	
		Markedly orientated cracking (<0.2 mm)[(1)] without leakage[a]		
	Joint deterioration	Staining or change in colour	–	
Level C *Slight deterioration* (6 % < $S_{w,rp} \leq 20$ %)	Visual or surface degradation defects	Small superficial craters	>10 and ≤50	
		Wear or scratches		
		Crushing or scaling of the borders		
		Change of shine and/or colour		
		Damp stains		
		Biological growth	≤30	
		Graffiti		
		Efflorescence		
	Cracking	Cracking with no predominant direction[a]	≤30	
		Markedly orientated cracking (>0.2 mm)[(2)] without leakage[a]		
	Joint deterioration	Without loss of filling material[a]	≤30	
		With loss of filling material[a]	≤10	
	Detachment	Loss of adherence	≤20	
		Swelling		

(continued)

Table 2.7 (continued)

Condition level	Defects		Area of cladding affected (%)	Illustrative example
Level D *Moderate degradation* (20 % < $S_{w,rp}$ ≤ 50 %)	Visual or surface degradation defects	Small superficial craters	>50	
		Wear or scratches		
		Crushing or scaling of the borders		
		Change of shine and/or colour		
		Damp stains		
		Biological growth	>30	
		Graffiti		
		Efflorescence		
	Cracking	Cracking with no predominant direction[a]	>30 and ≤ 50	
		Markedly orientated cracking (>1 mm)[3] without leakage[a]		
	Joint deterioration	Without loss of filling material[a]	>30 and ≤ 50	
		With loss of filling material[a]	>10 and ≤ 30	
	Detachment	Loss of adherence	>20	
		Swelling		
		Localized detachment	≤ 10	
Level E *Generalized degradation* ($S_{w,rp}$ ≥ 50 %)	Cracking	Cracking with no predominant direction[a]	>50	
		Markedly orientated cracking (>5 mm)[4]		
	Joint deterioration	Without loss of filling material	>50	
		With loss of filling material	>30	
	Detachment	Generalized detachment	>10	

[a]With leakage—the degradation level is increased by one
[1]Cracking, detectable at a distance greater than 5 m only if binoculars are used
[2]Tenuous cracking line, easily detectable at a distance greater than 5 m, using binoculars
[3]Well defined cracking visible from a distance of more than 5 m, without using binoculars
[4]Cracking characterized by a thick line in which a clear separation of the borders can be seen, from a distance of more than 5 m, with the aid of binoculars

According to Campante and Paschoal (2002), the main defects observed in ceramic claddings are efflorescence, detachment and cracking. In turn, Silvestre and de Brito (2009) classify the defects in four groups: (i) adhesion failure and/or detachment of ceramic claddings; (ii) cracking; (iii) deterioration of the cladding

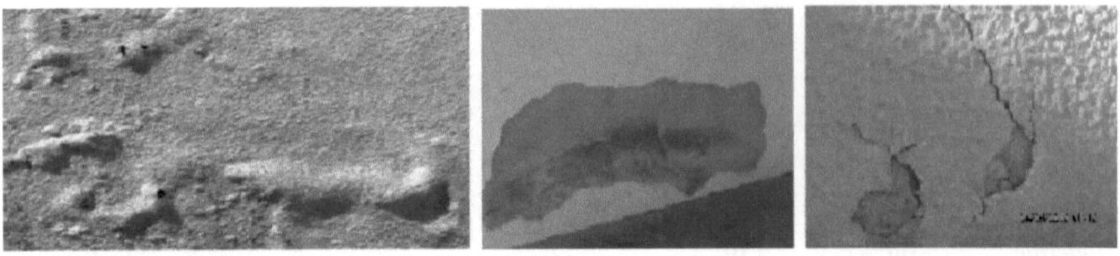

Fig. 2.11 Loss of adhesion or cohesion: debonding (*left*) and detachment in painted surfaces (*right*) (data sourced from Chai 2011)

tiling system, with direct consequences in its performance; (iv) defects of a strictly aesthetic nature. Besides these defects, Timellini and Palmonari (1989) mention the relevance of joints defects in the overall performance of this cladding system. Silvestre and de Brito (2007) also refer that there is a high incidence of defects in the material used to fill the joints, since this material is highly susceptible to the aggressive deterioration agents. Considering the literature, the defects observed in ceramic tiling systems are classified into four categories (Bordalo et al. 2011)

- **Visual defects** (Fig. 2.12)—which affect the visual appearance of the cladding, although they do not usually determine the end of the service life of ceramic claddings;
- **Cracking** (Fig. 2.13)—which is divided in three subcategories, taking into account the characteristics of each type of cracking:

Fig. 2.12 Visual or surface degradation defects in ceramic tiling systems (from *left* to the *right*): localized colour change, parasitic vegetation, efflorescence, crushing or scaling of the borders and small superficial craters (data sourced from Bordalo 2008)

Fig. 2.13 Cracking defects in ceramic tiling systems (from *left* to the *right*): cracked glazing, markedly orientated cracking and with no predominant direction (data sourced from Bordalo 2008)

Fig. 2.14 Joint deterioration defects in ceramic tiling systems (from *left* to the *right*): cracking of the filling material, cracking between the filling material and the edges of the ceramic tile and loss of filling material (data sourced from Bordalo 2008)

Fig. 2.15 Detachment defects in ceramic tiling systems: debonding (*left*) and detachment of ceramic tiles (data sourced from Bordalo 2008)

 - Glazing cracking, looking like a spider's web, which is a surface problem being usually caused by the ageing of the tiles or by the difference between the thermal expansion coefficients of the glazing and of the inner body of the tile;
 - Cracking with no predominant direction, which is usually more superficial and affects large portions of the cladding surface;
 - Markedly orientated cracking, which is usually local, deep and wide;

- **Joint deterioration defects** (Fig. 2.14)—which are usually related with the deterioration of the filling material; they compromise the overall performance of ceramic tiles, since joints are responsible for the system's tightness and ability to absorb deformations;
- **Detachment** (Fig. 2.15)—this is the most serious defect on ceramic tiling due to the hazardous consequences that may occur if ceramic elements fall from the façade and the cost of repairs (Lo 2002). Loss of adherence, swelling and detachment are the main defects present in this group.

Table 2.7 presents the proposed degradation levels defined for ceramic tiling systems, according to the defects previously described, their extent in the façade and the severity of their appearance.

2.5.3 Degradation of Façade Claddings

Sjöström (1985) states that service life prediction should include the definition of a mathematical model to quantify the degradation actually observed. Gaspar (2009) refers that it is possible to assess the overall level of degradation of any construction using a quantitative index that reflects the overall performance of the element under analysis. Therefore, Gaspar (2009) and Gaspar and de Brito (2008, 2011) proposed the definition of a numerical index, called as severity of degradation (S_w), to quantify the overall degradation of the façades claddings.

The severity of degradation is given by the ratio between the weighted degraded area and a reference area, equivalent to the total cladding area having the highest possible level of degradation as shown in Eq. (2.1). The weighted degraded area is given by the product of the façade area affected by different groups of defects, by a weighting factor related to the severity of each detected defect—k_n, and a weighting factor that reflects the relative weight of each defect on the overall degradation of the façade—$k_{a,n}$. This method, originally developed for rendered façades, established a general framework model, allowing its subsequent adaptation to the specific case of stone claddings (Silva et al. 2011), ceramic cladding systems (Bordalo et al. 2011), painted surfaces (Chai et al. 2014) and ETICS (Ximenes et al. 2015).

$$S_w = \frac{\Sigma\left(A_n \times k_n \times k_{a,n}\right)}{A \times \Sigma\left(k_{\max}\right)} \tag{2.1}$$

where S_w represents the severity of degradation of the façade, as a percentage, A_n the area of the cladding affected by a defect n, in m^2, k_n the defects' "n" multiplying factor, as a function of its condition (between 1 and 4), $k_{a,n}$ the weighting coefficient corresponding to the relative importance of each defect obtained through cost of repair or risk ($k_{a,n} \in R^+$) (if no instructions are provided, it is assumed $k_{a,n} = 1$), k the weighting factor equal to the highest condition level, and A the total area of the cladding, in m^2.

2.5.3.1 Relative Weighting of Defects

Different defects that appear in the façade claddings during their life cycle have different levels of severity. Indeed, although two defects can affect the same extent of cladding, e.g. localized staining and detachment, they represent different damage types, with distinct levels of risk and type of repair work required. The application of weighting coefficients depending on the severity of the defects allows obtaining an indicator of the overall degradation of façades more appropriate to the actual state of decay of the inspected facades. Therefore, in this study the defects are weighted based on the fulfilment of performance criteria or minimum requirements (safety and watertightness), their susceptibility to generate new defects and essentially based on their repair cost. The latter variable seems to be a preponderant parameter in the perspective of decision-makers. According to Rudbeck (2002), the cost is a fundamental aspect, indeed even more relevant than durability, to guide decision-making process throughout the different phases of the construction life cycle.

In the next sections, the weighting factors are established based on the cost of repair of the different defects, using present-day values, at the moment when repair is needed. The cost of repair is calculated as the ratio between the sum of the costs of each operation within the required intervention and the cost of replacing the cladding. Repair costs are estimated based on available literature, expert opinion and cost simulators, in order to adjust the model to the Portuguese market.

Natural Stone Claddings

Table 2.8 presents the weighting coefficients for natural stone claddings. The replacement of stone plates entails costs that easily equal or surpass the cost of a new cladding due to the need to remove of the old cladding.

However, a cladding's replacement is not a consensual decision; in fact, in some cases, owners choose to fasten the stone plates detached from the support or to change the fastening system, rather than replace the entire cladding. Also regarding loss of integrity of the stone elements, in some cases, owners opt for clogging the degraded stone with epoxy resins, instead of replacing them. Therefore, the costs of maintenance/repair present some subjectivity, since they depend on the acceptance criteria of users and owners.

Rendered Façades

Regarding rendered façades, the distinction between the three groups of defects is based on the cost of repair, ranging from simple cleaning, to remove stains and soot, repair of cracks or, in more serious situations, to repair and substitute areas affected by the loss of cohesion of mortar or detached. Table 2.9 presents the weighting coefficients adopted for rendered façades, revealing, as expected, a correlation between the cost of repair of the different defects and the severity of these defects empirically perceived.

Painted Surfaces

For painted surfaces, the quantification of the weighting coefficients is made iteratively, based on the analysis of different scenarios; the scenario with the best results is selected regarding its ability to translate the physical reality observed during the fieldwork. In this case, more qualitative criteria are adopted based on the study performed by Chai et al. (2014). A growing hierarchy of the defects "staining and colour changes", "cracking", "chalking" and "loss of adhesion" is adopted as shown in Table 2.10. When comparing this scenario with a neutral scenario

Table 2.8 Weighting coefficients ($k_{a,n}$) for natural stone claddings (data sourced from Silva et al. 2011)

Defect	Performance criteria Requirements		Possibility of generating new anomalies	Repair operation (cost in €/m²)	Ratio between repair cost and replacement cost[a] (%)	Weighting coefficient $k_{a,n}$
	Safety	Watertightness				
Visual or surface degradation	○○	●○	●○	Cleaning (11.7 €/m²)	13	0.13
Joints						
Degradation of filling material	●○	●○	●●	Joint repair (23.4 €/m²)	25	0.25
Loss of filling material				Replacement of the joint material in cladding directly adhered to the substrate involves some risks, and may damage the natural stone	100	1.0
Bond-to-substrate	●●	●●	●●	Replacement of stone plates always costs at least as much as executing a new cladding, and may cost more because of having to remove the degraded original cladding	120	1.2
Loss of integrity	●●	●●	●●	Repairing loss-of-integrity anomalies may involve only surface repair (epoxy resins or equivalent) or replacement of the stone plate	100	1.0

○○—No correlation; ●○—probable correlation; ●●—high correlation

[a]The cost of building a vertical granite cladding façade with a cementitious adhesive is around 93 €/m²

Table 2.9 Weighting coefficients ($k_{a,n}$) for rendered façades (data sourced from Gaspar and de Brito 2011)

Degradation condition	Stains		Cracking		Loss of adhesion	
B	$k_{a,n} = 0.12$	2.50 €/m²	$k_{a,n} = 0.95$	20.50 €/m²	$k_{a,n} = 1.53$	33.00 €/m²
C	$k_{a,n} = 0.53$	11.50 €/m²	$k_{a,n} = 0.95$	20.50 €/m²	$k_{a,n} = 1.53$	33.00 €/m²
D	$k_{a,n} = 0.53$	11.50 €/m²	$k_{a,n} = 1.12$	24.00 €/m²	$k_{a,n} = 1.53$	33.00 €/m²
E	$k_{a,n} = 0.53$	11.50 €/m²	$k_{a,n} = 1.53$	33.00 €/m²	$k_{a,n} = 1.53$	33.00 €/m²

Table 2.10 Weighting coefficients ($k_{a,n}$) for painted surfaces (data sourced from Chai et al. 2014)

Defect	Stains/colour change	Cracking	Chalking	Loss of adherence
$k_{a,n}$	0.25	1.00	1.00	1.50

(corresponding to the same severity for all defects—$k_{a,n} = 1$), the following conclusions can be drawn: (i) without weighting, the values of the severity of degradation of painted surfaces present a high scatter; (ii) the values of the severity of degradation without weighting are generally higher than the values with weighting, confirming the hypothesis that the model without weighting portrays a pessimistic view of the physical reality; (iii) the degradation model with weighting varies between 0 and 100 %, thus producing results easier to interpret.

Ceramic Claddings

Concerning adherent ceramic claddings, the same criteria applied to stone claddings is adopted, giving particular relevance to the repair cost associated to the different defects that affect the ceramic tiling systems (Table 2.11).

2.5.3.2 Definition of the End of Service Life of Façade Claddings

As already referred, the definition of the end of service life is not an easy task, depending on often subjective and context dependent acceptance criteria, which are not easily established by "pure" scientific methods. In this study, the maximum acceptable degradation level is established based on the careful analysis of the degradation condition of the claddings analysed, so that coherent values can be obtained for the Portuguese context.

From the analysis of each case study analysed (considering the four claddings under analysis, a sample of 718 case studies is evaluated), a critical degradation value was adopted equal to a severity of degradation of 20 %, beyond which the probability of no longer fulfilling the essential requirements is too high. In fact, in the analysis performed, the adoption of a maximum degradation level of 10 % seems too conservative to establish the end of the service life of the claddings

Table 2.11 Weighting coefficients ($k_{a,n}$) for painted surfaces (data sourced from Bordalo et al. 2011)

Defect	Performance criteria Requirements		Possibility of generating new anomalies	Repair operation (cost in €/m^2)	Ratio between repair cost and replacement cost[a]	Weighting coefficient $k_{a,n}$
	Safety	Watertightness				
Visual or surface degradation	○○	●○	●○	Cleaning (13.09 €/m^2)	18	0.18
Cracking	●●	●●	●●	The repair of cracking in ceramic tiles may involve different types of intervention, ranging from superficial clogging of the affected material to replacement of the degraded tiles.	100	1.0
Joints						
Degradation of filling material	●○	●○	●●	Joint repair (23.4 €/m^2)	32	0.32
Loss of filling material				Replacement of the joint material in cladding directly adhered to the substrate involves some risks, and may damage the natural stone	100	1.0
Loss of adhesion	●●	●●	●●	Replacement of ceramic tiles always costs at least as much as executing a new cladding, and may cost more because of having to remove the degraded original cladding	120	1.2

○○—No correlation; ●○—probable correlation; ●●—high correlation

[a]The cost of building a ceramic tiling system substantially varies according to the cost of the tile chosen. Assuming a cladding constituted by stoneware tiles with 40 × 40 cm, with a cost of 19 €/m^2, an execution cost around 75.57 €/m^2 is estimated

Fig. 2.16 Case study that illustrates the end of the service life of stone claddings (with a severity of degradation around 20 %) (data sourced from Silva 2009)

Fig. 2.17 Case study that illustrates the end of the service life of rendered façades (with a severity of degradation around 20 %) (data sourced from Gaspar 2009)

analysed. In turn, the adoption of a maximum degradation level of 30 or 40 % seems too high, revealing a generalized degradation state. Nevertheless, the maximum degradation level adopted in this study may vary according to the level of demand of the owners or users or even the social context of the building. Figures 2.16, 2.17, 2.18 and 2.19 illustrate the degradation condition that corresponds to the end of the service life of stone claddings, rendered façades, painted surfaces and ceramic claddings, respectively.

Fig. 2.18 Case study that illustrates the end of the service life of painted surfaces (with a severity of degradation around 20 %) (data sourced from Chai et al. 2014)

Fig. 2.19 Case study that illustrates the end of the service life of ceramic claddings (with a severity of degradation around 20 %) (data sourced from Bordalo 2008)

2.6 Characterization of the Samples Analysed

In this study, the samples analysed are as homogeneous as possible regarding the age of the claddings, their characteristics and the degradation agents to which they are subjected. All the case studies that present high degradation levels at an early stage, apparently caused by gross design or execution errors or due to vandalism were excluded, since their degradation is caused by discrete phenomena, which are not possible to predict or model. Concerning the age of the claddings, the sample covers a large range of ages, homogeneously distributed. Figure 2.20 shows the distribution of the samples according to the age of the façades analysed.

For rendered façades, the age of the case studies (meaning the time since the last major repair or maintenance intervention) ranges between 0 and 57 years. 31 % of the sample present ages lower than 5 years and are therefore within the warranty period of buildings in Portugal (this information is relevant since some early defects are corrected during the warranty period). The majority of the rendered façades (77 % of the sample) has less than 20 years. Between years 20 and 60, the sample is distributed with relative uniformity, with around four buildings in each year (however, there are intervals of less than 5 years, without any façade inspected). For painted surfaces, the claddings analysed present ages between 0 and 18 years,

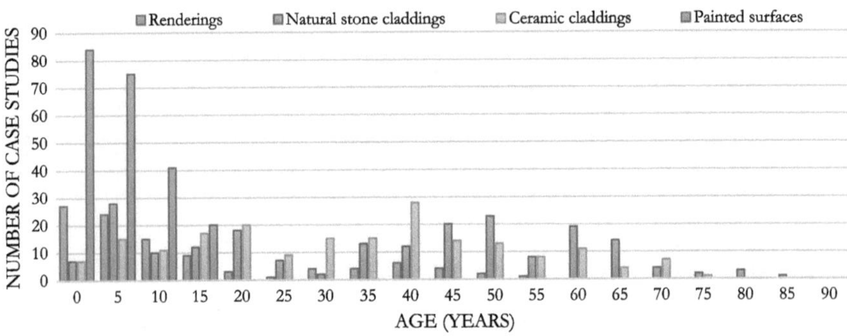

Fig. 2.20 Distribution of the samples according to the age of the façades analysed

with a high incidence of cases with less than 5 years (55 % of the sample). Regarding natural stone claddings, the sample analysed presents ages between 1 and 89 years. Ceramic tiling systems present ages between 1 and 77 years. In these two types of claddings, the majority of the sample has ages between 20 and 60 years (52 % of the sample for stone claddings and 61 % for ceramic claddings), and only 8 % of the sample has ages below 5 years. These data provide some indications regarding the durability of the types of cladding analysed. The samples corresponding to rendered façades and painted surfaces present a smaller range of ages, i.e. these samples have a lower average age when compared with stone or ceramic claddings. Since this study does not contemplate maintenance actions, it is not easy to obtain case studies of renderings or paintings with advanced ages that have not yet been subjected to any intervention.

Figure 2.21 shows the distribution of the severity of degradation of the samples analysed according to their age. Rendered façades are those that have higher degradation indexes, however, for this type of cladding, it has been considered the possibility of overlapping of the defects, which implies that the severity of degradation can be higher than 100 %. For the other types of claddings, the overlapping of defects was not considered, thus normalizing the severity of degradation index, assuming that in the worst case scenario the severity of degradation does not exceed 100 %. This methodological distinction is inherent to the fact that the samples analysed have been collected in previous studies; in fact, this book is the result of an extensive work, in constant progression, whose methodology, concerning the collection of data and the quantification of the degradation, has been evolving over time. By the analysis of the results of natural stone claddings, ceramic claddings and painted surfaces, the following conclusions can be drawn: (i) painted surfaces are those that have higher degradation indexes in earlier ages, with an average severity of degradation of 14.8 % and a maximum value 72 %; (ii) adherent ceramic claddings present an average severity of degradation of 13.1 %, with a maximum value of 61 %; (iii) finally, natural stone claddings present the lowest degradation indexes, with an average severity of degradation of 8.5 % and a maximum value of 43 %. A brief characterization of the samples analysed is presented in the next sections.

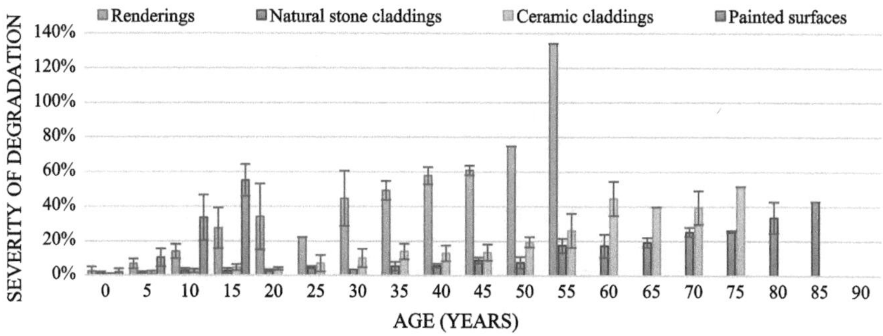

Fig. 2.21 Distribution of the severity of degradation obtained for the claddings analysed according to their age

2.6.1 Natural Stone Claddings

Table 2.12 presents the characterization of the sample analysed for natural stone claddings. During the collection of the sample, case studies were selected randomly and then filtered in order to obtain a representative sample of the types of natural stone found in Portuguese claddings. According to a study performed by Esteves

Table 2.12 Characterization of the sample analysed for natural stone claddings

Characteristics of natural stone claddings		% of case studies (number of cases)	Average ages (standard deviation), in years	Average S_w (minimum/maximum)
Type of stone	Limestone	35 % (72)	41.8 (25.5)	9.7 % (0 %/37 %)
	Granite	27 % (54)	23.9 (13.3)	3.3 % (0 %/12 %)
	Marble	38 % (77)	43.8 (20.9)	11.0 % (0 %/43 %)
Colour	Light	66 % (134)	42.0 (23.9)	10.0 % (0 %/43 %)
	Dark	34 % (69)	29.7 (17.1)	5.6 % (0 %/37 %)
Type of finishing	Smooth	47 % (96)	29.9 (19.5)	5.7 % (0 %/30 %)
	Rough	53 % (107)	44.9 (22.9)	11.1 % (0 %/43 %)
Dimension of stone plates	Medium sizes	64 % (129)	31.7 (21.6)	6.1 % (0 %/28 %)
	Large sizes	36 % (74)	48.5 (20.2)	12.7 % (0 %/43 %)
Thickness of stone plates	Less than 2.5 cm	76 % (154)	35.4 (19.7)	7.2 % (0 %/43 %)
	≥2.5 cm	24 % (49)	45.2 (28.9)	12.5 % (0.1 %/29 %)
Type of cladding	Integral/partial	30 % (61)	21.0 (21.8)	5.0 % (0 %/25 %)
	Bottom wall	70 % (142)	45.0 (18.9)	10.0 % (0 %/43 %)
Façade orientation	North	14 % (29)	29.3 (21.4)	6.6 % (1 %/37 %)
	NE/E/SE	43 % (88)	39.1 (21.5)	8.0 % (0 %/30 %)
	W/NW	25 % (51)	35.5 (21.9)	8.9 % (0 %/40 %
	S/SW	17 % (35)	44.8 (25.3)	10.9 % (0 %/43 %)
Distance from the sea	Less than 5 km	38 % (77)	38.4 (28.5)	11.1 % (0 %/43 %)
	More than 5 km	62 % (126)	37.4 (18.2)	7.0 % (0 %/40 %)
Wind-rain action	Moderate	84 % (170)	39.3 (23.2)	9.2 % (0 %/43 %)
	Severe	16 % (33)	30.1 (17.7)	5.0 % (1 %/14 %)
Exposure to damp	Low	59 % (119)	44.1 (19.7)	9.2 % (0 %/43 %)
	High	41 % (84)	28.8 (23.5)	7.6 % (0 %/37 %)
Type of use	Commerce and services	49 % (99)	35.6 (20.6)	7.6 % (0 %/28 %)
	Housing	51 % (104)	39.9 (24.3)	9.4 % (0 %/43 %)
Ease of inspection	Current	54 % (110)	44.2 (24.6)	11.1 % (0 %/43 %)
	Unfavourable	46 % (93)	30.3 (17.2)	5.4 % (1 %/37 %)

(2007), the overall consumption (production minus exportation plus importation) of natural stone in the Portuguese market in 2003 comprised 22.6 % marble and similar, 13.9 % granite and similar and 63.5 % pavement stone. By analysing only data common to both studies, i.e. admitting that the sample comprises only granite and marble, the study by Esteves (2007) leads to 62 and 38 % of marbles and granites, respectively. In this study, the sample analysed comprises a 59 % of marbles and 41 % of granites which was considered representative of the local context. The majority of the cases analysed has medium-sized stone plates (64 %) and thicknesses lower than 2.5 cm (76 %).

Regarding the orientation of façades, there is a predominance of case studies facing East, Southeast, and Northeast (43 %); the remaining cases are equally distributed by the other categories. Considering the environmental exposure conditions, the following conclusions can be drawn: (i) the majority of the case studies are located at more than 5 km from the sea (62 %); (ii) 84 % of the sample present a moderate exposure to wind-rain action. The sample analysed displays an almost uniform distribution in terms of exposure to damp, type of use and ease of inspection.

2.6.2 Rendered Façades

The characterization of the sample analysed for rendered façades is shown in Table 2.13. The sample analysed mainly consists of current cement renderings (60 %), followed by lime–cement renderings (21 %), single-layer renderings (13 %) and renderings with crushed marble (6 %). Lime–cement renderings correspond to older buildings (with a higher average age), thus presenting the highest values of the severity of degradation (48.6 %). The majority of the sample analysed correspond to housing buildings. Regarding the colour of the façade, white renderings represent 19 % of the sample; light colours are the most common (48 %). Concerning building geometry, the sample is well distributed between compact buildings (without protruding and exceptional construction elements) and irregular buildings.

The majority of rendered façades is located in buildings with eave's and ground floor protection, and without platbands or balconies copings. The majority of the rendered façades analysed presents medium design level (65 %) and only a small percentage of façades (6 %) presents a superior design level (corresponding to more recent buildings, with small average degradation indexes). The sample is relatively well distributed in terms of façades' orientation. The majority of the façades analysed are located at more than 5 km of the sea (77 %), in urban areas, close to traffic routes, thus presenting normal or unfavourable exposure conditions to the pollution sources. The majority of the case studies present favourable conditions of exposure to damp (66 %). Finally, the sample is distributed with relative uniformity for the variables façade protection level and ease of inspection.

Table 2.13 Characterization of the sample analysed for rendered façades

Characteristics of rendered façades		% of case studies (number of cases)	Average ages (standard deviation), in years	Average S_w (minimum/maximum)
Render type	Lime-cement renderings	21 % (21)	32.0 (17.0)	48.6 % (1 %/134 %)
	Current cement renderings	60 % (60)	11.6 (11.3)	13.9 % (0 %/58 %)
	Renderings with crushed marble	6 % (6)	9.1 (10.6)	8.5 % (1 %/32 %)
	Single-layer renderings	13 % (13)	7.0 (4.5)	7.9 % (0 %/23 %)
Colour	White	19 % (19)	8.7 (5.5)	12.5 % (0 %/49 %)
	Light colours	48 % (48)	18.5 (17.0)	25.4 % (0 %/134 %)
	Dark colours	33 % (33)	14.0 (14.3)	16.9 % (0 %/63 %)
Building geometry	Compact	41 % (41)	14.7 (14.4)	18.0 % (0 %/62 %)
	Irregular	59 % (59)	15.4 (15.3)	21.6 % (0 %/134 %)
Eaves' protection	Without protection	47 % (47)	12.3 (13.6)	17.0 % (0 %/134 %)
	With protection	53 % (53)	17.6 (15.7)	22.8 % (0 %/83 %)
Platbands copings	Without copings	79 % (79)	16.2 (15.4)	22.2 % (0 %/134 %)
	With copings	21 % (21)	11.1 (12.1)	12.1 % (0 %/67 %)
Balcony copings	Without copings	66 % (66)	16.5 (16.3)	23.4 % (0 %/134 %)
	With copings	34 % (34)	12.6 (11.4)	13.8 % (0 %/52 %)
Ground floor protection (socle)	Without protection	17 % (17)	10.7 (13.0)	14.6 % (0 %/67 %)
	With protection	83 % (83)	16.1 (15.2)	21.2 % (0 %/134 %)
Detailing/design level	Inferior	29 % (29)	14.3 (13.0)	20.2 % (0 %/63 %)
	Medium	65 % (65)	15.9 (16.1)	21.1 % (0 %/134 %)
	Superior	6 % (6)	10.5 (9.8)	9.2 % (1 %/32 %)
Façade orientation	E/SE	25 % (25)	13.5 (13.9)	17.2 % (0 %/64 %)
	N/NE	34 % (34)	17.8 (16.2)	26.1 % (0 %/134 %)
	W/NW	18 % (18)	14.2 (15.1)	19.2 % (0 %/83 %)
	S/SW	23 % (23)	13.8 (14.1)	15.1 % (0 %/63 %)
Distance from the sea	Less than 5 km	23 % (23)	18.1 (14.1)	26.4 % (0 %/83 %)
	More than 5 km	77 % (77)	14.3 (15.1)	18.2 % (0 %/134 %)

(continued)

Table 2.13 (continued)

Characteristics of rendered façades		% of case studies (number of cases)	Average ages (standard deviation), in years	Average S_w (minimum/maximum)
Exposure to damp	Unfavourable	11 % (11)	12.3 (5.2)	19.2 % (5 %/49 %)
	Normal	23 % (23)	22.5 (17.8)	34.0 % (0 %/134 %)
	Favourable	66 % (66)	13.1 (14.2)	15.4 % (0 %/64 %)
Distance from pollution sources	Unfavourable	24 % (24)	14.4 (13.2)	19.8 % (0 %/61 %)
	Normal	70 % (70)	16.1 (15.8)	21.3 % (0 %/134 %)
	Favourable	6 % (6)	6.7 (5.4)	7.8 % (0 %/20 %)
Façade protection level	Without protection	46 % (46)	15.8 (15.4)	22.7 % (0 %/64 %)
	Normal situation	24 % (24)	14.2 (15.8)	18.1 % (0 %/134 %)
	With protection	30 % (30)	14.9 (13.8)	17.7 % (0 %/67 %)
Type of property	Private	82 % (82)	15.1 (15.5)	19.8 % (0 %/134 %)
	Public sector	14 % (14)	18.4 (12.2)	26.4 % (5 %/63 %)
	Commerce and services	4 % (4)	7.0 (0.8)	5.3 % (1 %/9 %)
Ease of inspection	Normal	59 % (59)	19.6 (16.4)	26.5 % (0 %/134 %)
	Unfavourable	41 % (41)	8.9 (9.1)	11.0 % (0 %/58 %)

2.6.3 Painted Surfaces

Table 2.14 shows the characteristics of the sample analysed for painted surfaces. Plain paints and elastic membranes are the most common type of paint, representing three-fourth of the sample. Silicate and silicone paints are the less common type of paint (only 4 % of the sample). In the sample analysed, only 4 % of paintings present dark colours.

Most of the buildings analysed present a compact volume (81 %), heights below 15 m (73 %) and are home to services or commerce (67 %). The sample is relatively homogeneous according to façades' orientation, distance from the sea and exposure to damp. Finally, in the sample analysed there is a prevalence of case studies with current exposure to pollution sources.

2.6.4 Ceramic Tiling Systems

The characterization of the sample of adherent ceramic claddings is presented in Table 2.15. In this sample, there is a predominance of glazed tiles (86 %) with light colours (59 %), with sizes of less than 20 cm (83 %). In most cases, the cladding is

Table 2.14 Characterization of the sample analysed for painted surfaces

Characteristics of painted surfaces		% of case studies (number of cases)	Average ages (standard deviation), in years	Average S_w (minimum/maximum)
Type of paint	Plain paints	37 % (81)	6.6 (4.0)	12.5 % (0 %/60 %)
	Elastic membranes	38 % (83)	5.5 (3.9)	10.3 % (0 %/69 %)
	Silicate and silicone paints	4 % (9)	8.1 (5.8)	25.8 % (2 %/72 %)
	Textured paint	21 % (47)	10.1 (4.8)	24.6 % (0 %/68 %)
Colour	White	30 % (66)	6.5 (3.7)	11.9 % (0 %/69 %)
	Light	66 % (144)	7.1 (4.7)	15.3 % (0 %/72 %)
	Dark	4 % (10)	8.5 (6.9)	25.8 % (0 %/63 %)
Type of finishing	Smooth	56 % (124)	6.3 (4.1)	12.9 % (0 %/72 %)
	Rough	44 % (96)	7.8 (5.0)	17.3 % (0 %/69 %)
Building geometry	Compacta	81 % (179)	7.4 (4.5)	15.7 % (0 %/72 %)
	Irregular	19 % (41)	5.2 (4.4)	10.7 % (0 %/69 %)
Façade orientation	E/SE	29 % (65)	8.1 (4.4)	17.3 % (0 %/72 %)
	N/NE	19 % (41)	6.9 (4.1)	12.5 % (0 %/63 %)
	W/NW	31 % (68)	6.3 (4.7)	13.5 % (0 %/68 %)
	S/SW	21 % (46)	6.4 (4.6)	15.1 % (0 %/69 %)
Wind-rain action	Severe	33 % (72)	6.2 (3.8)	12.1 % (0 %/63 %)
	Moderate	45 % (99)	5.5 (4.1)	10.4 % (0 %/72 %)
	Low	22 % (49)	11.2 (3.9)	27.5 % (2 %/68 %)
Distance from the sea	Less than 5 km	52 % (114)	5.4 (3.8)	10.1 % (0 %/69 %)
	More than 5 km	48 % (106)	8.7 (4.7)	19.8 % (0 %/72 %)
Exposure to damp	High	53 % (116)	5.4 (3.7)	10.1 % (0 %/69 %)
	Low	47 % (104)	8.7 (4.7)	20.0 % (0 %/72 %)
Distance from pollution sources	Current	79 % (174)	6.6 (4.2)	13.2 % (0 %/69 %)
	Unfavourable	21 % (46)	8.5 (5.4)	20.7 % (0 %/72 %)
Type of use	Commerce and services	67 % (72)	6.7 (3.5)	13.0 % (0 %/68 %)
	Housing	33 % (148)	7.1 (5.0)	15.6 % (0 %/72 %)
Ease of inspection	Current	73 % (161)	7.2 (4.8)	15.8 % (0 %/72 %)
	Unfavourable	27 % (59)	6.3 (3.6)	12.0 % (0 %/63 %)

applied over a masonry support (54 %), not presenting peripheral joints (85 %) but with peripheral protection (52 %). Regarding the façades orientation, the number of case studies analysed for each category is evenly distributed. Finally, most of the

Table 2.15 Characterization of the sample analysed for ceramic tiling systems

Characteristics of ceramic claddings		% of case studies (number of cases)	Average ages (standard deviation), in years	Average S_w (minimum/maximum)
Type of surface	Glazed	86 % (167)	36.3 (18.8)	14.2 % (0 %/61 %)
	Not glazed	14 % (28)	24.0 (15.6)	6.5 % (1 %/19 %)
Colour	Light	59 % (115)	38.2 (18.7)	15.5 % (0 %/61 %)
	Dark	41 % (80)	29.3 (18.0)	9.6 % (0 %/52 %)
Tiles size	L $\leq$ 20 cm	83 % (162)	37.9 (17.9)	14.7 % (0 %/61 %)
	L > 20 cm	17 % (33)	18.0 (14.2)	5.4 % (0 %/21 %)
Substrate	Masonry	54 % (106)	43.1 (17.3)	19.2 % (1 %/61 %)
	Concrete	46 % (89)	24.3 (15.3)	5.8 % (0 %/32 %)
Peripheral joints	No	85 % (166)	34.6 (18.2)	13.0 % (0 %/61 %)
	Yes	15 % (29)	34.1 (22.6)	14.1 % (0 %/48 %)
Peripheral protection	No	48 % (94)	32.8 (19.7)	12.4 % (0 %/61 %)
	Yes	52 % (101)	36.2 (18.0)	13.8 % (0 %/52 %)
Façade orientation	E/SE	25 % (48)	34.4 (18.3)	12.1 % (0 %/48 %)
	N/NE	30 % (59)	36.2 (19.1)	14.2 % (0 %/58 %)
	W/NW	24 % (46)	29.9 (16.6)	9.4 % (0 %/50 %)
	S/SW	22 % (42)	37.3 (21.2)	16.9 % (0 %/61 %)
Distance from the sea	Less than 5 km	71 % (139)	38.1 (19.7)	16.1 % (0 %/61 %)
	More than 5 km	29 % (56)	25.6 (13.0)	5.7 % (0 %/26 %)
Wind-rain action	Severe	27 % (53)	41.3 (20.3)	19.3 % (0 %/61 %)
	Moderate	50 % (97)	29.8 (18.1)	10.2 % (0 %/50 %)
	Low	23 % (45)	36.7 (16.3)	12.2 % (1 %/44 %)
Exposure to damp	High	43 % (84)	37.3 (20.1)	17.4 % (0 %/61 %)
	Low	57 % (111)	32.4 (17.7)	9.9 % (0 %/45 %)
Ease of inspection	Current	65 % (127)	38.6 (16.9)	14.9 % (0 %/61 %)
	Unfavourable	35 % (68)	26.9 (20.0)	9.8 % (0 %/48 %)

case studies are located at less than 5 km from the sea (71 %), with moderate exposure to wind-rain action (50 %) and low exposure to damp (57 %).

2.7 Conclusion

The degradation of buildings is a complex phenomenon that depends on many factors and severely affects the built heritage. Degradation begins as soon as constructions are put into use, due to ageing and the effect of environmental agents; it is also directly affected by an incipient culture of maintenance of the building somewhat prevalent in many countries—in which repair actions are only taken after the occurrence of defects—which leads to the buildings' premature loss of performance. The service life prediction of buildings and their components has paramount importance for the concept of a sustainable environment, enabling a more rational use of resources. In particular, buildings may present a better performance during their life cycle, thus reducing their financial and environmental costs. The knowledge of durability and service life of building components is also crucial for maintenance policies—for it allows planning in a technically informed manner the timely occurrence of maintenance investments—which are indispensable in the context of the present huge demand to rehabilitate.

In recent decades, several international standards and studies have been put forward, which intend to establish methodologies to assess the durability of buildings as well as predict their service life. The main approaches to the problem can be divided in deterministic, probabilistic and engineering methods. These methods are useful to stakeholders in the construction sector, who should decide on the best approach to the problem, taking into account the available data, the desired result, the complexity of each approach and the advantages/disadvantages inherent of each one of the service life prediction methods. Generally speaking, more complex approaches (stochastic or computational models) lead to more rigorous models; however, often does it seem appropriate to sacrifice some accuracy in exchange for a greater applicability and simplicity of the model—and especially in the case of exterior façades claddings.

In the definition of service life prediction models, one of the main factors to be taken into consideration is the maximum acceptable degradation level, which establishes the end of service life of the construction elements. However, as referred to in this chapter, this theoretical limit is somewhat subjective and may vary according to owners and users demands; and these depend on several factors such as the social and economic context of societies. Additionally, the requirements of users and owners permanently change over time, thus requiring continual (re)investment in buildings.

In this chapter, a model is proposed to quantify the overall degradation of existing façades claddings in real life exposure conditions. This model is based on the determination of a numerical index (referred to as severity of degradation), based on the extent of the façades degradation, the severity of the defects detected and their cost of repair. The proposed methodology is based on the fieldwork study of claddings degradation condition, analysed in service conditions, based on an extensive survey, using visual inspections only. This model was successfully applied to the identification and quantification of the degradation of various types of claddings (renderings, ETICs, natural stone claddings, adherent ceramic tiles and painted surfaces). This index (S_w) is used as a reference value (observed on field) that may benchmark and validate the results from the models proposed in the following chapters. In spite of its limitations, this approach presents some advantages, since: (i) it can be easily complemented, at any time, with new data; and (ii) it is easy to apply, even for other types of non-structural elements, allowing their adjustment to other realities (e.g. it can be applied in other countries and building contexts, by simply changing the weighting coefficients and making them meet the costs practiced or the conditions observed in those countries or contexts).

References

Aarseth LI, Hovde PJ (1999) A stochastic approach to the factor method for estimating service life. In: 8th DBMC international conference on durability of building materials and components; Vancouver, Canada, pp 1247–1256

Abraham TH (2002) (Physio)logical circuits: the intellectual origins of the McCulloch-Pitts neural networks. J Hist Behav Sci 38(1):3–25

AIJ (1993) The English edition of the principal guide for service life planning of buildings. Architectural Institute of Japan, Tokyo

Aikivuori AM (1999) Critical loss of performance—what fails before durability. In: 8th International conference on durability of buildings materials and components, Vancouver, Canada, pp 1369–1376

ASTM E632 (1990) Standard practice for developing accelerated tests to aid prediction of the service life of building components and materials. In: Annual book of ASTM standards, section 4: construction, vol 04.07. Building seals and sealants; fire standards; building constructions. Easton, USA. American Society for Testing and Materials, p 1078

Augenbroe GLM, Park C-S (2002) Towards a maintenance performance toolkit for GSA, interim report submitted to GSA, Georgia Institute of Technology, Atlanta, USA

Balaras A, Droutsa K, Argiriou AA, Asimakopoulos DN (2000) EPIQR surveys of apartment buildings in Europe. Energy Build 31(2):111–128

Balaras A, Droutsa K, Dascalaki E, Kontoyiannidis S (2005) Deterioration of European apartment buildings. Energy Build 37(5):515–527

Barberousse H, Ruot B, Yéprémian C, Boulon G (2007) An assessment of façade coatings against colonisation by aerial algae and cyanobacteria. Build Environ 42(7):2555–2561

Bone S, Heard H, Horsfall D (1989) Defects in buildings. Department of Environment, PSA Directorate of Building Development, HMSO, London

Bonshor R, Bonshor L (2001) Cracking in buildings. BRE, London

Bordalo R (2008) Service life prediction of adherent ceramic tiling systems. Master thesis in Civil Engineering, Instituto Superior Técnico, University of Lisbon, Lisbon (in Portuguese)

Bordalo R, de Brito Jorge, Gaspar P, Silva A (2011) Service life prediction modelling of adhesive ceramic tiling systems. Build Res Inf 39(1):66–78

Bourdeau L (1999) Sustainable development and the future of construction: a comparison of visions from various countries. Build Res Inf 27(6):355–367

Bovea MD, Díaz-Albo E, Gallardo A, Colomer FJ, Serrano J (2010) Environmental performance of ceramic tiles: improvement proposals. Mater Des 31(1):35–41

Brand S (1997) How buildings learn: what happens after they're built?, 1st edn. Phoenix Illustrated, London

Brandt E, Rasmussen M (2002) Assessment of building conditions. Energy Build 34(2):121–125

BSI (1992) Guide to durability of buildings and building elements, products and components, BS 7543. British Standards Institution, London

Campante E, Paschoal J (2002) Durability of facades with ceramic coverings—Why they fail'. In: 9th International conference on durability on building materials and components, 2002, Brisbane, Australia, pp 17–21

Camuffo D (1995) Physical weathering of stones. Sci Environ 167(1–3):1–14

Chai C (2011) Service life prediction of painted surfaces on exterior walls. Master thesis in Civil Engineering, Instituto Superior Técnico, University of Lisbon, Lisbon (in Portuguese)

Chai C, de Brito J, Gaspar P, Silva A (2014) Predicting the service life of exterior wall painting: techno-economic analysis of alternative maintenance strategies. J Constr Eng Manage 140 (3):04013057

Chew MYL (1999) Factors affecting ceramic tile adhesion for external cladding. Constr Build Mater 13(5):293–296

Chew MYL, Ping TP (2003) Staining of facades. World Scientific Publishing, Singapore, p. 160

Chew M (2005) Defect analysis in wet areas of buildings. Constr Build Mater 19(3):165–173

Clifton JR (1993) Predicting the service life of concrete. ACI Mater J 90(6):611–617

Cowan P (1963) Studies in the growth, change and ageing of buildings. Trans Bartlett Soc 1:55–84

CSA S478-95 (2001) (Canadian Standards Association) Guideline on durability in buildings. CSA, Canada, pp 9–17

Daniotti B, Spagnolo SL (2008) Service life prediction tools for buildings' design and management. In: 11th DBMC, International conference on durability of building materials and components, Istanbul, Turkey, 2008, T72

Daniotti B, Spagnolo SL, Paolini R (2008) Factor method application using factors' grid. In: 11th International conference on durability of building materials and components, Istanbul, Turkey, 2008, T41

Davies G, Szigeti F (1999) Are facilities measuring up? Matching building capacities and functional needs. In: CIB W078 workshop on information technology in construction 1999, Vancouver, Canada, pp 1856–1866

Design guide to refurbishment (1999) Building Maintenance and Management Centre, Tokyo, Japan (in Japanese)

DeSimone LD, Popoff F (1998) Eco-efficiency. In: The business link to sustainable development, 2nd edn. MIT Press, USA, p 280

Donca G, Mihăilă I, Ganea M, Hirle D, Nica M (2007) Maintenance role in life cycle management. Ann Oradea Univ Fascicle Manage Technol Eng 6(16):2158–2163

Edvardsen C, Mohr L (2000) Designing and rehabilitating concrete structures: probabilistic approach (DuraCrete). In: 5th CANMET/ACI, international conference on durability of concrete, 2000, pp 1192–1208

Emídio F, de Brito J, Gaspar P, Silva A (2014) Application of the factor method to the estimation of the service life of natural stone cladding. Constr Build Mater 66:484–493

Esteves LAR (2007) Portuguese natural stone. The future is paved today. Master thesis in Management and Industrial Strategy, Instituto Superior de Economia e Gestão, University of Lisbon, Lisbon (in Portuguese)

European Organisation for Technical Approvals (EOTA) (1999) Assumption of working life of construction products in guideline for European Technical Approvals and Harmonized Standards. December 1999. Guidance Document 002

Faber MH, Gehlen C (2002) Probabilistischer Ansatz zur Beurteilung der Dauerhaftigkeit von bestehenden Stahlbetonbauten. Beton und Stahlbetonbau 97(8):421–429

Flanagan R, Norman G, Meadows J, Robinson G (1989) Life cycle costing: theory and practice. BSP Professional Books, Oxford

Flores-Colen I, de Brito J (2010) A systematic approach for maintenance budgeting of buildings façades based on predictive and preventive strategies. Constr Build Mater 24(9):1718–1729

Flores-Colen I, de Brito J, Freitas VP (2008) Condition assessment of facade rendering though in situ testing. In: 11th DBMC international conference on durability on building materials and components 2008, Istanbul, Turkey, Paper T71

Flourentzou F, Brandt E, Wetzel C (1999) MEDIC—A method for predicting residual service life and refurbishment investments budgets. In: 8th International conference on durability of buildings materials and components, Vancouver, Canada, pp 1280–1288

Frangopol DM, Kallen M-J, Noortwijk JM (2004) Probabilistic models for life-cycle performance of deteriorating structures: review and future directions. Prog Struct Mat Eng 6(4):197–212

Freitas VP, Sousa M, Abrantes V (1999) Survey of the durability of facades of 4000 dwellings in northern Portugal. In: 8th International conference on the durability of building materials and components, Ottawa, Canada, pp 1040–1050

Freitas VP, Corvacho H, Quintela M, Delgado JMPQ (2013) Durability assessment of adhesive systems for bonding ceramic tiles on façades: the research and the practice. In: Durability of building materials and components, building pathology and rehabilitation, vol 3. Springer, Berlin, pp 173–205

Frohnsdorff GJ, Martin JW (1996) Towards prediction of building service life: the standards imperative. In: 7th International conference on durability of building materials and components. Stockholm, Sweden, pp 1417–1428

Galbusera MM, de Brito J, Silva A (2014) Application of the factor method to the prediction of the service life of ceramic external wall claddings. J Build Perform Constr Facil, pp 19–29. doi:10.1016/j.conbuildmat.2014.05.045

Gaspar P (2009) Service life of constructions: development of a method to estimate the durability of construction elements. Application to renderings of current buildings. Doctor thesis in sciences of engineering, Instituto Superior Técnico, Technical University of Lisbon, Portugal (in Portuguese)

Gaspar P, de Brito J (2005) Mapping defect sensitivity in external mortar renders. Constr Build Mater 19(8):571–578

Gaspar PL, de Brito J (2008) Quantifying environmental effects on cement-rendered facades: A comparison between different degradation indicators. Build Environ 43(11):1818–1828

Gaspar PL, de Brito J (2011) Limit states and service life of cement renders on façades. Mater Civil Eng 23(10):1393–1404

Guide to Condition Assessment for Refurbishment (1993). Building Maintenance & Management Centre, Tokyo, Japan, 1993 (in Japanese)

Haagenrud SE (2004) Factors causing degradation. Guide and bibliography to service life and durability research for buildings and components. In: Joint CIB W80/RILEM TC 140—Prediction of service life of building materials and components, pp 2.1–2.105

Haapio A, Viitaniemi P (2008) How workmanship should be taken into account in service life planning. In: 11th International conference on durability of building materials and components, Istanbul, Turkey, T45

Hansen EJP, Ekman T, Hansen KK (1999) Durability of cracked fibre reinforced concrete structures exposed to chlorides. In: 8th International conference on the durability of building materials and components, Ottawa, Canada, pp 280–289

Hed G (1999) Service life planning of building components. In: 8th International conference on durability of building materials and components, Vancouver, Canada, pp 1543–1551

Ho DCW, Lo SM, Yiu CY, Yau LM (2004) A survey of materials used in external wall finishes in Hong Kong. Hong Kong Surveyor 15(2):7–11

Hovde PJ (2000) Factor methods for service life prediction: a state-of-the-art. Draft Report, Norwegian University of Science and Technology, Trondheim, Norway

Hovde PJ (2002) The factor method for service life prediction from theoretical evaluation to practical implementation. In: 9th International conference on durability of buildings materials and components, Brisbane, Australia, Paper 232

Hovde P (2004) Factor methods for service life prediction. In: CIB W080/RILEM 175 SLM: Service life methodologies prediction of service life for buildings and components, task group: performance based methods of service life prediction 2004, Trondheim, Norway, pp 1–51

HQAL (2000) Housing quality assurance law. Centre for better living, Tokyo, Japan (in Japanese)

Iselin DG, Lemer AC (eds) (1993) The fourth dimension in building: strategies for minimizing obsolescence. National Research Council, Building Research Board, National Academy Press, Washington, DC

ISO 15686-1 (2011) Buildings and constructed assets—service life planning—Part 1: general principles and framework. International Organization for Standardization, Switzerland

John VM, Sjöström C, Agopyan V (2002) Durability in the built environment and sustainability in developing countries. In: 9th International conference on durability of building materials and components, Brisbane, Australia, Paper 11

Kooymans R, Abbott J (2006) Developing an effective service life asset management and valuation model. J Corp Real Estate 8(4):198–212

Kus H (2002) Service life of external renders. In: XXX IAHS world congress on housing, vol III. Coimbra, pp 1875–1882

Kus H, Carlsson T (2003) Microstructural investigations of naturally and artificially weathered autoclaved aerated concrete. Cem Concr Res 33(9):1423–1432

Lacasse MA, Sjöström C (2004) Recent advances in methods for service life prediction of buildings materials and components—an overview. In: CIB World Building Congress, Canada, pp 1–10

Lair J (2003) Failure modes and effect analysis and service life prediction. Intermediary report (D4-C2-jl-01 draft 2), IEA task 27 (project C2: failure mode analysis); CSTB, France, pp 166–212

Lee N, Bennett J, Jones M, Marston N, Kear G (2008) A durability assessment tool for the new zealand building code. In: 11th International conference on durability of building material and components, Istanbul, Turkey, T45

Leira B, Lindgård J, Nesje A, Sund E, Sægrov S (1999) Degradation analysis by statistical methods. In: 8th International conference on durability of building materials and components, Vancouver, Canada, pp 1436–1446

Liang MT, Wu JH, Liang CH (2001) Multiple layer fuzzy evaluation for existing reinforced concrete bridges. J Infrastruct Syst 7(4):144–159

Lo Y (2002) Delamination of external wall finishes of housing. In: XXX IAHS world congress on housing—housing construction—an interdisciplinary task. Coimbra, Portugal, pp 1571–1576

Long AE, Henderson GD, Montgomery FR (2001) Why assess the properties of near-surface concrete? Constr Build Mater 15(2–3):65–79

Lounis Z, Lacasse MA, Vanier DJ, Kyle BR (1998) Towards standardization of service life prediction of roofing membranes. In: Wallace TJ, Rossiter Jr WJ (eds) Roofing research and standards development, vol 4. American Society for Testing and Materials (ASTM STP 1349)

Mansur AAP, Nascimento OL, Vasconcelos WL, Mansur HS (2008) Chemical functionalization of ceramic tile surfaces by silane coupling agents: polymer modified mortar adhesion mechanism implications. Mater Res 11(3):293–302

Marteinsson B (2003) Assessment of service lives in the design of buildings—development of the factor method. Licentiate thesis, KTH's Research School—HiG, Centre of Built Environment, University of Gävle, Sweden

Marteinsson B, Jónsson B (1999) Overall survey of buildings—performance and maintenance. In: 8th International conference on the durability of building materials and components, Vancouver, Canada, pp 1634–1654

Mateus R, Bragança L, Koukkari H (2008) Sustainability assessment and rating of Portuguese buildings. In: World sustainable conference (SB08), Melbourne, Australia, pp 959–966

Mc Duling J, Horak E, Cloete C (2008) Service life prediction beyond the 'factor method'. In: 11th International conference on durability of building materials and components, Istanbul, Turkey, T42

Meola C, Maio RD, Roberti N, Carlomagno GM (2005) Application of infrared thermography and geophysical methods for defect detection in architectural structures. Eng Fail Anal 12 (6):875–892

Moreno SH (2012) The method by factors to estimate service life in buildings projects according to norm ISO 15686. Manage Res Pract 4(4):5–11

Moser K (1999) Towards the practical evaluation of service life—illustrative application of the probabilistic approach. In: 8th International conference on durability of buildings materials and components, Vancouver, Canada, pp 1319–1329

Moser K (2003) Engineering design methods for service life planning—state of the art. In: International workshop on management of durability in the building process (WMDBP 2003). Politecnico di Milano, Milan, Paper 40

Moser K (2004) Engineering design methods for service life prediction. In: CIB W080/RILEM 175 SLM: Service life methodologies prediction of service life for buildings and components, task group: performance based methods of service life prediction, Trondheim, Norway, pp 52–95

Moser K, Edvardsen C (2002) Engineering design method for service life prediction. In: 9th International conference on the durability of building materials and components 2002, Brisbane, Australia, Paper 222

National Statistics Institute (INE) (2001) National statistics—census. http://www.ine.pt/prodserv/quadro/mostraquadro (in Portuguese)
Nireki T (1996) Service life design. Constr Build Mater 10(5):403–406
Norvaišienė R, Miniotaitė R, Stankevičius V (2003) Climatic and air pollution effects on building facades. Mater Sci 9(1):102–105
NP EN ISO 4628-1 (2005) Paints and varnishes—evaluation of degradation of coatings—designation of quantity and size of defects, and of intensity of uniform changes in appearance—Part 1: General introduction and designation system, Portuguese Quality Institute, Lisbon, Portugal, p 8
NP EN ISO 4628-2 (2005) Paints and varnishes—evaluation of degradation of paint coatings—designation of intensity, quantity and size of common types of defect—Part 2: Designation of degree of blistering, Portuguese Quality Institute, Lisbon, Portugal, p 16
NP EN ISO 4628-4 (2005) Paints and varnishes—evaluation of degradation of coatings—designation of quantity and size of defects, and of intensity of uniform changes in appearance—Part 4: Assessment of degree of cracking, Portuguese Quality Institute, Lisbon, Portugal, p 20
NP EN ISO 4628-5 (2005) Paints and varnishes—evaluation of degradation of coatings—designation of quantity and size of defects, and of intensity of uniform changes in appearance—Part 5: Assessment of degree of flaking, Portuguese Quality Institute, Lisbon, Portugal, p 11
NP EN ISO 4628-7 (2005) Paints and varnishes—evaluation of degradation of coatings—designation of quantity and size of defects, and of intensity of uniform changes in appearance—Part 7: Assessment of degree of chalking by velvet method, Portuguese Quality Institute, Lisbon, Portugal, p 8
NS 3422 (1994) Specification texts for operation, maintenance and renewal of buildings and civil engineering works. Norges Standardiseringsforbund, Oslo, Norway
Paulo PV, Branco F, de Brito J (2014) Buildings life: a building management system. Struct Infrastruct Eng Maintenance Manage Life Cycle Des Perform 10(3):388–397
Pearce D (2003) The social and economic value of construction. In: The construction industry's contribution to sustainable development. NCRISP, Davis Langdon Consultancy, London
Re Cecconi F (2002) Performance leads the way to service life prediction. In: 9th International conference on durability of buildings materials and components, Brisbane, Australia, Paper 213
Rikey M, Cotgrave A (2005) The context of maintenance. In: Construction technology. The technology of refurbishment and maintenance, vol. 3. Macmillan Palgrave, New York, pp 50–56
Rimestad L (1998) The use of field failure data in accelerated testing. In: Safety and reliability, Hansen GK, Sandtorv HA (eds) Balkema, Rotterdam, pp 1209–1216
Ross SM (1996) Stochastic processes, 2nd edn. John Wiley & Sons, New York
Rudbeck C (2002) Service life of building envelope components: making it operational in economical assessment. Constr Build Mater 16(2):83–89
Sarja A (2005) Generic limit state design of structures. In: 10th International conference on durability of building materials and components, Lyon, France, TT3-161
Shohet I, Laufer A (1996) Exterior cladding methods: a technoeconomic analysis. J Constr Eng Manage 122(3):242–247
Shohet IM, Paciuk M (2004) Service life prediction of exterior cladding components under standard conditions. Constr Manage Econ 22(10):1081–1090
Shohet I, Rosenfeld Y, Puterman M, Gilboa E (1999) Deterioration patterns for maintenance management—a methodological approach. In: 8th International conference on durability of buildings materials and components, Vancouver, Canada, pp 1666–1678
Siemes T, Edvardsen C (1999) Duracrete: service life design for concrete structures. In: 8th International conference on durability of buildings materials and components, Vancouver, Canada, pp 1343–1356

Silva A (2009) Service life prediction of natural stone walls cladding, Master thesis (in Portuguese). Instituto Superior Técnico, Lisbon, Portugal
Silva A, de Brito Jorge, Gaspar P (2011) Service life prediction model applied to natural stone wall claddings (directly adhered to the substrate). Constr Build Mater 25(9):3674–3684
Silva A, Dias JLR, Gaspar PL, de Brito J (2013) Statistical models applied to service life prediction of rendered façades. Autom Constr 30:151–160
Silva A, Gaspar PL, de Brito J (2014) Durability of current renderings: a probabilistic analysis. Autom Constr 44:92–102
Silvestre J, de Brito J (2007) Statistical analysis of defects of tiles' joints. Análisis estadístico de los defectos de juntas cerámicas. Materiales de Construcción, Instituto de Ciencias de la Construcción Eduardo Torroja, Madrid, Spain 57(285):85–92
Silvestre J, de Brito Jorge (2009) Ceramic tiling inspection system. Constr Build Mater 23 (2):653–668
Sjöström C (1985) Overview of methodologies for prediction of service life. In: Problems in service life prediction of building and construction materials NATO ASI series, vol 95, pp 3–20
Sjöström C, Davies H (2005) Built to last: service life planning. ISO Focus 2(11):13–15
Slaughter ES (2001) Design strategies to increase building flexibility. Build Res Inf 29(3):208–217
Soronis G (1996) Standards for design life of buildings: utilization in the design process. Constr Build Mater 10(7):487–490
Straub A (2003) Using a condition-dependent approach to maintenance to control costs and performances. Facil Manage 1(4):380–395
Takata S, Kimura F, Van Houten F, Westkamper E, Shpitalni M, Ceglarek D, Lee J (2004) Maintenance: changing role in life cycle management. CIRP Ann 53(2):643–655
Talon A, Boissier D, Chevalier J-L, Hans J (2005) Temporal quantification method of degradation scenarios based on FMEA. In: 10th International conference on durability of building materials and components, Lyon, France, TT4-139
Timellini G, Palmonari C (1989) Ceramic floor and wall tile performance and controversies. EdiCer, Sassuolo
Van Winden C, Dekker R (1998) Rationalization of building maintenance by Markov decision models: a pilot case study. J Oper Res Soc 49(9):928–935
Vanier DJ (1999) Why industry needs asset management tools. In: Innovations in urban infrastructure seminar of the APWA international public works congress, Denver, USA, pp 11–25
Watt DS (1999) Building pathology—principles and practice, 1st edn. Blackwell Science Ltd., Blackwell Publishing Company, London
Wekesa BW, Steyn GS, Otieno FAO (2010) The response of common building construction technologies to the urban poor and their environment. Build Environ 45(10):2327–2335
Wetzel A, Zurbriggen R, Herwegh M (2010) Spatially resolved evolution of adhesion properties of large porcelain tiles. Cement Concr Compos 32(5):327–338
Wyatt D (2005) The contribution of FMEA and FTA to the performance review and auditing of service life design of constructed assets. In: 10th International conference on durability of building materials and components, Lyon, France, TT4-206
Ximenes S, de Brito J, Gaspar PL, Silva A (2015) Modelling the degradation and service life of ETICS in external walls. Mater Struct 48:2235–2249
Zhang X, Gao H (2011) Determining an optimal maintenance period for infrastructure systems. Comput Aided Civil Infrastruct Eng 27(7):543–554

Chapter 3
Deterministic Models

3.1 Simple Regression Analysis

3.1.1 Theoretical Foundations

Regression analysis is one of the most used statistical techniques in exogenous, predictor or explanatory, responsible for the behaviour of the dependent variable (Draper and Smith 1998; Montgomery and Ruger 2003). In other words, regression techniques are usually applied to describe the relationship between variables. When regression analysis is used in predictive models, the main objective is to explain a given reality, trying to predict the behaviour of a dependent variable based on the knowledge of one or more independent variables. However, in some cases, a regression analysis is not used as a predictive or explanatory model, but instead is applied to fit a mathematical or regression equation to the data available. In fact, King (1986) argues that the purpose of the regression analysis is basically to measure the effects of the independent variables on the explanation of the behaviour of the dependent variable. Hagquist and Stenbeck (1998) report that in this case the approach to the regression model should focus on addressing questions such as how variable x affects variable y or how the effect of x on y is modified by z. Applying this approach to service life prediction, the questions raised are how degradation (y) evolves over time (x) and how the characteristics of the facade coatings (z) influence this variation.

Using a simple regression analysis it is possible to analyse the degradation evolution of façades claddings. Degradation evolution is the loss of performance demonstrated by an element during its service life and expresses the progress of the loss of the material's capacity to answer the requisites demanded (Bordalo et al. 2011). This loss of performance can be expressed graphically for each sample under analysis by a cloud of points whose abscissas represent the "age" of the buildings and ordinates represent the "severity of degradation" (Gaspar 2009). The methodology applied in this book is based on the survey of the degradation

A. Silva et al., *Methodologies for Service Life Prediction of Buildings*,
Green Energy and Technology, DOI 10.1007/978-3-319-33290-1_3

condition of the façade coatings during an extensive field work. For this, various cases are analysed in real in-use conditions and with different degradation indexes. It is then possible to obtain, for each case study, a coordinate (*x*, *y*), where *x* is the age of the buildings or their elements (age is considered as the period of time from the last corrective action until the inspection date) and *y* represents the observed degradation. Knowing the coordinates of all the case studies and using a simple regression analysis, it is possible to identify the average degradation curve of the element (obtaining the regression function or trend that best fits the cloud of points that composes the sample) (Shohet et al. 2002). These degradation functions or curves are usually associated to specific degradation mechanisms (Shohet et al. 1999; Florentzou et al. 1999). Shohet et al. (1999) suggest four typical deterioration patterns due to the effects of the degradation agents: a linear pattern associated with degradation agents that act permanently; a concave pattern associated with biological agents, whose deterioration develops rapidly at an early stage, but tends to slow down over time; a convex pattern associated with the physical and chemical phenomena, which act slowly initially but whose action is felt cumulatively (in these cases, façades with higher degradation levels present a greater probability of their degradation increasing at a faster rate); an "S"-shaped pattern associated with a degradation phenomenon whose intensity changes over time (this pattern corresponds to a façade that tends to show premature defects that apparently stabilize over time but accelerate again near the end of the service life).

3.1.1.1 Description

Wooldridge (2009) refers that the main purpose of a regression analysis is that *y* and *x* are two variables that represent a population. Therefore, a regression analysis is used to explain *y* as a function of *x* or study the way in which the variability of *y* varies with the variability of *x*. Consequently, in simple linear regression analysis, the relationship between the dependent variable (*y*) and the independent variable (*x*) is given by Eq. (3.1).

$$y = B_0 + B_1 \cdot x + \varepsilon \tag{3.1}$$

where *y* represents the dependent variable, B_0 and B_1 the regression coefficients, *x* the independent variable and ε the model random errors (it can also be seen as a "noise" factor, which reflects the effects of other factors that influence the behaviour of the dependent variable but which are not covered by the analysis (Sykes 1992).

In a regression analysis, the regression coefficients B_0 and B_1 are obtained by the least squares method. According to Abdi (2010), the least squares method is one of the most widely used methods in statistics. In this method, a possible line (or curve, in nonlinear regression) is superimposed on the data, resulting in a different set of estimated errors given by the difference between the observed values and the values estimated by the possible lines/curves. The regression analysis then chooses, among all possible lines/curves, the one that minimizes the sum of the squares of the estimated errors (Sykes 1992).

The general solution of the differential equation that minimizes the sum of the squares of the estimated errors is given by Eq. (3.2) (Myers 1990).

$$(X'X)\hat{B} = X'y \Leftrightarrow \hat{B} = (X'X)^{-1}X'y \tag{3.2}$$

where $\hat{B}$ represents the vector estimation of the model coefficients, X the matrix of the n observations of X_i ($i = 1, 2, \ldots, p$) (if the constant parameter is included in the model, the matrix's first column should has the value 1, allowing determining coefficient B_0) and y is the vector of the n observations of the dependent variable.

Montgomery and Peck (2001) present a linearization of this matrix system, allowing, for a simple linear regression analysis, obtaining analytically the values of the coefficients B_0 and B_1, as shown in Eqs. (3.3) and (3.4).

$$B_1 = \frac{\sum x_i y_i - \bar{y}\sum x_i}{\sum x_i^2 - \bar{x}\sum x_i} \tag{3.3}$$

$$B_0 = \bar{y} - B_1\bar{x} \tag{3.4}$$

where B_0 and B_1 represent the regression coefficients, y_i the observed values of the dependent variable, $\bar{y}$ the average of the observed values of the dependent variable, x_i the observed values of the independent variable, $\bar{x}$ the average of the observed values of the independent variable, $\Sigma x_i y_i$ the sum of the multiplication between variables x and y, Σx_i the sum of the values of the variable x and Σx_i^2 the sum of the squared values of variable x.

In regression analysis, the Pearson's product moment correlation coefficient (r) and the coefficient of determination (R^2) are the most widely used indexes to evaluate the fitting of the regression model to the data. The Pearson's product moment correlation coefficient (r) is a statistical parameter that allows analysing the linear dependence between variables, knowing the relationship between them and the magnitude and direction of this relationship (either positive or negative) (Higgins 2006). The Pearson's correlation coefficient (r) between two variables can be determined, as seen in Eq. (3.5).

$$r_{xy} = \frac{S_{xy}}{S_x S_y} \quad \text{or} \quad r_{xy} = \frac{\sum x_i y_i - \frac{\sum x_i \sum y_i}{n}}{\sqrt{\left(\sum x_i^2 - \frac{\sum x_i^2}{n}\right)\left(\sum y_i^2 - \frac{\sum y_i^2}{n}\right)}} \tag{3.5}$$

where r_{xy} represents Pearson's correlation coefficient, S_{xy} is the sample covariance between x and y; S_x the standard deviation of sample for variable x; S_y the standard deviation for variable y, $\Sigma x_i y_i$ the sum of the multiplication between x and y, Σx_i^2 the square of the sum of the values of x; Σy_i^2 the square of the sum of the values of y; and n the number of cases in the sample.

The values of the Pearson's correlation coefficient (r) range between −1 (negative linear correlation) and 1 (positive linear correlation). Values −1 and 1 represent a perfect correlation between the dataset and the regression trend. A correlation coefficient equal to zero reveals that there is no relationship between the two variables. In absolute value, higher values of the correlation coefficient reveal a better fitting of the model to the data. Piaw (2006) proposes a scale which relates the value of the Pearson's correlation coefficient (r) with the quality of the correlation between variables. According to this scale, an absolute value of r equal to 1 reveals a perfect correlation; an absolute value of r between 0.9 and 1 reveals a very strong correlation; an absolute value of r between 0.7 and 0.9 reveals a strong correlation; an absolute value of r between 0.3 and 0.7 reveals an average correlation; an absolute value of r between 0 and 0.3 reveals a weak correlation; and a value of r equal to 0 reveals a null correlation.

Additionally, the coefficient of determination can be used to evaluate the goodness-of-fit of the model to the sample. Zar (1999) states that the coefficient of determination (R^2) measures the proportion of the variability of y that is attributable to the dependence of y of all x_i, as defined by the adjustment of the model to the data. In other words, this coefficient measures the proportion of the variability of the dependent variable that can be explained by the regression model obtained (Berry and Feldman 1985; Bowerman et al. 2005)—Eq. (3.6).

$$R^2 = \frac{\sum_{i=1}^{n}(\hat{y}_i - \bar{y})^2}{\sum_{i=1}^{n}(y_i - \bar{y})^2} = 1 - \frac{\sum_{i=1}^{n}(y_i - \hat{y}_i)^2}{\sum_{i=1}^{n}(yi - \bar{y})^2} = \frac{\text{SSR}}{\text{SST}} = 1 - \frac{\text{SSE}}{\text{SST}} \tag{3.6}$$

where R^2 is the coefficient of determination, y_i the observed values of the dependent variable, $\bar{y}$ the average of the observed values of the dependent variable, $\hat{y}_i$ the values predicted by the regression model, SSR the regression sum of squares (SST) (the variance explained by the regression of y on x) given by $\sum_{j=1}^{n}\left(\hat{Y}_j - \bar{Y}\right)^2$, SSE the residual or error SST, which represents the residual variance or the variance due to the errors (not explained by the regression model), given by $\sum_{j=1}^{n}\left(Y_j - \hat{Y}_j\right)^2$, SST the total SST, which is the variance of the dependent variable, given by $\sum_{j=1}^{n}\left(Y_j - \bar{Y}\right)^2$; the total variability is the sum of the variance explained by the model (SSR) and the variance explained by the errors (SSE).

The coefficient of determination always varies between 0 and 1. Similar to the correlation coefficient, when R^2 is equal to 1, it means that there is a perfect correlation and that the sample data lie exactly on the regression line. Conversely, when R^2 is equal to 0, it means that the correlation is zero, indicating that x_i is completely useless in the explanation y_i (Cottrell 2003). As expected, when the variance explained by the model (SSR) is close to the values of SST, the ratio

SSR/SST tends to 1, which implies a perfect adjustment to the data (R^2 = 1); on the other hand, if the value of SSE is close to the value of SST, then the ratio SSE/SST tends to 1, which implies a null adjustment, since R^2 tends to zero. Maroco (2007) reveals that the use of R^2 as the only parameter to evaluate the model's goodness-of-fit may be subjective as it varies with the field of research; in fact, in exact sciences, it is considered that a model with a good potential should reveal a R^2 higher than 0.9 while in social sciences (related to more complex phenomena) it is considered that a model with a R^2 higher than 0.5 already presents an acceptable fit of the model to the data.

3.1.2 Application to Façade Claddings

Empirically, it is assumed that a coating whose age is 0, which implies that it has been put into use at that instant, should not show any degradation. However, design and execution conditions can influence the degradation of façade claddings even before they are put into use (Viitaniemi and Haapio 2008). The execution phase is one of the most critical periods in the buildings' life cycle. Some of the decisions taken during this phase may lead to various defects, which change the natural degradation pattern of coatings. However, design errors are not easily detectable when the coatings are applied. Moreover, even in the absence of design errors, in some situations, façades installation occurs much earlier than the official date of construction (documented in Municipalities), which explains the eventual existence of deterioration when the building starts formally its service life. Therefore, in real situations, in instant $x = 0$, the degradation can be different from zero ($y \neq 0$). In fact, the different claddings within the sample reveal different degradation ratios. In some cases (either due to the characteristics of the coatings, the exposure conditions or a combination of both), the degradation does not manifests itself in the early stages of the façades' life cycle, or during this period is almost imperceptible, and is not detected by visual inspection. Therefore, it seems appropriate to adopt an average degradation curve, in which the degradation phenomenon appears to stabilize for an initial period of time, and evolves more rapidly thereafter. Thus, it seems relevant that the proposed regression models include the constant (B_0), contemplating these situations.

In the analysis of the degradation evolution over time, empirical knowledge suggests an increase in the severity of degradation values with age. However, this relationship is not necessarily linear, and two main situations may occur:

- A nonlinear relationship between age and severity of degradation; in this case, the use of a simple nonlinear regression (e.g. 2nd or 3rd degree polynomial curve) would result in a coefficient of determination much higher than that obtained with a linear trend line;

- In linear regression, the variable age may not be able to explain 100 % of the variability of the severity of degradation, which implies that some of the variability of the severity of degradation has to be explained by other factors; thus, the use of a linear or nonlinear multiple regression, encompassing other variables in addition to age, which allows obtaining a model closer to the reality observed during fieldwork.

3.1.2.1 Natural Stone Claddings

In this book, 203 natural stone claddings are analysed. The development of the cladding's degradation over time is initially studied using a graphical procedure and a simple linear regression line is adjusted to the scatter of points of the sample (average degradation curve), whose abscissas measure the variable "age" and whose ordinates measure the variable "severity"—Fig. 3.1. The limits that establish a 95 % confidence interval are also represented and the case studies that fall outside of the defined confidence interval are represented in red. A coefficient of determination (R^2) of 0.639 is obtained, revealing a correlation between the dataset and the degradation curve. In this case, 63.9 % of the variability of the severity of degradation of stone claddings is explained by the model and the remaining 36.1 % must be explained by other factors, which are not contemplated in this analysis.

By using a linear trend to describe the loss of performance of stone claddings over time it is assumed that the degradation process is constant over time, which is not necessarily true. Due to the complexity of the degradation phenomena, it is found that the development of degradation actions varies over time. Therefore, using this model may lead to the underestimation or overestimation of the façades condition during their life cycle.

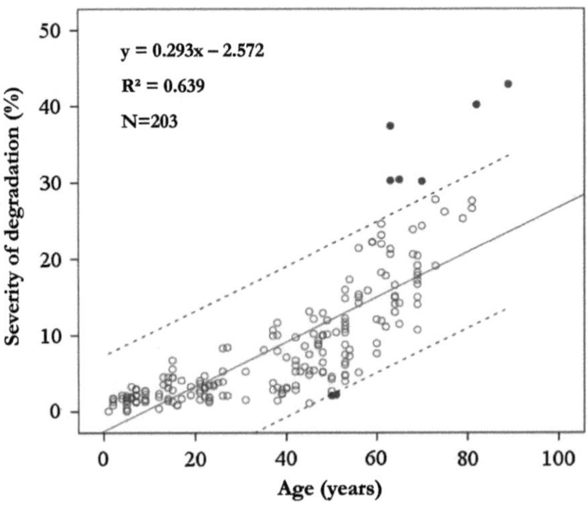

Fig. 3.1 Degradation evolution of stone claddings over time (linear regression analysis)

Thus it seems reasonable to adjust the degradation curves to the cloud of points that represents the sample using a nonlinear simple regression analysis. In this study, a polynomial regression analysis is applied, where the relationship between variables is given by the general expression (3.10).

$$y = B_0 + B_1 \cdot x + B_2 \cdot x^2 + B_3 \cdot x^3 + \cdots + B_n \cdot x^n + \varepsilon \tag{3.10}$$

where y represents the dependent variable, B_0, B_1, B_2 and B_n the regression coefficients, x the independent variable and ε the model random errors.

Figure 3.2 presents a simple nonlinear regression analysis applied to the whole sample, adjusting a second-degree polynomial (on the left) and a third-degree polynomial (on the right) to the case studies. In both analyses, a stronger correlation is obtained between the model and the sample with a coefficient of determination (R^2) of 0.769 and 0.775 for second- and third-degree polynomials lines, respectively.

For stone claddings, the third-degree polynomial curve is the one that best fits the cloud of points that comprise the sample. This model leads to a coefficient of determination 13.6 % higher than the one obtained using a linear regression. In this case, the significant increase achieved justifies the adoption of a more complex model. The third-degree polynomial trend resembles a convex-shaped pattern of degradation, which according to Shohet et al. (1999) corresponds to a physical and chemical deterioration phenomenon, which acts slowly in the initial stage of the façades life cycle, but whose action is felt cumulatively over time. In this deterioration pattern, the higher the age, the higher probability of a more rapid increase in the degradation.

Table 3.1 presents the estimated service life obtained by simple regression models (linear and polynomial), according to the three levels of maximum allowable degradation limit (10, 20 and 30 %). For a maximum acceptable degradation level of 10 %, the estimated service life of stone claddings varies between 43 and 52 years. Analysing the condition of façades and according to the estimated service life obtained, it seems that the adoption of a maximum degradation level of 10 %

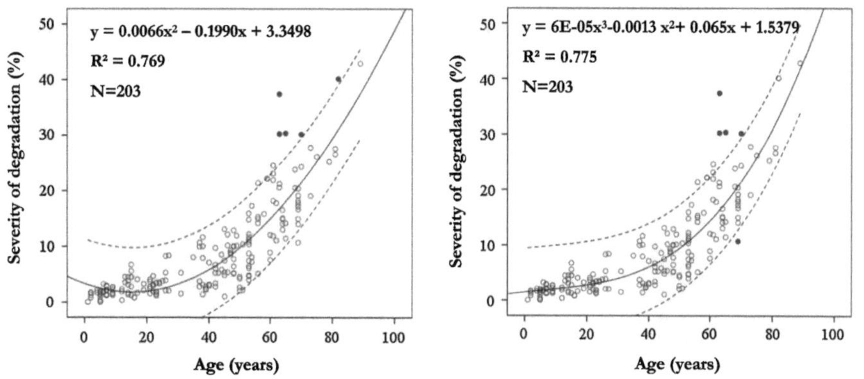

Fig. 3.2 Degradation evolution of stone claddings over time (non-linear regression analysis)

Table 3.1 Estimated service life of stone claddings according to the maximum allowable degradation level

	Simple linear regression (R^2 = 0.639)			Simple polynomial regression, 2nd degree (R^2 = 0.768)			Simple polynomial regression, 3rd degree (R^2 = 0.775)		
	10 %	20 %	30 %	10 %	20 %	30 %	10 %	20 %	30 %
Estimated service life (years)	43	77	111	51	68	81	52	69	80

seems too conservative to establish the end of the service life of stone claddings. For a maximum degradation level of 20 %, the estimated service life obtained varies between 68 and 77 years and for a maximum degradation level of 30 %, the estimated service life obtained varies between 80 and 111 years. A maximum acceptable level of degradation of 30 % appears to be somewhat too high, since the degradation of the claddings is too widespread. Thus, a maximum degradation level of 20 % is considered in this book, which is found to be more adequate in the description of the degradation phenomenon of stone claddings, based on the sample analysed during fieldwork. It is important to notice that this theoretical limit is hard to define and that it must be set with caution since it depends mostly on the acceptance criteria of the building owners/promoters.

Figure 3.3 shows the degradation evolution according to the type of natural stone. In this study, three categories related to the type of stone are considered: granite, limestone and marble. There is a good correlation between the sample and the curve for marble façades (R^2 = 0.837). Limestone claddings show a medium correlation (R^2 = 0.691) and granite claddings present a low coefficient of determination (R^2 = 0.204), mostly due to the fact that in the sample analysed there is a lack of case studies aged over 51 years and therefore it is not possible to define with

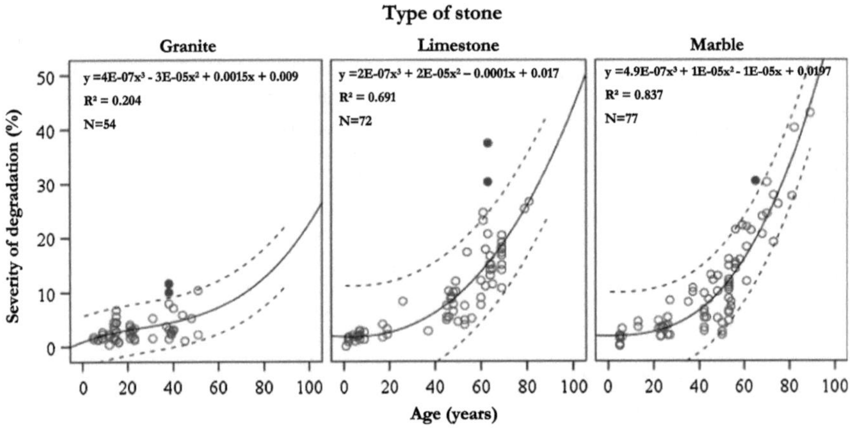

Fig. 3.3 Degradation evolution according to the type of natural stone

clarity the pattern of degradation of this type of stone. Based on the degradation curves, it is found that granite claddings are the most durable, with longer service lives (notwithstanding the small statistical significance of the obtained curve), nearly 25 % higher than limestone claddings and about 31.3 % higher than marble claddings.

Figure 3.4 presents the degradation curve of natural stone claddings according to the colour of the stone plates. Two categories are considered: light colours, comprising white, rose, yellow and similar colours; and dark colours, for example, black, grey and red. Both light and dark colours exhibit a good correlation between the field results and the degradation curve, with a coefficient of determination of 0.828 and 0.744, respectively. In the sample analysed, light colours show higher expected service lives, despite the fact that there is no information regarding a direct relationship between the colour and degradation of stone claddings in the literature on the theme.

Concerning the type of finishing, the degradation curves obtained reveal a good correlation between the data and the model (Fig. 3.5). However, for a maximum degradation level of 20 % and a polynomial trend line, it is not possible to establish a relationship between the type of finishing and service life, since the estimated service lives for both curves are similar. If a linear trend line is used, rough finished claddings present lower service lives, which can be explained by the greater area of stone exposed to adverse atmospheric conditions.

Figure 3.6 presents the degradation curves according to the area of stone plates, revealing a high coefficient of determination (higher than 0.7) for both categories. Claddings with larger plates reach the end of their service life 9.7 % sooner than

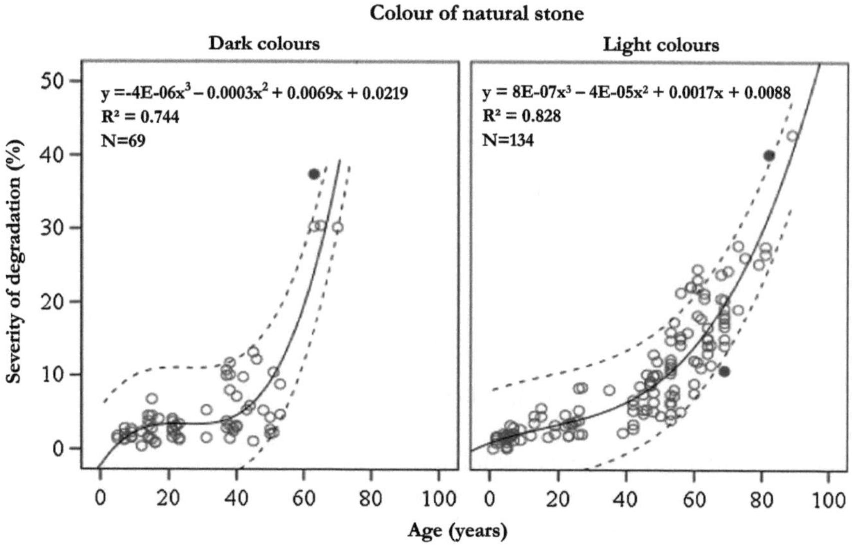

Fig. 3.4 Degradation evolution according to the colour of natural stone

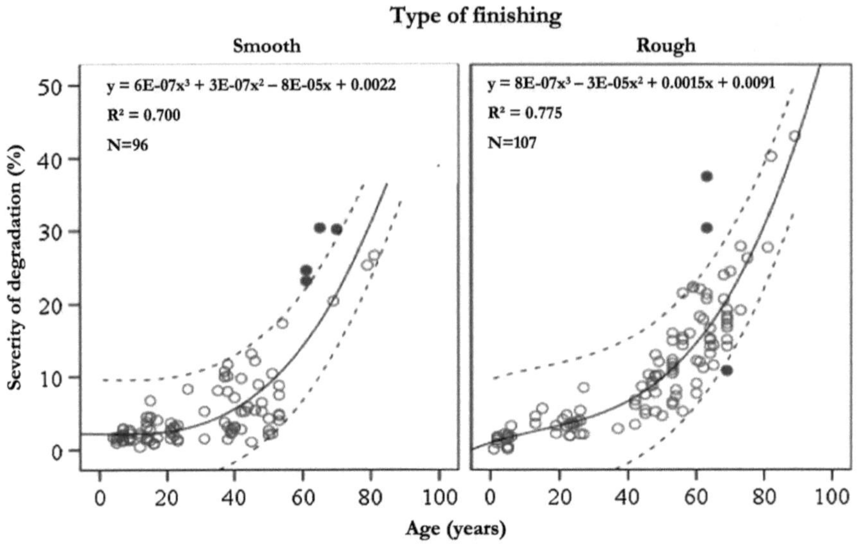

Fig. 3.5 Degradation evolution according to the type of finishing

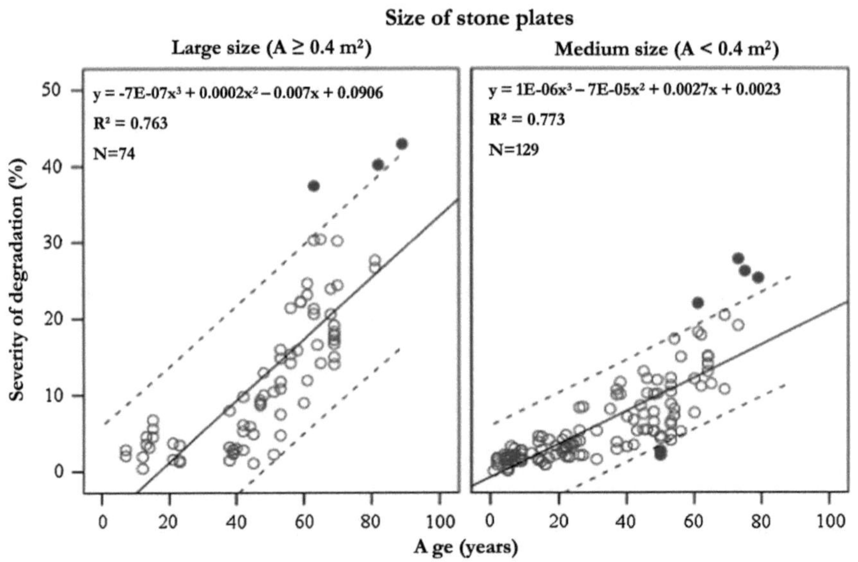

Fig. 3.6 Degradation evolution according to the area of the stone plates

claddings with medium-size plates. Silva et al. (2012) refer that larger plates have a lower relative area of joints and consequently a higher concentration of stresses; furthermore, larger plates are also thicker/heavier per unit area of façade, having a larger contact area between the plate and the substrate, making it more difficult to

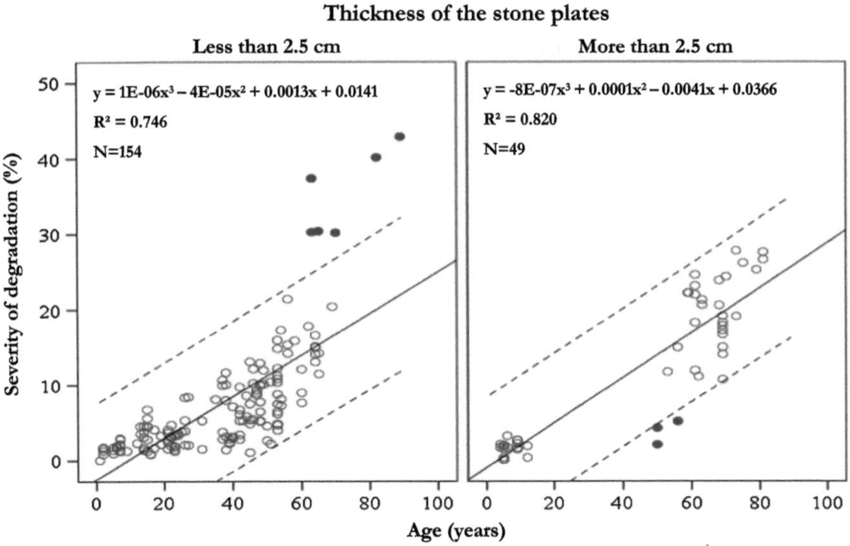

Fig. 3.7 Degradation evolution according to the thickness of the stone plates

guarantee the adhesion of the cladding. Therefore, in the sample analysed, larger stone plates directly adhered to the substrate are more prone to cracking and detachment from the substrate.

Figure 3.7 shows the degradation evolution of stone claddings according to the thickness of the stone plates. In this case, two categories are considered: stone plates less than 2.5 cm thick and stone plates more than 2.5 cm thick. A linear degradation curve is adopted for the thickness of stone plates, since the use of a polynomial trend leads to a curve without physical sense, i.e. for a given period of time façades degradation condition improves before deteriorating again. Both curves reveal high correlation values. Stone claddings less than 2.5 cm thick present lower service lives (about 2.9 %) than thicker claddings.

Concerning the location of the cladding on the façade, two groups are proposed: claddings higher up the façade (full or partial cladding) and those that are applied only on the ground floor of the building (at socle level). The degradation curves (Fig. 3.8) reveal a high correlation for the elevated cladding (R^2 = 0.866) and for cladding at socle level (R^2 = 0.730). The latter have a longer service life (4.3 % higher than claddings in higher levels), which may be explained by the easier access for maintenance actions and the protection against environmental weathering agents (although they are also more prone to vandalism).

Figure 3.9 presents the degradation evolution of stone claddings according to the orientation of the façade. The sample is uniformly distributed by the four categories considered. The degradation curves reveal a medium correlation level between the field data and the curve for the façades facing North and NE/E/SE, and a very good correlation for façades located W/NW and S/SW. According to Gaspar (2009),

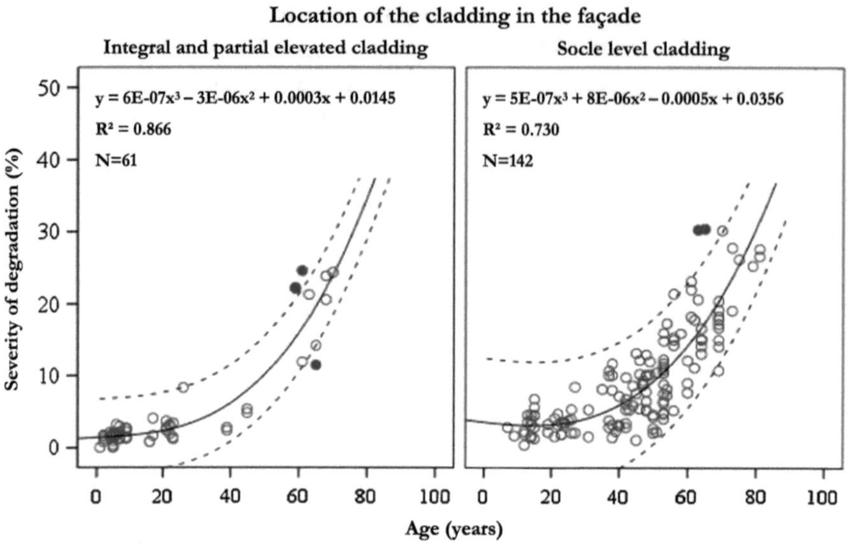

Fig. 3.8 Degradation evolution according to the location of the cladding in the façade

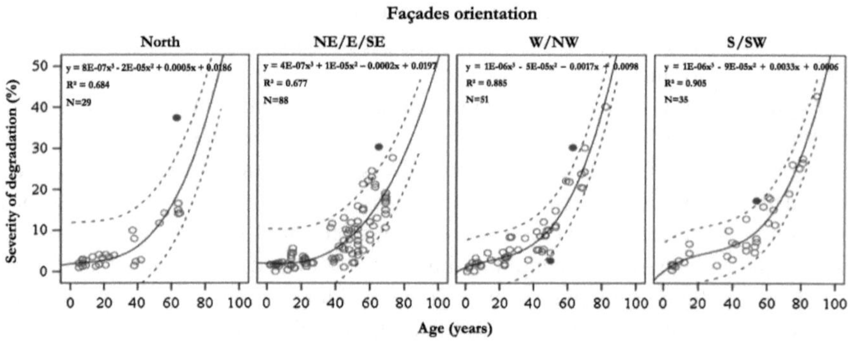

Fig. 3.9 Degradation evolution according to façades orientation

North and West are the most aggressive quadrants in Portugal: North is wetter and colder and the prevailing winds come from the west, with a greater probability of simultaneous occurrence of wind and rain. In the sample analysed, the claddings facing West are those with lower service lives; on the contrary, claddings facing South are the ones with higher service lives.

Figure 3.10 shows the degradation curves according to distance from the sea. Two categories are considered: façades located at more than 5 km (generally located in consolidated areas, protected from the effect of wind transporting marine salts) and façades located in coastal areas, less than 5 km from the sea. Claddings in coastal areas reach sooner the end of their service lives, with a decrease of 7 % of

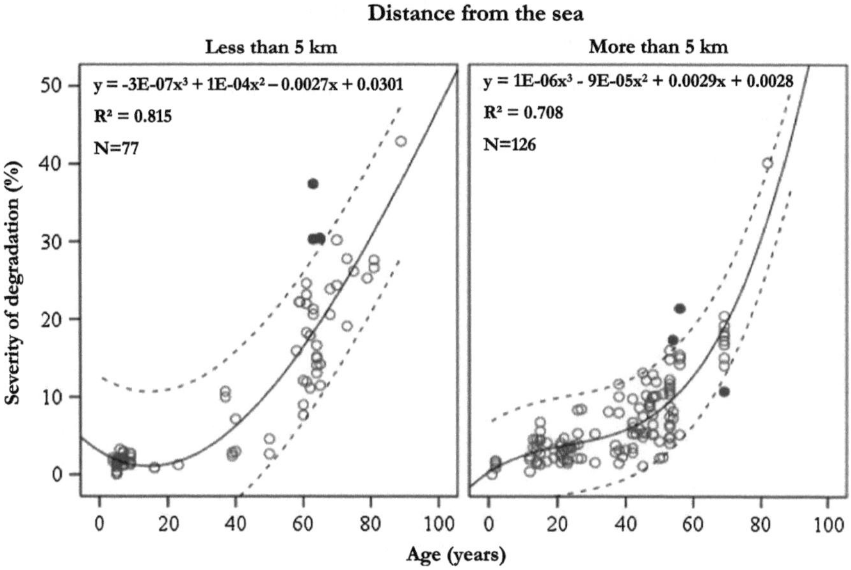

Fig. 3.10 Degradation evolution according to the distance from the sea

their expected service life when compared with claddings located at more than 5 km. In fact, various studies (Seinfeld and Pandis 1998; Pitts and Pitts 2000) reveal that in coastal regions, façades are exposed to the effect of (usually strong) winds, carrying salts—mainly chlorides and sulphates—which attack with greater or lesser severity all construction materials, promoting their degradation.

Concerning the combined effect of wind and rain action (Fig. 3.11), two categories are considered: moderate (buildings located in an urban area and protected by nearby buildings or vegetation from the influence of prevailing winds) and severe exposure. Both degradation curves reveal a good correlation between the sample and the model. The second-degree polynomial curve leads to an identical estimated service life for both situations. Analysing a less restricted degradation level (30 %), it is possible to detect that claddings with severe exposure are more prone to degradation, with a lower estimated service life.

Figure 3.12 presents the degradation curves according to exposure to damp. This study distinguishes between two situations: low and high exposure to damp. It is considered that buildings with high exposure to damp are close to damp sources (e.g. in coastal areas, near rivers or wetlands), in areas with high relative humidity, and are not protected by contiguous buildings and consequently are subjected to prevailing winds.

The degradation curves obtained are statistically significant, with a good correlation between the curve and the dataset (with a coefficient of determination higher than 0.8). Claddings with high exposure to damp are more prone to suffer higher degradation levels, showing a decrease in their estimated service lives around 11.3 % when compared to claddings with low exposure to damp.

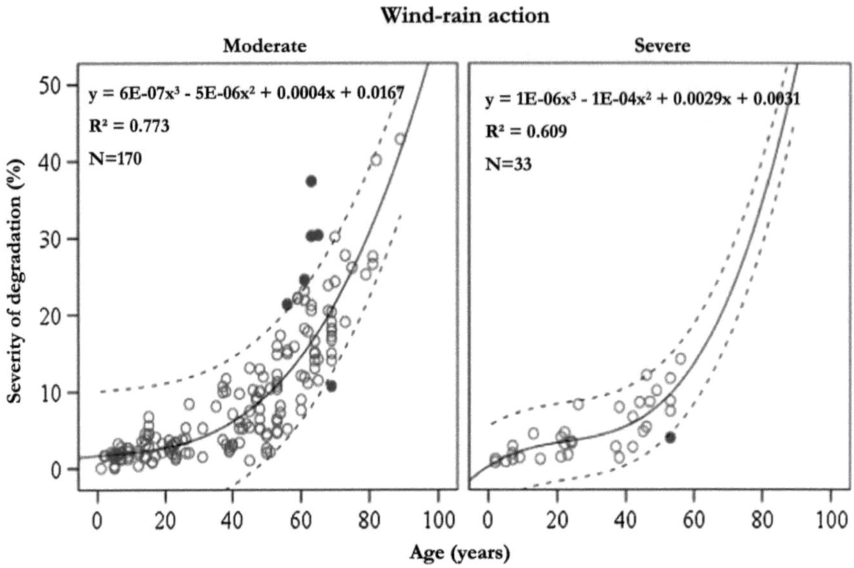

Fig. 3.11 Degradation evolution according to exposure to wind-rain action

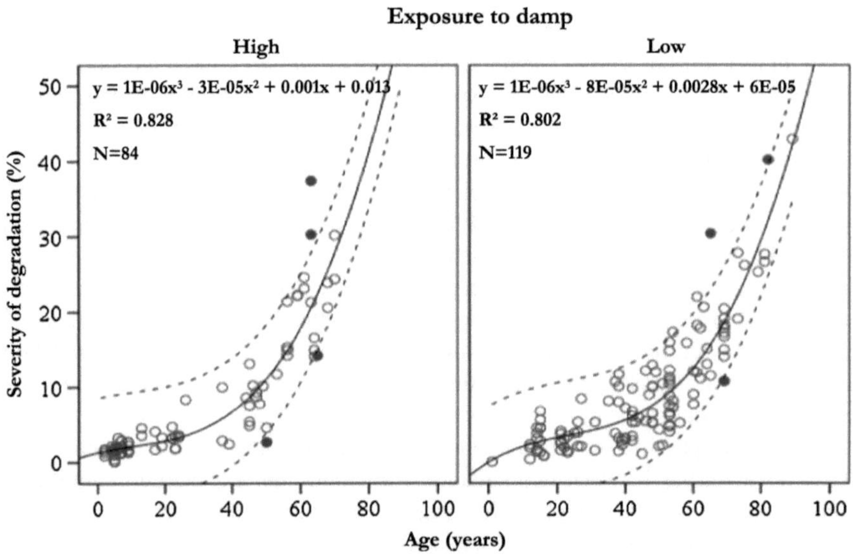

Fig. 3.12 Degradation evolution according to exposure to damp

Regarding the use of the building (Fig. 3.13), a distinction is made between (a) buildings accommodating services and commerce and (b) residential buildings. The degradation curves obtained present a coefficient of determination around 0.7, revealing a good correlation between the sample and the model. In this case, it is

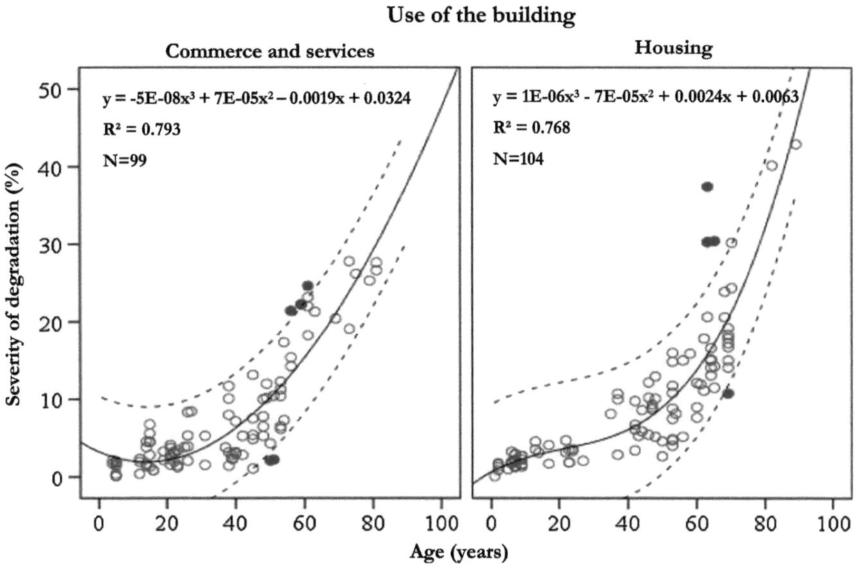

Fig. 3.13 Degradation evolution according to the use of the building

not possible to draw clear conclusions, since there is an intersection between the degradation curves of the two categories. Analysing a maximum degradation level of 20 %, it is concluded that residential buildings seem to be slightly more durable; however, if this limit were raised to 30 %, the results would be different and cladding systems in buildings accommodating services and commerce would last longer than claddings in residential buildings. Once again, it must be stressed that the theoretical limit that establish the end of service life of claddings depends significantly on the expectations of each individual and that it is not constant over time. Along these lines, Gaspar (2009) has shown that, except for high-end residential buildings, commercial and corporate buildings tend to have more demanding performance requirements (for reasons related to the image conveyed to customers) than current residential buildings; and therefore the former have shorter service life cycles on claddings.

Figure 3.14 presents the degradation evolution for natural stone claddings according to the height of the building. Concerning this parameter, two situations are identified: low buildings (up to three storeys high) and current buildings (more than 3 floors). Both degradation curves are statistically significant, revealing that claddings in low buildings are more durable, showing an increase of their estimated service lives of about 13 % when compared with taller buildings. This conclusion is coherent with several studies referring that the taller the building the greater its susceptibility to weather effects (Westberg et al. 2001).

Table 3.2 summarizes the results obtained (and previously discussed) for the degradation curves, using a linear and a polynomial trend line and according to two maximum acceptable degradation levels.

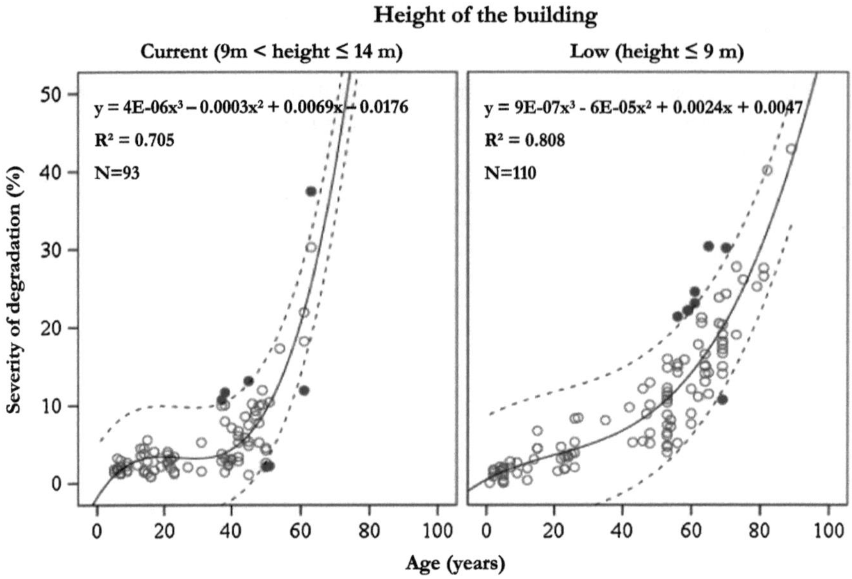

Fig. 3.14 Degradation evolution according to the height of the building

3.1.2.2 Rendered Façades

Table 3.3 presents the estimated service life of rendered façades according to different values of maximum allowable degradation levels. For a maximum degradation level of 20 %, the estimated service life varies between 16 and 18 years. In turn, for a maximum degradation level of 30 %, the estimated service life varies between 22 and 25 years. Shohet and Paciuk (2004) obtained similar results concerning the service life of rendered façades; adopting two levels of performance, the authors obtained for a more demanding performance level an estimated service life of 15 years (with a range between 12 and 19 years) and for a lower requirement level (situations where building's owners wish to minimize maintenance costs), an estimated service life of 23 years (with a range between 19 and 27 years).

The analysis of the degradation curves allows estimating the service life of rendered façades according to their characteristics (Table 3.4). The results obtained reveal that, on average, using a nonlinear regression model in the case of rendered façades produces a statistical significance increase of only 1.7 % when compared with a linear model. Therefore, this significance increase is practically residual. Additionally, the degradation curves defined with a linear trend produce a model easier to handle (from a mathematical viewpoint) to predict the service life of façade coatings. In some cases, the degradation curves obtained with a second-degree polynomial trend present two main disadvantages: (i) some degradation curves defined with few case studies have no physical meaning; (ii) the estimated service life using a second-degree polynomial trend, reveals inconsistent results for some

Table 3.2 Estimated service life of stone claddings according to their characteristics

Stone claddings characteristics		Number of case studies	Simple linear regression R^2	Estimated service life (years) 20 %	30 %	Polynomial regression (3rd degree) R^2	Estimated service life (years) 20 %	30 %
Type of stone	Limestone	72	0.627	83	123	0.691	72	86
	Marble	77	0.680	69	96	0.837	66	77
	Granite	54	0.203	234	359	0.204	96	111
Colour	Light	134	0.692	77	111	0.828	70	81
	Dark	69	0.427	82	117	0.744	61	67
Type of finishing	Smooth	96	0.523	89	130	0.700	69	80
	Rough	107	0.652	74	106	0.775	69	80
Area of stone plates	Medium size	129	0.660	96	142	0.773	72	82
	Large size	74	0.640	67	92	0.763	65	78
Thickness of the stone plates	Less than 2.5 cm	154	0.541	82	119	0.746	67	77
	>2.5 cm	49	0.790	71	104	0.820	69	90
Location of the cladding	Integral/partial	61	0.777	75	110	0.866	66	76
	Bottom wall	142	0.588	75	104	0.730	69	80
Orientation	North	88	0.587	87	126	0.677	66	75
	NE/E/SE	29	0.597	78	113	0.684	70	82
	W/NW	51	0.701	68	97	0.885	64	74
	S/SW	35	0.705	73	104	0.905	72	82
Distance from the sea	≤5 km	77	0.751	66	96	0.815	66	80
	>5 km	126	0.549	96	136	0.701	71	80
Wind-rain action	Severe	33	0.565	126	190	0.609	68	77
	Moderate	170	0.648	75	136	0.773	68	80
Exposure to damp	High	84	0.728	70	102	0.828	63	71
	Low	119	0.626	78	109	0.802	71	81
Use of the building	Commerce and services	99	0.668	77	110	0.792	68	81
	Housing	104	0.614	78	113	0.768	69	79
Height	Current	93	0.407	97	143	0.705	60	66
	Low	110	0.689	74	106	0.808	69	80

features of the rendered facades, e.g. the façades protection level, where the estimated service life of façades with unfavourable conditions is higher than the estimated service life of renders with protection; using a linear regression trend, this inconsistency does not happens. Thus, for rendered façades and analysing the results obtained using a simple linear regression (for a maximum degradation level of 20 %), it seems that, in physical terms, the results obtained are consistent:

Table 3.3 Estimated service life of rendered façades according to the maximum allowable degradation level

	Simple linear regression (R^2 = 0.856)			Simple polynomial regression, 2nd degree (R^2 = 0.868)			Simple polynomial regression, 3rd degree (R^2 = 0.881)		
	10 %	20 %	30 %	10 %	20 %	30 %	10 %	20 %	30 %
Estimated service life (years)	9	16	22	10	18	25	8	16	25

Table 3.4 Estimated service life of rendered façades according to their characteristics

			Simple linear regression			Polynomial regression (2nd degree)		
Renderings characteristics		Number of case studies	R^2	Estimated service life (years)		R^2	Estimated service life (years)	
				20 %	30 %		20 %	30 %
Render type	Current cement renderings	60	0.813	17	25	0.821	16	24
	Lime–cement renderings	21	0.813	16	22	0.831	18	25
	Renderings with crushed marble	6	0.977	20	29	0.977	20	29
	Single-layer renderings	13	0.488	18	26	0.494	24	–
Colour	White	19	0.806	13	18	0.837	14	17
	Light	48	0.846	16	22	0.880	21	27
	Dark	33	0.934	17	25	0.934	17	24
Height	Low ($h \leq 9$ m)	13	0.883	13	19	0.885	13	18
	Current ($9 < h \leq 14$ m)	50	0.841	17	23	0.870	21	28
	Alto ($h > 14$ m)	37	0.884	16	24	0.888	17	25
Building shape	Irregular	41	0.887	15	21	0.899	17	24
	Compact	59	0.872	17	25	0.873	16	25
Detailing/design level	Inferior	29	0.854	15	22	0.864	13	20
	Medium	65	0.860	16	22	0.882	20	26
	Superior	6	0.982	20	28	0.996	17	27
Eaves' protection	Without protection	47	0.872	14	20	0.904	18	24
	With protection	53	0.876	16	23	0.876	16	23
Ground floor protection (socle)	Without protection	17	0.956	15	21	0.956	15	22
	With protection	83	0.843	16	22	0.857	18	25
Platbands copings	Without copings	79	0.848	15	22	0.857	17	24
	With copings	21	0.912	17	25	0.942	21	28
Balcony copings	Without copings	34	0.874	15	21	0.881	17	23
	With copings	66	0.864	19	28	0.866	18	27

(continued)

Table 3.4 (continued)

Renderings characteristics		Number of case studies	Simple linear regression R^2	Estimated service life (years) 20 %	30 %	Polynomial regression (2nd degree) R^2	Estimated service life (years) 20 %	30 %
Façades orientation	North	15	0.869	17	22	0.944	25	30
	Northeast	19	0.814	15	22	0.815	14	21
	East	14	0.774	15	23	0.776	15	22
	Southeast	11	0.927	17	24	0.939	15	21
	South	13	0.922	17	24	0.974	22	28
	Southwest	10	0.839	19	28	0.875	13	20
	West	15	0.915	14	20	0.921	15	21
	Northwest	3	0.991	19	28	1.000	11	14
Distance from the sea	≤3.5 km	23	0.837	14	21	0.837	14	21
	>3.5 km	77	0.861	16	23	0.886	20	27
Exposure to damp	Unfavourable	11	0.606	13	18	0.683	15	18
	Current	66	0.874	16	21	0.922	20	26
	Favourable	23	0.915	18	25	0.915	17	25
Distance from pollution sources	Unfavourable	24	0.880	15	22	0.886	14	21
	Current	70	0.849	16	22	0.873	20	27
	Favourable	6	0.974	16	23	0.983	15	21
Façade protection level	Without protection	46	0.829	15	21	0.862	19	25
	Normal situation	24	0.930	16	23	0.930	15	22
	With protection	30	0.928	17	25	0.931	17	25

- It is confirmed that current cement renderings have an estimated service life around 5.9 % higher than lime–cement renderings; Kus et al. (2004) refer that lime–cement renderings are less permeable to water, retarding the water penetration into the material but retaining the water for longer periods in its interior, promoting the occurrence of various defects. In turn, according to the sample analysed, single-layer renderings present higher durability than current cement rendering or lime–cement renderings. Single-layer renderings correspond to industrial mortars, with a high variety of compositions; the composition of this type of mortar is always well controlled by being pre-dosed in factory, giving them an advantage relative to renders dosed in situ. Finally, according to the sample analysed, renderings with crushed marble are the most durable ones; this finding can be explained by the fact that the buildings that have this type of rendering also possess a superior design level (Gaspar 2009);
- Concerning the façade's colour, white and light colours present lower estimated service lives than renderings with dark colours. According to Gaspar (2009), the method used to evaluate the degradation condition (visual inspections) of

rendered façades can explain, to a certain extent, the results obtained. In fact, in light coloured façades the detection of defects is easier, contrary to what happens with darker colours that tend to mask problems;

- In the sample analysed, current buildings have an estimated service life higher than tall buildings; however, both have an estimated service life higher than lower buildings. There are a number of possible explanations to this result. To begin with, lower buildings represent only 13 % of the sample (only 13 case studies, which not allow obtaining unequivocal conclusions about the degradation trend of rendered façades in this type of building); furthermore—in the sample studied—this type of buildings tend to have a smaller number of dwellers and tend to be older, and therefore the users' financial ability to pay for frequent maintenance on the façades is more limited; finally in isolated detached houses there is a higher exposed area of the façade, relative to the building envelope when compared to apartment buildings;
- Regarding buildings' shape, two categories are adopted: (i) compact buildings, without protruding elements or indentations, with a regular architecture; (ii) irregular buildings, with prominent and exceptional elements in the building façade. Façades in buildings with compact shape have an estimated service life about 11.8 % higher than façades in building with irregular shape, which somehow also reinforces the results obtained relative to the building height;
- Concerning the design level, façades in buildings with a superior level of detail reveal a better performance level, with an estimated service life about 20 % higher than façades in buildings with a medium level of detail and around 25 % higher than façades in buildings with an inferior level of detail;
- Rendered façades in buildings with eaves' protection have an estimated service life around 12.5 % higher than façades in buildings without protection;
- Regarding the buildings' ground floor protection (socle), renderings in buildings with protection have a better performance, with an estimated service life around 6.3 % higher than renderings in unprotected façades;
- Façades in buildings without copings deteriorate faster (with an estimated service life approximately 21.1 and 11.8 % lower, respectively, than façades with balconies and platbands copings);
- In Portugal, as mentioned before, North and West quadrants are the most aggressive; in the sample analysed, renderings facing West have the lowest estimated service life (14 years); on the other hand, renderings faced Southwest and Northwest are the ones with better performance;
- Rendered façades located in coastal areas—at less than 3.5 km from the sea—present an estimated service life around 12.5 % lower than façades at more than 3.5 km (especially if the latter are located in consolidated areas, which tend to mitigate the effects of the wind that carries marine salts);
- Concerning exposure to damp, façades with favourable exposure conditions are those with better performance, reaching the end of their service life about 11.1 % later than renderings with current exposure conditions and about 27.8 % later than renderings with unfavourable exposure conditions;

- Regarding distance from pollution sources, three situations are considered: (i) favourable conditions, for buildings located outside urban centres, in an environment without pollution sources; (ii) current conditions, for buildings located in urban areas with moderate traffic; (iii) and unfavourable situations, for buildings located in urban centres, close to main roads. The difference between the three categories considered are not significant; however, renderings with favourable conditions present an estimated service life about 6.3 % higher than renderings subjected to unfavourable conditions;
- In rendered façades, the protection level is directly related to the exposure to wind-driven rain; the simultaneous occurrence of wind and rain causes the wetting of the façade and directly affects the durability of the façade coatings—especially those made of porous materials—enhancing the appearance of more or less serious defects such as moisture staining, runoff staining, efflorescence, biological colonization, occurrence of thermophoresis phenomena, which in turn may lead to cracking or a combination of defects (Franke et al. 1998; Blocken and Carmeliet 2004; Briggen et al. 2009); in this study, three categories are adopted concerning the protection level of the rendered façades: (i) protected façades, in low-rise buildings, in compact urban environments, protected from the prevailing winds by the surrounding buildings, the topography or vegetation; (ii) current situation, in mid-rise buildings in current urban contexts, protected from the action of prevailing winds; and (iii) façades without protection, located in tall buildings or in open fields/open spaces, street corner buildings or located on crossroads. The results show that façades with protection have an estimated service life 5.9 and 11.8 % higher than façades in current situations and without protection, respectively.

3.1.2.3 Painted Surfaces

The estimated service life of painted surfaces according to three maximum allowable degradation levels (10, 20 and 30 %) is analysed in Table 3.5. In painted surfaces, for a maximum degradation level of 10 %, an estimated service life that varies between 6 and 7 years is obtained. For a maximum degradation level of 20 %, the estimated service life varies between 9 and 10 years and, for a maximum

Table 3.5 Estimated service life of painted surfaces according to the maximum allowable degradation level

	Simple linear regression (R^2 = 0.820)			Simple polynomial regression, 2nd degree (R^2 = 0.886)			Simple polynomial regression, 3rd degree (R^2 = 0.886)		
	10 %	20 %	30 %	10 %	20 %	30 %	10 %	20 %	30 %
Estimated service life (years)	6	9	11	7	10	12	7	10	12

degradation level of 30 %, the estimated service life varies between 11 and 12 years.

Table 3.6 presents the estimated service life of painted surfaces according to their characteristics. In this study, a maximum degradation level of 20 % is adopted to establish the end of the service life of painted surfaces. This maximum deterioration index corresponds to a compromise level of demand between a level too restrictive (10 %), which would mean repainting of the façade every 6th or 7th year,

Table 3.6 Estimated service life of painted surfaces according to their characteristics

			Simple linear regression			Polynomial regression 3rd degree)		
Painted surfaces characteristics		Number of case studies	R^2	Estimated service life (years)		R^2	Estimated service life (years)	
				20 %	30 %		20 %	30 %
Type of product	Smooth paints	79	0.844	8.9	12.0	0.930	9.9	11.9
	Plastic membranes	84	0.819	8.5	11.6	0.907	9.4	11.5
	Silicate and silicone paints	9	0.907	6.7	9.1	0.981	9.4	11.2
	Textured paints	48	0.760	8.8	11.5	0.812	9.9	12.3
Colour	White	66	0.803	8.9	12.0	0.910	9.7	11.8
	Light	144	0.829	8.5	11.4	0.892	9.7	11.9
	Dark	10	0.807	6.7	9.8	0.813	4.7	8.3
Type of finishing	Smooth	124	0.815	8.5	11.5	0.819	9.7	11.7
	Rough	96	0.927	8.6	11.5	0.869	9.8	12.2
Surface preparation	Paint over rendering	31	0.814	9.2	11.4	0.822	9.7	11.7
	Repainting over previous paint coat	17	0.638	9.3	11.7	0.680	9.7	11.7
Building shape	Compact	179	0.809	8.6	11.6	0.875	9.7	12.0
	Irregular	41	0.875	7.7	10.5	0.945	9.0	10.9
Façades orientation	E/SE	64	0.795	8.9	11.7	0.863	9.9	12.1
	N/NE	41	0.736	9.5	13.0	0.759	9.9	12.6
	W/NW	66	0.869	8.1	10.9	0.945	9.4	11.3
	S/SW	34	0.876	7.7	10.5	0.936	9.0	11.2
Wind-rain action	Severe	72	0.749	9.0	12.6	0.863	9.5	11.4
	Moderate	98	0.856	8.1	10.9	0.940	9.7	11.8
	Low	37	0.781	9.4	11.7	0.800	10.0	12.2
Distance from the sea	≤5 km	114	0.832	8.4	11.5	0.914	9.4	11.5
	>5 km	106	0.794	8.7	11.5	0.863	9.8	12.0

(continued)

Table 3.6 (continued)

Painted surfaces characteristics		Number of case studies	Simple linear regression			Polynomial regression 3rd degree)		
			R^2	Estimated service life (years)		R^2	Estimated service life (years)	
				20 %	30 %		20 %	30 %
Exposure to damp	High	116	0.833	8.5	11.5	0.915	9.4	11.5
	Low	104	0.792	8.7	11.5	0.861	9.9	12.0
Distance from pollution sources	Current	174	0.801	8.6	11.6	0.870	9.7	11.8
	Unfavourable	46	0.844	8.3	11.1	0.912	9.8	12.1
Use of the building	Commerce and services	72	0.739	8.7	11.6	0.804	9.5	11.5
	Housing	148	0.845	8.4	11.4	0.915	9.8	12.0
Height	Low	161	0.837	8.4	11.2	0.895	9.8	12.0
	Current	59	0.736	9.3	13.0	0.861	9.6	11.3

a goal that cannot be fulfilled in practice—despite some legal recommendations of maintenance cycles of 8 years—and a maximum degradation level of 30 %, where the coatings' degradation is already widespread, generating a negative impact on the perception of the building stock.

The service life of painted façades is estimated using a third-degree polynomial trend, with a very good correlation between the degradation curve and the dataset and consequently with a high statistical significance. As mentioned before, this study is based on data collected through visual inspections of façades in in-service conditions; for this reason, some features of the coatings cannot be accurately measured. Specifically, the type of paint applied is not easy to determine (additional diagnosis techniques would be necessary, involving for example chemical analyses, outside the scope of this study). Because of this limitation, in the present book, four generic categories are adopted to describe the type of paint: (i) smooth paints, which include traditional (water-based) and non-traditional (e.g. with incorporation of nanoparticles) smooth paints; (ii) textured paints (water-based paints with sand in its formulation, which gives them some resistance and a rough finishing); (iii) elastic membranes, which correspond to water-based paints with acrylic and styrene-acrylic resins; (iv) and silicones and silicates paints, the first of which are emulsion paints with siloxane resins and the second ones are silicate dispersion paints.

From the analysis of the estimated service lives according to the characteristics of the painted surfaces, the following conclusions are obtained:

- In the sample analysed, smooth and textured paints are those that have higher estimated service lives, about 5 % higher than silicones/silicates paints and elastic membranes;

- Temperature variations promote the presence of defects and the loss of adhesion of painted surfaces. The surface temperature of the coatings depends on the colour of the pigments used in paint formulation; Teo et al. (2005) refer that light coloured coatings with smooth finishing can better reflect UV radiation than dark coloured and rough finished coatings; the results obtained confirm this degradation pattern, since in the sample analysed painted surfaces with dark colours have estimated service lives approximately 51.5 % lower than white-painted surfaces or those in light colours;
- Painted surfaces with rough finishing present estimated service lives about 1 % higher than painted surfaces with smooth finishing. This conclusion is confirmed by the results obtained by Garrido et al. (2012), which predict the service life of 100 painted surfaces in old buildings and conclude that textured paints are more durable than smooth finished paints. This result can be explained to a certain extent by the presence of sand in rough finished paints, which gives them higher strength and durability than traditional smooth paints;
- Regarding surface preparation, two situations are considered: paint over rendering and repainting over previous paint coat. The evaluation of this characteristic is complex, since more often than not it is impossible to identify through visual inspection only which of the two situations occurs; thus, it is necessary to consult the archives from Municipalities or from the construction firms responsible for the execution of the work. Through such methodology it was possible to identify the surface preparation conditions in 48 case studies only. Unfortunately, the analysis of this sample does not allow obtaining significant differences between the estimated service life values for the two categories analysed;
- Concerning the buildings' shape, painted surfaces located in buildings with irregular shape have estimated service lives around 7 % lower than façades in buildings with compact shape; generally, the existence of protruding elements promotes the retention of water and dirt, leading to defects in the façade;
- Painted surfaces in buildings facing South are those with lower estimated service lives, as suggested by the study of Garrido et al. (2012), since for the studied sample South-oriented buildings are exposed to UV radiation for longer periods of time, which significantly affects the painted surfaces by accelerating its degradation process; in the sample analysed, there is a prevalence of defects concerning colour and brightness changes, cracking and loss of adhesion in façades in buildings facing South and West; on the other hand, painted surfaces in buildings facing North and East are those with higher estimated service lives;
- Painted surfaces with severe exposure to wind-driven rain action have an estimated service life around 2 % lower than coatings with moderate exposure and about 5 % lower than coatings with low exposure;
- Painted surfaces in coastal areas show a decrease of their estimated service life of around 4.1 % relative to coatings located at more than 5 km from the sea;
- Coatings with low exposure to damp present an estimated service life 5.1 % higher than coatings with high exposure;

- Contrarily to expectations, painted surfaces with unfavourable exposure conditions to pollutant sources have a slightly higher service life (around 1 %) than painted surfaces with current exposure to pollution; these results can perhaps be explained by the synergy between different degradation agents, since in the sample analysed the majority of painted surfaces with unfavourable exposure to pollutants are located in urban areas, at more than 3.5 km from the sea and with lower exposure to damp, which leads to reducing the degradation indexes of these coatings;
- Regarding the use conditions, in the sample analysed, painted surfaces in residential buildings have estimated service lives 3.1 % lower than painted surfaces located in buildings with services and commerce, the reasons for these results are identical to the ones presented on the section of the stone claddings;
- Painted surfaces in lower buildings present estimated service lives around 2 % higher than those in tall buildings, perhaps because according to Teo et al. (2005) these buildings are more exposed to solar radiation and more subjected to the effects of wind and rain action.

3.1.2.4 Ceramic Tiling Systems

Table 3.7 presents the estimated service life of ceramic tiling systems according to three maximum allowable degradation levels (10, 20 and 30 %). In this case, the criteria described for the other claddings previously analysed are adopted. A maximum degradation level of 20 % is assumed to establish the end of the service life of ceramic tiling systems. For this degradation level, an estimated service life that varies between 46 and 50 years is obtained.

Table 3.8 shows the estimated service life values of ceramic tiling systems according to their characteristics. In an overall analysis, it is found that a third-degree polynomial trend is the best model to describe the loss of performance of ceramic claddings over time. Assuming a maximum degradation level of 20 %, the following conclusions are obtained:

- Regarding the tiles' surface, Casasola et al. (2012) refer that the glaze provides the ceramic tiles with aesthetic properties, making them less porous and more resistant (chemically and mechanically); unfortunately, in the sample analysed, only 28 case studies are of unglazed ceramic tiles, with an average age of

Table 3.7 Estimated service life of ceramic tiling systems according to the maximum allowable degradation level

	Simple linear regression (R^2 = 0.659)			Simple polynomial regression, 2nd degree (R^2 = 0.765)			Simple polynomial regression, 3rd degree (R^2 = 0.767)		
	10 %	20 %	30 %	10 %	20 %	30 %	10 %	20 %	30 %
Estimated service life (years)	29	46	63	36	49	59	36	50	60

Table 3.8 Estimated service life of ceramic tiling systems according to their characteristics

Renderings characteristics		Number of case studies	Simple linear regression R^2	Estimated service life (years) 20 %	30 %	Polynomial regression (3rd degree) R^2	Estimated service life (years) 20 %	30 %
Surface	Glazed	167	0.673	46	62	0.774	49	59
	Not glazed	28	0.465	77	116	–	–	–
Colour	Light	115	0.701	45	59	0.773	49	58
	Dark	80	0.605	55	80	0.734	54	63
Size	$L \leq 20$ cm	162	0.655	46	62	0.757	50	60
	$L > 20$ cm	33	0.612	58	86	0.656	44	50
Substrate	Masonry	106	0.654	44	59	0.724	49	59
	Concrete	89	0.488	70	102	0.613	52	59
Peripheral joints	With joint	29	0.793	45	64	0.867	52	62
	Without joint	166	0.638	46	63	0.753	50	59
Peripheral protection	With protection	101	0.683	47	63	0.785	50	60
	Without protection	94	0.637	46	63	0.752	50	59
Façades orientation	North	58	0.631	47	66	0.705	49	60
	South	41	0.793	41	55	0.872	47	57
	East	40	0.635	49	68	0.748	53	61
	West	43	0.556	51	71	0.681	51	60
	Northwest	3	0.760	47	66	–	–	–
	Southeast	8	0.394	56	76	–	–	–
	Southwest	1	–	–	–	–	–	–
	Northeast	1	–	–	–	–	–	–
Distance from the sea	≤5 km	139	0.661	45	61	0.757	50	60
	>5 km	56	0.435	72	105	0.501	53	61
Wind-rain action	Severe	53	0.646	42	58	0.743	49	59
	Moderate	97	0.689	49	68	0.806	50	60
	Low	45	0.582	51	69	0.673	52	62
Exposure to damp	High	84	0.689	42	57	0.774	48	58
	Low	111	0.644	52	72	0.770	53	62
Height	High	68	0.694	46	65	0.822	51	60
	Current	127	0.641	46	62	0.732	50	60

24 years, which makes it impossible to define a degradation curve based on a third-degree polynomial that properly expresses the degradation of this type of coating; therefore, it is not possible to obtain unequivocal conclusions regarding the influence of the tiles' surface on the degradation of ceramic claddings; in further analysis performed in this book and using other methodologies, this question will be addressed again;

- Ceramic claddings with dark colours present higher estimated service lives (around 9.3 %) than those with light colours; this result should be analysed with caution, since dark colours tend to mask some defects, making the inspection of this type of claddings more difficult;
- In this study, two categories are considered concerning the size of the tiles: large tiles ($L > 20$ cm) and small tiles ($L \leq 20$ cm). Large tiles are more prone to suffer from detachment, and BS 5385: 1991 recommends the use of metal fasteners when ceramic tiles larger than 20 cm are used; in this study, ceramic claddings with small-size tiles present an estimated service life 11.1 % greater than claddings with large tiles;
- In the application of ceramic claddings, it is essential to ensure an adequate preparation of the substrate, as well as the selection of materials compatible with each other, ensuring the adhesion and durability of the coating system; in the sample analysed, ceramic tiles applied over masonry substrates present an estimated service lives around 5.8 % lower than those applied over concrete substrates; it is important to notice that ceramic claddings applied over masonry substrates correspond to older buildings (with an average age of 43 years) while those over concrete substrates correspond to newer buildings (with an average age of 24 years);
- Peripheral joints allow accommodating the deformation of the ceramic tiles system and, as expected, claddings without peripheral joints present lower estimated service lives (with a decrease of 3.8 % relative to claddings with joints);
- Regarding peripheral protection, in the sample analysed there are no significant differences between the estimated service life for claddings with and without protection;
- Various authors (Ho et al. 2005; Yiu et al. 2007) have demonstrated the influence of environmental agents on the degradation of ceramic tiles; according to Zhu and Guan (1997), the degradation of ceramic claddings occurs mainly due to cyclical environmental actions such as heating/cooling, wetting/drying or freeze/thaw; some authors (Guan et al. 1997; Yiu et al. 2007) also analyse the importance of solar radiation and temperature gradient on the deterioration of ceramic claddings; in Portugal, the façades facing South are those more exposed to solar radiation; in the sample analysed, ceramic claddings in buildings facing South are those with lower estimated service lives;
- Ceramic claddings in coastal areas tend to present higher degradation rates with estimated service lives lower than claddings located in inland areas; claddings located at less than 5 km from the sea show a decrease in its service life of 5.7 % compared to claddings located more than 5 km from the coast;

- A greater aggressiveness of the exposure conditions to wind-rain action leads to a decrease of the estimated service life of ceramic claddings; claddings with low exposure present an estimated service life around 3.8 % higher than those with moderate exposure and around 5.8 % higher than those with severe exposure;
- Concerning exposure to damp, claddings with low exposure levels present an increase of their service life of 9.4 % relative to those with high exposure to damp;
- Regarding the height of the building, there are no significant differences between tall and current buildings; however, tall buildings tend to present slightly higher estimated service lives.

3.2 Multiple Linear Regression Analysis

3.2.1 Theoretical Foundations

Multiple linear regression analysis can be seen as an extension of the simple linear analysis. Contrary to what occurs with the simple regression (that only includes a single explanatory variable), multiple linear regression incorporates more than one independent variable (Satapathy et al. 2009). Wooldridge (2009) refers that multiple regression analysis allows obtaining more efficient models, since it allows introducing a large number of variables that contribute to the explanation of the dependent variable. According to Sykes (1992), mathematics, unlike humans, is able to visualize more than three dimensions, referring that multiple linear regression is able to handle an arbitrarily large number of independent variables. This perception raises the question of how to select the variables that actually contribute to the model's explanation. This is particularly relevant in service life prediction, where often there are multiple variables that affect the durability of the building components. Multiple linear regression allows identifying, among all independent variables included in the model, those that best contribute to explain the variability of the dependent variable. In this respect this methodology answers to the concertos of a number of authors that argue that an effective service life prediction methodology should be able to distinguish the significant variables from those less significant (Martin et al. 1996).

In multiple linear regression, the relationship between the dependent and the independent variables is given generically by Eq. (3.11).

$$y = B_0 + B_1 \cdot x_1 + B_2 \cdot x_2 + \cdots + B_k \cdot x_k + \varepsilon \tag{3.11}$$

where y represents the dependent variable, B_0, B_1 and B_k the regression coefficients, x_1, x_2 and x_k the independent variables and ε the model random errors.

There are different methods for the appropriate selection of independent variables (Samal et al. 2008). In this study, a stepwise technique is used, which is a step-by-step selection method that uses a sequential procedure to determine the variables to be included in the model. In the stepwise method, the variables that do not comply with the selection criteria are excluded. Thus, the basic assumptions of the regression analysis are tested and the effects of multicollinearity, i.e. pseudo-independent variables that are interrelated and thus jeopardize the multiple regression analysis are eliminated (Leung et al. 2001). However, statistics are not always a good replacement for logical understanding and knowledge of the scientific bases in the area under analysis (Maroco 2007; Freedman 2010). More important than obtaining good statistical results is the careful examination of the results obtained, evaluating whether they agree with what is to be expected both theoretically and physically. Sometimes, when the researcher has a wide range of potential explanatory variables, it is necessary to have a previous knowledge concerning how the variables are related to each other, having established beforehand a hierarchy of variables that seem more relevant to the explanation of the phenomena under analysis (Williams 2004; Morad et al. 2007).

3.2.1.1 Statistical Inference of Multiple Linear Regression Model

As in simple regression analysis, in multiple linear regression, the Pearson's product moment correlation coefficient (r) and the coefficient of determination (R^2) can be used to evaluate the goodness-of-fit of the model to the data. However, it is important to notice that the determination coefficient increases as the number of independent variables grows (D'Urso and Santoro 2006; Taylor 2009). Therefore, the adjusted R^2 (adjusted determination coefficient) is also used, which only increases if the addition of a new variable leads to an improvement in the adjustment of the regression model to the data—Eq. (3.12).

$$R_a^2 = 1 - \frac{(1 - R^2)(n - 1)}{(n - 1 - p)} \tag{3.12}$$

where n represents the number of cases in the sample and p the number of independent variables.

In the selection of variables (using a stepwise technique), it is necessary to evaluate whether the independent variables are predictors of the dependent variable (degradation severity) through the model's significance test. The significance of the model may be tested evaluating two hypotheses:

- Null hypothesis (H_0): $B_1 = B_2 = \cdots = B_k = 0$;
- Alternative hypothesis (H_1) : $\exists i : B_i \neq 0$ (i.e. there is at least one i for which B_i is different from zero).

If the null hypothesis (H_0) is true, then none of the independent variables included in the model explains the variability of the dependent variable. On the other hand, if the alternative hypothesis (H_1) is true, then at least one of the variables analysed significantly contributes to explaining the dependent variable. Box and Draper (1987) refer that essentially all models are wrong, but some are useful. This means that all mathematical models have some uncertainty and some degree of imprecision associated to them. In the analysis of model's significance, two errors may occur:

- Type I error: when the null hypothesis is rejected, being true;
- Type II error: when the null hypothesis is accepted, being false.

The statistical tests are therefore adapted to control type I error through the level of significance (α), which is the probability of rejecting the null hypothesis, when it is true. Using as example the law enforcement terms, researchers aim to avoid type I errors, since when there is a reasonable doubt, the jury must decide in favour of the defendant (i.e. it is better that ten guilty persons escape than that one innocent suffers, as referred by English jurist William Blackstone). Davis et al. (2014) refer that in a courtroom, the jury needs to establish a reference value for the evaluation of guilt. The authors also state that if the jury decides that they must be 95 % sure of their verdict (which corresponds to an alpha level of 5 %), then if a mistake is made, the jury can wrongly convict an innocent 5 % of the time and release a guilty person 95 % of the time.

In the literature, there is some controversy associated with the statistical significance of the regression analysis though. According to a number of authors (Royall 1997; Hubbard and Bayarri 2003; Gigerenzer et al. 2004), it is not possible to precisely define what "statistically significant" means, because the literature on the subject is also mixed up concerning this concept. Currently, the scientific literature suggests that for a given level of significance or probability of type I error (α) H_0 should be rejected if the p-value is less than or equal to α (Rice 1989). According to Dallal (2007), 5 % is considered the default significance level, and the term "statistically significant" is usually associated with p-value ≤ 0.05. The author also states that there are different theories to explain the use of a p-value equal to 5 %; however, they all refer to Fisher and his study published in 1925—"Statistical Methods for Research Workers". In the end, the definition of p-value is not consensual (Altman 1991; Ziliak and McCloskey 2009). Therefore, each researcher must evaluate, for each specific situation, which is the p-value that makes more sense to obtain the best regression model.

To test the hypotheses regarding the significance of the model, it is necessary to analyse the total variability of the dependent variable (y) that is given by the total SST. As discussed before (in the explanation of the coefficient of determination—R^2), the total SST is given by the sum of the variance explained by the model (SSR) and the variance due to the errors (SSE). This leads to the relationship expressed in Eq. (3.13) (Sirkin 2006).

$$\sum_{i=1}^{n}(y_i-\bar{y})^2=\sum_{i=1}^{n}(\hat{y}_i-\bar{y})^2+\sum_{i=1}^{n}(y_i-\hat{y}_i)^2 \quad \text{i.e. SST} = \text{SSR} + \text{SSE} \tag{3.13}$$

where y_i is the observed values of the dependent variable, $\bar{y}$ the average of the observed values of the dependent variable, $\hat{y}_i$ the values predicted by the regression model, SSR the regression SST (the variance explained by the regression of y on x) given by $\sum_{j=1}^{n}\left(\hat{Y}_j-\bar{Y}\right)^2$, SSE the residual or error SST, which represents the residual variance or the variance due to the errors (not explained by the regression model), given by $\sum_{j=1}^{n}\left(Y_j-\hat{Y}_j\right)^2$, and SST the total SST, which is the variance of the dependent variable, given by $\sum_{j=1}^{n}\left(Y_j-\bar{Y}\right)^2$; the total variability is the sum of the variance explained by the model (SSR) and the variance explained by the errors (SSE).

To evaluate the model's significance, the F-test can also be used. The F-test is given by the ratio between the variance explained by the regression model (SSR) and the variance explained by the model's errors (SST residuals), as expressed in Eq. (3.14).

$$F=\frac{\text{SSR}/p}{\text{SSE}/(n-p-1)}=\frac{\text{MSR}}{\text{MSE}} \tag{3.14}$$

where MSR are the mean squares of the regression model and MSE are the mean squares of the errors.

If the null hypothesis (H_0) is true, the F-test has a Fisher–Snedecor distribution. Therefore, to evaluate the statistical significance of the model (for a significance level α), the value of F is compared with the critical value of F (for $F(\alpha;\ p;\ n-p-1)$, where n represents the number of samples and p the number of explanatory variables included in the regression model), given in the Fisher–Snedecor tables, which relate the numerator (v_1) and the denominator (v_2) degrees of freedom for a given level of significance; in this study 5 %. If F is higher than the critical value, the null hypothesis ($B = 0$) is neglected, i.e. the model has a statistically significant explanation capacity.

All the estimations associated with the F-test, and the p-value related to this test, are given in the table of analysis of variance (ANOVA) (Fisher 1935). In fact, the analysis of variance (ANOVA) is a powerful tool, widely used in statistical inference (Marin-Galiano and Kunert 2006; Leon and Zhu 2008). The ANOVA table allows evaluating whether the regression model is statistically significant. However, when the null hypothesis is rejected, it only means that there is at least one independent variable that influences the variability of the dependent variable, and it does not identify these variables. Therefore, it is important to identify and estimate the coefficients of the mathematical expression resulting from the multiple linear regression analysis performed. The linear regression coefficients (B) of the independent variables represent the alteration undergone by the dependent variable whenever the independent variable changes by one unit. Also, it is important to evaluate some statistical parameters associated with these coefficients, such as:

- Standard error—standard errors associated with the regression coefficients;
- Standardized coefficients (Beta)—obtained by normalizing all the variables included in the analysis; the greater the value of these coefficients the greater the *t* values and the smaller the significance values;
- *t*-test—used to determine whether each regression coefficient is useful to predict/explain the dependent variable; the *t*-test can be estimated as indicated in Eq. (3.15), where B_i represents the regression coefficient of the variable *i* and $S(B_i)$ the standard error associated with the regression coefficient B_i;

$$t_i = \frac{B_i}{S(B_i)} \tag{3.15}$$

- The *t*-test is represented by a *t*-Student distribution (Fisher 1925), symmetrical around zero, which is why the *t* values can be positive and negative. The *t* values of the regression are then compared with the critical value of *t* (from a *t*-Student distribution table) and if their absolute value is greater the model is considered well adjusted;
- Finally, it is necessary to evaluate whether, for all independent variables, the significance value is lower than the threshold *p*-value (5 %); the stepwise method guarantees that only the variables that comply with this condition are included in the regression model.

3.2.1.2 Verification of the Assumptions of the Model

In the definition of a regression model, it is recommended to examine whether the assumptions inherent in the estimation of the model parameters are accomplished (Wallentin and Ågren 2002). Belsley et al. (1980) introduce the expression "regression diagnostics", proposing several methods to assess the existence of influential points and collinearity. Meloun and Militký (2001) refer that the diagnosis of the regression model should include the analysis of three crucial points: (i) regarding the data, evaluate the quality of the sample according to the model's demand; (ii) concerning the model, evaluate the adjustment of the model to the dataset; (iii) regarding the method, ensure that the assumptions inherent to the estimation of the regression model are accomplished.

Multicollinearity

In various studies—in biology, social sciences and econometrics—the independent variables tend to relate with each other and with other variables related to the dependent variable but not included in the model. This effect of inter-relationship between variables is usually called multicollinearity (Bell and Wang 1997). Multicollinearity is a major concern in multiple linear regression analysis (Bayhan and Bayhan 1998; Spanos and McGuirk 2002; Wooldridge 2009), since if the independent variables are interrelated the multiple regression analysis is jeopardized (Leung et al. 2001; Pires et al. 2008). To detect the presence of multicollinearity effects, the variance inflation factor (VIF) can be used, which is determined for all the independent variables of the model, as indicated in Eq. (3.16) (Hsieh and Lavori 2000).

$$\mathrm{VIF}(B_i) = \frac{1}{1 - R_i^2} \tag{3.16}$$

The existence of multicollinearity implies that the regression coefficients (B_i) does not correctly reflect the effect of the independent variable i in the dependent variable, since the value of B_i is influenced by the existence of other explanatory variables in the model (Kutner et al. 2005). In the estimation of VIF, R_i^2 is the coefficient of determination of the regression analysis between the independent variable i and the remaining independent variables included in the model (Leclerc and Pireaux 1995). The higher the value of R_i^2, the greater is the influence of the remaining independent variables in explaining of the independent variable i, which implies that there is a correlation between the independent variables. Thus, the higher the value of R_i^2, the higher the VIF value and the greater the multicollinearity effect. Various authors (Judge et al. 1985; Montgomery et al. 2001) report that VIF values above 5 indicate the existence of multicollinearity. Moreover, other authors (Marquardt 1970; Myers 1990; Johnson and Wichern 2007; Hair et al. 2007) indicate that the existence of multicollinearity can be diagnosed for VIF values higher than 10 (VIF values exceeding 10 implies a R_i^2 value of 0.9).

Another indicator of multicollinearity is the bivariate correlation matrix between independent variables. In the literature, there is no consensual threshold from which it may be possible to predict model estimation problems due to collinearity between independent variables. Maroco (2007) refers that bivariate correlations between the independent variables higher than 0.75 ($|r| > 0.75$) lead to collinearity problems. On the other hand, Ethington (2010) indicates that bivariate correlations between independent variables above 0.7 can lead to collinearity issues. Farrar and Gauber (1967) refer that to avoid multicollinearity problems, the value of correlation between independent variables should be limited to 0.8 or 0.9. However, the correlation matrix only provides an accurate indication of the presence of an inter-correlation between independent variables when the model contains only two independent variables. When there are more than two variables, the bivariate correlation matrix does not provide unequivocal conclusions (Mason and Perreault 1991).

Another method to evaluate the presence of multicollinearity is the analysis of the eigenvalues (λ) of the correlation matrix between the independent variables. Hocking (2003) suggests that an eigenvalue below 0.1 indicates the existence of moderate collinearity, while eigenvalues below 0.05 indicate the existence of high collinearity. Moreover, the condition index k (that is given by the ration between the maximum eigenvalue and the minimum eigenvalue - $k = \lambda_{\max}/\lambda_i$) can be used in multicollinearity diagnosis. If the values of the condition index are greater than 15, there is a moderate risk of collinearity; values between 30 and 100 reveal a moderate to strong dependency between independent variables; and indices higher than 100 reveal serious collinearity problems (Belsley et al. 1980; Lin 2007).

Analysis of Model Residuals

The analysis of the residuals (model errors)—i.e. the difference between the predicted and observed values—is an informative tool to examine the adequacy of the regression model to the observed data (Beebe et al. 1998; Deschepper et al. 2006). There are four assumptions regarding the model residuals that should be verified (Rauch 1997; Aguirre-Basurko et al. 2006): (i) the mean of the residuals should be zero; (ii) statistical independence of the residuals, i.e. the assumption that errors are independent of one another, implying that subjects are responding independently (Stevens 2009); (iii) homoscedasticity, i.e. the variance of the residuals should be constant; (iv) normality of the residuals distribution. Breaking these assumptions can lead the investigator to incorrect conclusions regarding the regression model (the model could be inefficient or seriously biased) (Jarque and Bera 1980), and lead to an increased risk of Type I error (Berry and Feldman 1985; Tabachnick and Fidell 1996; Keith 2015).

Usually, the evaluation of these assumptions is performed through graphical analysis. The normal *P-P* plot can be used to find whether the random variable "residuals" have a normal distribution. If the distribution are normal ($e_j \sim N(\mu, \sigma^2)$), all the points in the graph are located around a line (McRoberts 2009). The average and the variance of the residuals can be analysed based on the graph of the residuals *versus* the value of the independent variables. To verify the assumptions mentioned before, the residuals should be randomly distributed in an uniform bandwidth around the value 0 (without any pattern)—$E(e_j) \approx 0$ (Atkinson 1985; Velilla 1998). Concerning the residuals variance, there are two possible distinct situations: (i) the variance of the error term is constant (homoscedasticity); (ii) the variance of the error term is not constant (heteroscedasticity). The verification of homoscedasticity is usually based on the residual plots analysis (Dette 2002). When the residuals' distribution shows a given pattern (increasing or decreasing), it indicates that the variance of residuals is not constant, thus breaking one of the assumptions of the regression model, compromising the statistical power of the analysis and distorting the results (Osborne and Waters 2002).

One of the most exact techniques of testing the existence of a correlation was proposed by Durbin and Watson (1971), who refer the need to compare the result provided by the summary table of the model [where the result is obtained as presented in Eq. (3.17)] with an upper (d_U) and a lower (d_L) limit. These limits are given in the tables of Savin and White (1977), whose values depend on the significance level, the number of independent variables (excluding the constant) and the number of case studies in the sample (Silva et al. 2012). The Durbin–Watson statistics parameter provided in the summary table of the model is therefore used to check for self-correlation between the residual/errors of the regression model. An autocorrelation between the model residuals implies that the error committed at instant t influences the error committed at instant $t + 1$, i.e. there is a self-correlation of errors over time.

$$d = \frac{\sum_{i=2}^{n} (\hat{\varepsilon}_i - \hat{\varepsilon}_{i-1})^2}{\sum_{i=1}^{n} \varepsilon_i^2} \tag{3.17}$$

where d represents the Durbin–Watson test value, $\sum_{i=2}^{n} (\hat{\varepsilon}_i - \hat{\varepsilon}_{i-1})^2$ the sum of the squared differences of successive errors and $\sum_{i=1}^{n} \varepsilon_i^2$ is the sum of the square of the errors. If $d < d_L$, then the null hypothesis is rejected (exists dependence); if $d > d_U$, then the null hypothesis is not rejected, i.e. there is independence between the errors; if $d_L < d < d_U$, then the test is inconclusive.

Detection of Outliers and Influential Cases

In the literature there are various definitions for outlier. According to Kendall and Buckland (1982), outliers are values of a sample of n observations that are significantly different from the remaining data and raise the question of whether they are from a different population or the sampling technique is faulty. Johnson and Wichern (2007) refer that an outlier is an observation that appears inconsistent compared to the other cases in the dataset and Barnet and Lewis (1994) state that an outlier is an observation that is markedly deviated from the other observations of the sample.

Draper and Smith (1998) refer that outliers should not be automatically removed but instead should be analysed with some caution. Indeed, it is interesting to understand why a well-adjusted model to the dataset cannot represent the behaviour of a particular point. El-Basyouny and Sayed (2010) state that there are different approaches to deal with outliers; one of the possible strategies is ignore the behaviour of these cases, including them in the model. The authors argue that:

(i) The distinction between "true" outliers (to be excluded) and "false" outliers (to include in the model) it is not clear in practice;
(ii) Each time that an outlier is removed, it is necessary to perform a new regression analysis, which can lead to new outliers, thus initiating a vicious cycle;
(iii) If the identified outliers represent a small percentage of the total sample (e.g. 5 %), it is acceptable to include them in the model, in particular in cases where the researcher has no absolute certainty whether such cases are really abnormal.

In some situations, a case study cannot be considered as an outlier but instead the model is unable to deal with or represent this particular case. Outliers are often considered as an error and aberrant data, but in some cases they may bring important information concerning the phenomena under analysis (Ben-Gal 2005). In fact, a number of authors refer that the first step in outlier detection is the identification and the prior knowledge of what is normal or abnormal (Williams et al. 2002; Liu et al. 2004; Hodge and Austin 2004).

There different methods that can be applied in the detection of outliers. Usually, this detection is associated with the calculation of residuals (Evans 1999). Scatterplots of residuals *versus* independent variables can be used to identify outliers (Larsen and McCleary 1972). In this analysis, the outliers correspond to points located away from the centre and the majority of the dataset (Serdahl 1996), i.e. outliers are extreme observations, whose residuals are considerably higher, in absolute value, than the residuals of the other observations (Maroco 2007).

Sawalha and Sayed (2006) refer that outliers should only be deleted if they have significant effects on the model fit, leading to incorrect conclusions. Wheeler et al. (2010) refer that in outlier diagnosis different statistical indicators can be used, such as: the leverage values; Cook's distance (Cook 1977; Cook and Weisberg 1982); and DfFITS and DfBetas values. All these indicators are related to the hat matrix ($H = (X'X)^{-1}X'$). The diagonal values of this matrix (h_{ii}) give an estimated influence of each case in the model fit and are designed as leverage values (Yaffee 2004; Jensen 2010). Maroco (2007) reports that leverage values below 0.2 are acceptable, between 0.2 and 0.5 can be risky and values above 0.5 indicate the presence of a multivariate outlier. Hoaglin and Welsch (1978) consider that a leverage higher than $2p/n$ (where p is the number of independent variables and n the sample size) suggests the presence of an influential case or outlier (Belsley et al. 1980; Gunst and Eubank 1983). Velleman and Welsch (1981) suggest, for smaller samples, the value of $3p/n$ as the limit above which a case study can be considered an outlier. The other measures are extremely useful in the evaluation of influential cases, and all of them are related with the leverage value. Briefly, these parameters establish the relationship between the model with all dataset and the model after the exclusion of the influential case (Barrett and Gray 1992; Staniswalis 2006; Jahufer and Jianbao 2009). The detection of outliers using the DfFITS values uses the following rule: for medium-size samples, it is assumed that there is an outlier if $|\text{DfFITS}| > 1$; for large samples, it is assumed that there is an outlier if $|\text{DfFITS}| > 2\sqrt{p/n}$ (Neter et al. 1996). Belsley et al. (1980) suggest that observations with DfBetas higher than 2 or $2\sqrt{n}$ indicates influential cases. Regarding the Cook's distance (D_i), observations with a value higher than 1 affect the estimation of the regression parameters (Chatterjee et al. 2000). For samples greater than 15, MacDonald (2002) proposes that influential points are diagnosed in the following situations: $D_i > 0.7$ for $p = 2$ (one independent variable); $D_i > 0.8$ for $p = 3$ (two independent variables); and $D_i > 0.85$ for $p > 3$ (more than two independent variables).

3.2.2 *Application of Multiple Linear Regression to Façade Claddings*

In this section, a multiple linear regression is used to evaluate the influence of claddings characteristics (materials quality, design and execution level, use and maintenance conditions) and the environmental conditions (independent variables) that best contribute to explain the degradation of façade claddings (dependent variable). For this, the software SPSS (Statistical Package for Social Sciences) is used which, despite being more used in the social sciences and in the medical field, is also a powerful tool in the construction sector. The reliability of the service life prediction models proposed in this chapter depends significantly on the data collected during fieldwork. In this section, a multiple linear regression analysis is applied to the four types of claddings analysed. These models include all the variables with statistical significance in the explanation of the variability of the façades' degradation. In the selection and construction of models, the stepwise technique (described above) is used, and the statistical significance and validation of the assumptions of the regression are analysed.

3.2.2.1 Natural Stone Claddings

The first step to define the multiple linear regression models is to identify and quantify the characteristics that influence the durability of stone claddings. In the simple regression analysis, the variables more relevant to the degradation of stone claddings, and whose quantification is possible using visual inspections only, were identified. However, most of these features correspond to qualitative variables, including type and colour of the stone, type of finishing, area and thickness of stone plates, location of the cladding, orientation, distance from the sea, wind-rain action, exposure to damp, use and height of the building. The variables are quantified by the ratio between the estimated service life of each category for each characteristic and the average estimated service life for the whole sample (in this case, 69 years). Table 3.9 shows the quantification of the variables considered.

Once the variables have been identified and quantified, it is possible to proceed to the modelling itself. Table 3.10 presents the summary of the multiple linear regression models for stone claddings.

In this analysis, all model assumptions are validated, for a 5 % significance level and none of the case studies analysed is considered an outlier. The multiple regression model obtained includes as explanatory variables age, distance from the sea, area of the stone plates and exposure to damp. This model leads to a coefficient of determination (R^2) of 0.711, revealing that the four explanatory variables included in the model are able to explain 71.1 % of the variability of the severity of degradation of stone claddings.

Table 3.11 shows the ANOVA analysis (Analysis of variance) of the multiple linear regression model for stone claddings. Using a 5 % level of significance, and

Table 3.9 Quantification of the independent variables considered for stone claddings

Independent variables	Variables quantification independent variables			
Type of stone	Limestone: 1.04	Marble: 0.96		Granite: 1.39
Colour	Light colours: 1.01		Dark colours: 0.88	
Type of finishing	Smooth: 0.993		Rough: 0.991	
Area of stone plates	Medium size: 1.04		Large size: 0.94	
Thickness of the stone plates	Less than 2.5 cm: 0.97		Higher than 2.5 cm: 1.00	
Location of the cladding	Integral/partial: 0.96		Bottom wall cladding: 1.00	
Orientation	North: 0.96	NE/E/SE: 1.01	W/NW: 0.93	S/SW: 1.04
Distance from the sea	≤5 km: 0.96		>5 km: 1.03	
Wind-rain action	Moderate: 0.986		Severe: 0.984	
Exposure to damp	Low: 1.03		High: 0.91	
Use of the building	Housing: 1.00		Commerce and services: 0.99	
Height of the building	Current: 1.00		High: 0.87	

Table 3.10 Summary of the multiple linear regression analysis model obtained for stone claddings

Model[e]	R	R^2	Adjusted R^2	Standard error of the estimate
1	0.799[a]	0.639	0.637	0.050
2	0.830[b]	0.689	0.686	0.046
3	0.837[c]	0.701	0.696	0.046
4	0.843[d]	0.711	0.706	0.045

[a]Predictors: (constant), age (years)
[b]Predictors: (constant), age (years), distance from the sea
[c]Predictors: (constant), age (years), distance from the sea, area of stone plates
[d]Predictors: (constant), age (years), distance from the sea, area of stone plates, exposure to damp
[e]Dependent variables: degradation severity (%)

Table 3.11 ANOVA table of multiple linear regression model obtained for stone claddings

Model[b]		Sum of squares	Degrees of freedom (df)	Mean square	F	Sig.
6	Regression	0.987	4	0.247	122.060	0.000[a]
	Residual	0.400	198	0.002		
	Total	1.388	202			

[a]Predictors: (constant), age (years), distance from the sea, area of the stone plates, exposure to damp
[b]Dependent variables: degradation severity (%)

with 4 numerator degrees of freedom and 198 denominator degrees of freedom, the critical value of F is −2.434 (obtained through the Fisher–Snedecor tables), significantly lower than the value obtained in the ANOVA analysis (122.06). This result confirms that the null hypothesis ($B = 0$) can be rejected, i.e. the model has a statistically significant explanation capacity, and there is at least one statistically significant independent variable.

After verifying the model's assumptions, the next step is to determine the coefficients of the mathematical expression resulting from the multiple linear regression analysis performed—Table 3.12.

The collinearity diagnosis for the explanatory variables included in the model reveals that the VIF is always lower than 5, thereby indicating that there is no risk of multicollinearity between the variables included in the model. Furthermore, all the variables included in the model are useful in the prediction of the dependent variable (claddings' degradation), since the significance values presented in Table 3.12 are always lower than the adopted significance level ($\alpha = 0.05$). Also, the critical value of t (from a t-Student distribution table, for a sample of 203 case studies and for a significance level of 5 %) is 1.645, significantly lower (in absolute value) than the t values obtained for the independent variables included in the model.

Based on this model, it can be said that the degradation severity depends on the age of the façades (A), the distance from the sea (S), the area of the stone plates (AS) and the exposure level to damp (D). Based on Eq. (3.18), it is possible to estimate the degradation severity (S_w) of stone claddings based on the four variables considered in the model.

$$S_w = 0.003 \cdot A - 0.429 \cdot S - 0.195 \cdot \mathrm{AS} - 0.174 \cdot D + 0.772 \tag{3.18}$$

To apply this mathematical formulation, the explanatory variables must be replaced (according to the claddings characteristics) by their numerical value, presented in Table 3.9.

The four variables included in the model present different levels of relevance for the explanation of the variability of the degradation of stone claddings. Age is clearly the most relevant variable, contributing to explain 89.9 % of the variability of the degradation severity; distance from the sea contributes with 7 % for the predictive ability of the model and the remaining model variables contribute about 2 % for the explanation of the degradation of stone claddings.

From the analysis of the multiple linear regression model obtained, it is found that: (i) as age increases degradation also increases; (ii) claddings in coastal areas have higher degradation rates than claddings located at more than 5 km from the sea; (iii) claddings with medium-size stone plates have lower degradation rates than large-size stone plates; (iv) claddings with high exposure level to damp present higher degradation indexes. These results are coherent, expressing conveniently the phenomena under analysis. Assuming a maximum degradation level of 20 %, an average estimated service life of 77.2 years is obtained, with a standard deviation of 7.65 years.

Table 3.12 Regression coefficients of multiple linear regression model obtained for stone claddings

Model		Non-standardized coefficients		Standardized coefficients	t	Sig.	95 % confidence interval for B		Collinearity statistics	
		B	Standard error	Beta			Lower bound	Upper bound	Tolerance	VIF
4	(Constant)	0.772	0.119		6.477	0.000	0.537	1.007		
	Age (years)	0.003	0.000	0.794	18.179	0.000	0.003	0.003	0.763	1.310
	Distance from the sea	−0.429	0.097	−0.183	−4.416	0.000	−0.621	−0.238	0.852	1.174
	Area of stone plates	−0.195	0.069	−0.115	−2.809	0.005	−0.331	−0.058	0.869	1.150
	Exposure to damp	−0.174	0.063	−0.120	−2.741	0.007	−0.299	0.049	0.758	1.320

3.2.2.2 Rendered Façades

Regarding rendered façades, the same methodology was adopted. Table 3.13 presents the quantification of the independent variables considered for rendered façades (previously analysed in this chapter, in the simple regression analysis).

A preliminary analysis is performed to identify potential outliers. For a wider significance level (a 10 % significance level), an influential case is identified. Figure 3.15 shows the distribution of the residuals (standardized) in function of the values predicted by the model (standardized). In the graphical analysis it is clear that the residuals are distributed randomly around zero, not revealing any trend. There is, however, a data point that stands out (marked in red), which is apparently an outlier. The casewise diagnosis reveals that this case study has a standard deviation from the mean of 5.46, and thus can be considered an outlier.

Since this case corresponds to just an observation (1 %) it was decided to remove it. A new multiple linear regression analysis with only 99 case studies is

Table 3.13 Quantification of the independent variables considered for rendered façades

Independent variables	Variables quantification			
Render type	Lime–cement rendering 0.90	Single-layer rendering 1.25	Current cement rendering 1.18	Rendering with crushed marble 1.38
Colour	White: 0.88	Light colours: 0.96		Dark colours: 1.12
Height of the building	$h \leq 9$ m: 0.96	$9 < h \leq 14$ m: 1.02		$h > 14$ m: 1.10
Building shape	Compact: 1.10		Irregular: 0.94	
Detailing/design level	Inferior: 1.02	Current: 0.99		Superior: 1.41
Eaves' protection	No: 0.88		Yes: 1.06	
Ground floor protection (socle)	No: 0.99		Yes: 1.00	
Platbands copings	No: 0.92		Yes: 1.14	
Balcony copings	No: 0.94		Yes: 1.31	
Orientation	W: 0.88 NW: 1.19	S: 1.06 SW: 1.19	E: 0.94 SE: 1.06	N: 1.06 NE: 0.94
Distance from the sea	≤ 3.5 km: 0.95		>3.5 km: 1.02	
Exposure to damp	Unfavourable: 0.71		Current: 0.99	Favourable: 1.14
Distance from pollution sources	Unfavourable: 1.00		Current: 1.00	Favourable: 1.13
Façade protection level	Without protection: 0.92		Normal situation: 1.06	With protection: 1.13

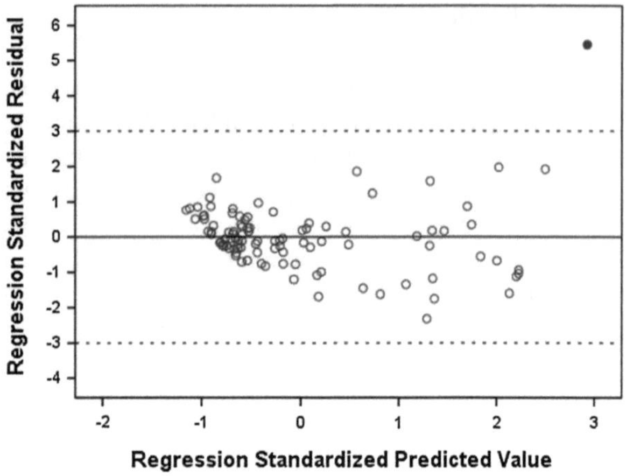

Fig. 3.15 Distribution of the residuals (standardized) in function of the values predicted by the model (standardized)

then performed. In this analysis, no outlier is identified, in accordance with the defined criteria.

Table 3.14 presents the summary of the model obtained for rendered façades with 99 case studies. The explanatory variables considered in the analysis are, in order of relevance, age, orientation, the presence of balcony copings, render type, exposure to damp and the façade protection level. These six variables are able to explain 90.5 % of the variability of degradation severity, and the remaining 9.5 % is due to other factors not contemplated in this model.

In Table 3.15, the ANOVA table of the multiple linear regression model for rendered façades is shown. It is concluded that the model has a statistically significant capacity, since the value of F (145.534) is much higher than the critical

Table 3.14 Summary of the multiple linear regression analysis model obtained for rendered façades

Model[g]	R	R^2	Adjusted R^2	Standard error of the estimate
1	0.936[a]	0.876	0.875	0.077
2	0.940[b]	0.884	0.882	0.074
3	0.943[c]	0.890	0.887	0.073
4	0.946[d]	0.895	0.891	0.072
5	0.949[e]	0.901	0.896	0.070
6	0.951[f]	0.905	0.898	0.069

[a]Predictors: (constant), age (years)
[b]Predictors: (constant), age (years), orientation
[c]Predictors: (constant), age (years), orientation, balcony copings
[d]Predictors: (constant), age (years), orientation, balcony copings, render type
[e]Predictors: (constant), age (years), orientation, balcony copings, render type, exposure to damp
[f]Predictors: (constant), age (years), orientation, balcony copings, render type, exposure to damp, façade protection level
[g]Dependent variables: degradation severity (%)

Table 3.15 ANOVA table of multiple linear regression model obtained for rendered façades

Model[b]		Sum of squares	Degrees of freedom (df)	Mean square	*F*	Sig.
6	Regression	4.157	6	0.693	145.534	0.000[a]
	Residual	0.438	92	0.005		
	Total	4.595	98			

[a]Predictors: (constant), age (years, orientation, balcony copings, render type, exposure to damp, façade protection level
[b]Dependent variables: degradation severity (%)

value (equal to 1.843, obtained by the Fisher–Snedecor tables for a 10 % significance level, six numerator degrees of freedom and 92 denominator degrees of freedom). The significance value obtained (0.000) is less than 10 % (the threshold *p*-value admitted), reveling that there is at least one statistically significant independent variable.

The coefficients of the mathematical expression resulting from the multiple linear regression analysis are presented in Table 3.16. In the case of rendered façades, the values of the *t* test for the independent variable coefficients, in absolute terms, is always above the critical value of *t* (equal to 1.291, obtained by interpolation using the *t*-Student distribution tables, for a 10 % significance level, and a sample of 99 cases), revealing that the model is well adjusted to the dataset. The collinearity diagnosis for the explanatory variables reveals that the VIF is always lower than 5, showing there is no risk of multicollinearity between the explanatory variables.

It is then possible to define a numerical equation (expression (3.19)) that represents the relationship between the degradation severity of rendered façades and the six explanatory variables considered.

$$S_w = 0.013 \cdot A - 0.166 \cdot O - 0.088 \cdot B - 0.2 \cdot R - 0.109 \cdot D - 0.152 \cdot P + 0.757 \tag{3.19}$$

where S_w represents degradation severity, A age of the renderings, O façades orientation, B presence of balconies copings, R render type, D exposure to damp, and P façade protection level. For the application of the mathematical equation it is necessary to replace the variables by their numerical value, presented in Table 3.13.

The results obtained are consistent, leading to the following conclusions: (i) as age increases degradation also increases; (ii) rendered façades with balcony copings have lower degradation indexes; (iii) façades with higher protection levels present lower values of degradation severity; (iv) renderings more exposed to damp show higher degradation levels; and (v) mortars considered in the simple regression analysis as more durable (e.g. renderings with crushed marble) present lower degradation values than less-durable renders (e.g. lime–cement renderings). For rendered façades and assuming a maximum degradation level of 20 %, an average estimated service life of 15.7 years is obtained, with a standard deviation of 3.02 years.

Table 3.16 Regression coefficients of multiple linear regression model obtained for rendered façades

Model		Non-standardized coefficients		Standardized coefficients	t	Sig.	95 % confidence interval for B		Collinearity statistics	
		B	Standard error	Beta			Lower bound	Upper bound	Tolerance	VIF
6	(Constant)	0.757	0.157		4.831	0.000	0.446	1.069		
	Age (years)	0.013	0.001	0.870	21.351	0.000	0.012	0.014	0.624	1.603
	Orientation	−0.166	0.074	−0.075	−2.247	0.027	−0.312	−0.019	0.939	1.065
	Balcony copings	−0.088	0.040	−0.072	−2.202	0.030	−0.167	−0.009	0.981	1.019
	Render type	−0.200	0.067	−0.123	−2.971	0.004	−0.334	−0.066	0.599	1.668
	Exposure to damp	−0.109	0.050	−0.072	−2.183	0.032	−0.208	−0.010	0.965	1.036
	Façade protection level	−0.152	0.081	−0.066	−1.888	0.062	−0.312	0.008	0.861	1.162

3.2.2.3 Painted Surfaces

Table 3.17 presents the quantification of the independent variables considered for painted surfaces. This numerical translation is given by the ratio between the estimated service life for characteristic of the sample and the average service life estimated for the whole sample (in the case of painted surfaces, 9.7 years).

The summary of the multiple linear regression model for painted surfaces is given in Table 3.18. In the construction of the multiple linear regression model, a stepwise technique is used, and all the assumptions regarding the model's significance, for a 5 % significance level, are validated. In this case, none of the case studies is an outlier or an influential case.

Table 3.17 Quantification of the independent variables considered for painted surfaces

Independent variables	Variables quantification			
Type of product	Smooth paints: 1.02 Silicates and silicone paints: 0.97		Plastic membranes: 0.97 Textured paints: 1.02	
Colour	White: 1.00	Light colours: 1.00		Dark colours: 0.49
Type of finishing	Smooth: 1.00		Rough: 1.01	
Building shape	Compact: 1.00		Irregular: 0.93	
Orientation	N/NE: 1.02	E/SE: 1.02	W/NW: 0.97	S/SW: 0.93
Distance from the sea	≤5 km: 0.97		>5 km: 1.01	
Wind-rain action	Low: 1.03	Moderate: 1.00	Severe: 0.98	
Exposure to damp	Low: 1.02	High: 0.97		
Distance from pollution sources	Current: 1.00	Unfavourable: 1.00		
Use of the building	Housing: 1.01	Commerce and services: 0.98		
Height of the building	Current: 1.00	High: 1.01		

Table 3.18 Summary of the multiple linear regression analysis model obtained for painted surfaces

Model[e]	R	R^2	Adjusted R^2	Standard error of the estimate
1	0.905[a]	0.820	0.819	0.073
2	0.911[b]	0.829	0.828	0.071
3	0.914[c]	0.836	0.833	0.070
4	0.916[d]	0.839	0.836	0.069

[a]Predictors: (constant), age (years)
[b]Predictors: (constant), age (years), orientation
[c]Predictors: (constant), age (years), orientation, colour
[d]Predictors: (constant), age (years), orientation, colour, distance from the sea
[e]Dependent variables: degradation severity (%)

Table 3.19 ANOVA table of multiple linear regression model obtained for painted surfaces

Model[b]		Sum of squares	Degrees of freedom (df)	Mean square	F	Sig.
4	Regression	5.369	4	1.342	279.260	0.000[a]
	Residual	1.033	215	0.005		
	Total	6.402	219			

[a]Predictors: (constant), age (years), orientation, colour, distance from the sea
[b]Dependent variables: degradation severity (%)

The variables with explanatory power in the variability of the degradation of painted surfaces are age, façade orientation, colour and distance from the sea. The model obtained has a coefficient of determination (R^2) of 0.839, thus revealing a high correlation between the model and the dataset.

Table 3.19 shows the ANOVA analysis for the multiple linear regression model for painted surfaces. The model is statistically significant, with a value of F(279.26) significantly higher than the critical value—2.401 (obtained through the Fisher–Snedecor tables, for a 5 % significance level, 4 numerator degrees of freedom of and 215 denominator degrees of the freedom).

In Table 3.20, the regression coefficients and the diagnosis of collinearity for the explanatory variables included in the model are presented. The results show that there is no risk of multicollinearity between variables in the model (VIF values below 5). The significance values for the regression coefficients are always less than 5 % (adopted significance level) indicating that all the variables included are statistically significant. Furthermore, the t values obtained for the independent variables included in the model are significantly higher (in absolute value) than the critical value of t (from a t-Student distribution table, for a sample of 220 case studies and a significance level of 5 %).

The degradation severity of painted surfaces can be determined based on the four independent variables considered, as shown in Eq. (3.20).

$$S_w = 0.035 \cdot A - 0.500 \cdot O - 0.140 \cdot C - 0.501 \cdot S + 1.029 \tag{3.20}$$

where S_w represents degradation severity, A age of the painted surfaces, O façades orientation, C colour of the painted surfaces, and S distance from the sea. For the application of the mathematical equation it is necessary to replace the variables by their numerical value, presented in Table 3.17.

In an overall analysis, the model is consistent, revealing that: (i) as age increases claddings' degradation also increases; (ii) façades facing North and East have lower degradation values than those oriented South and West; (iii) façades with dark colours, with higher solar absorption coefficient, show higher degradation levels; (iv) painted surfaces at more than 5 km from the sea show lower degradation indexes than those in coastal areas. For a maximum degradation level of 20 %, the model obtained conduct to an average estimated service life of 8.5 years, with a standard deviation of 0.7 years.

Table 3.20 Regression coefficients of multiple linear regression model obtained for painted surfaces

Model		Non-standardized coefficients		Standardized coefficients	t	Sig.	95 % confidence interval for B		Collinearity statistics	
		B	Standard error	Beta			Lower bound	Upper bound	Tolerance	VIF
4	(Constant)	1.029	0.293		3.511	0.001	0.451	1.606		
	Age (years)	0.035	0.001	0.934	31.529	0.000	0.033	0.037	0.856	1.169
	Orientation	−0.500	0.132	−0.105	−3.803	0.000	−0.760	−0.241	0.978	1.023
	Colour	−0.140	0.045	−0.087	−3.148	0.002	−0.228	−0.052	0.977	1.024
	Distance from the sea	−0.501	0.252	−0.059	−1.986	0.048	−0.999	−0.004	0.858	1.166

3.2.2.4 Ceramic Tiling Systems

Table 3.21 shows the numerical quantification of the variables considered in the service life prediction of ceramic tiling systems. The procedure mentioned before in the numerical translation of qualitative variables is used once again, in this case using the ratio between the estimated service life for each characteristic of the ceramic claddings and the average estimated service life for the whole sample (in the case of ceramic tiling systems, 50 years).

Table 3.22 shows the summary of the multiple linear regression analysis model for ceramic tiling systems. All the assumptions inherent to the construction of the model have been validated for a 5 % significance level. Regarding ceramic tiling systems, none of the case studies was excluded. The final model includes three explanatory variables: age, exposure to damp and tiles size. This model leads to a coefficient of determination (R^2) of 0.698, thus revealing that 69.8 % of the variability of the degradation severity can be explained by the variables included in the model.

Table 3.21 Quantification of the independent variables considered for ceramic tiling systems

Independent variables	Variables quantification			
Surface	Not glazed: 1.53	Glazed: 0.91		
Colour	Light colours: 0.97	Dark colours: 1.07		
Tiles size	$L \leq 20$ cm: 1.00	$L > 20$ cm: 0.88		
Substrate	Concrete: 1.03	Masonry: 0.97		
Peripheral joints	Yes: 1.03		No: 0.99	
Peripheral protection	Yes: 1.00		No: 0.99	
Orientation	N/NE: 0.99	E/SE: 1.05	W/NW: 1.01	S/SW: 0.94
Distance from the sea	≤5 km: 0.99		>5 km: 1.05	
Wind-rain action	Low: 1.04	Moderate: 1.00	Severe: 0.97	
Exposure to damp	Low: 1.05	High: 0.95		
Height of the building	Current: 1.00	High: 1.01		

Table 3.22 Summary of the multiple linear regression analysis model obtained for ceramic tiling systems

Model[d]	R	R^2	Adjusted R^2	Standard error of the estimate
1	0.812[a]	0.659	0.657	0.079
2	0.829[b]	0.687	0.684	0.076
3	0.836[c]	0.698	0.694	0.075

[a]Predictors: (constant), age (years)
[b]Predictors: (constant), age (years), distance from the sea
[c]Predictors: (constant), age (years), distance from the sea, tiles size
[d]Dependent variables: degradation severity (%)

Table 3.23 ANOVA table of multiple linear regression model obtained for ceramic tiling systems

Model[b]		Sum of squares	Degrees of freedom (df)	Mean square	*F*	Sig.
3	Regression	2.492	3	0.831	147,344	0.000[a]
	Residual	1.077	191	0.006		
	Total	3.569	194			

[a]Predictors: (constant), age (years), distance from the sea, tiles size
[b]Dependent variables: degradation severity (%)

The ANOVA analysis of the model is presented in Table 3.23. The model is statistically significant, with a value of *F*—147.34—significantly higher than the critical value—2.658 (obtained through the Fisher–Snedecor table for a 5 % significance level, with 3 numerator degrees of freedom and 191 denominator degrees of freedom).

Table 3.24 presents the regression coefficients and the diagnosis of collinearity for the explanatory variables included in the model. The VIF values show that there is no risk of multicollinearity among the variables included in the model. The significance values obtained for the regression coefficients are always less than 5 % (adopted significance level) indicating that all the variables included in the model are useful in the prediction of the degradation of ceramic tiling systems. The critical value of t (from a t-Student distribution table, for a sample of 195 case studies and for a significance level of 5 %) is 1.645, significantly lower (in absolute value) than the t values obtained for the independent variables included in the model. Equation (3.21) can be used to determine the degradation severity of adhered ceramic cladding according to the variables included in the model.

$$S_w = 0.006 \cdot A - 0.513 \cdot D - 0.345 \cdot T + 0.779 \quad (3.21)$$

where S_w represents degradation severity, A age of the ceramic claddings and D exposure to damp and T tiles size. For the application of the mathematical equation it is necessary to replace the variables by their numerical value, presented in Table 3.21.

The multiple regression model leads to consistent results: (i) degradation increases over time; (ii) claddings with unfavourable exposure conditions to damp tend to degrade more rapidly; (iii) ceramic claddings with large-size tiles tend to have higher degradation rates. For a significance level of 20 %, an average estimated service life of 46 years, with a standard deviation of 4.52 years, is obtained for adhered ceramic claddings.

Table 3.24 Regression coefficients of multiple linear regression model obtained for ceramic tiling systems

Model		Non-standardized coefficients		Standardized coefficients	*t*	Sig.	95 % confidence interval for *B*		Collinearity statistics	
		B	Standard error	Beta			Lower bound	Upper bound	Tolerance	VIF
3	(Constant)	0.779	0.182		4.277	0.000	0.521	1.425		
	Age (years)	0.006	0.000	0.833	19.227	0.000	0.005	0.007	0.832	1.202
	Exposure to damp	−0.513	0.110	−0.190	−4.661	0.000	−0.946	−0.348	0.854	1.172
	Tiles size	−0.345	0.130	−0.130	−2.655	0.009	−0.671	−0.138	0.765	1.308

3.2.3 Analysis of Causal Effects Between Explanatory Variables

Causal inference can be used to measure the relationship between two variables, defining how the variation of one variable is a cause of the variation in other variable (Moore 1997). Blalock Jr. (1964) refers that correlation analysis cannot be directly used to establish causality, since correlations only measure the degree to which several variables vary together. However, according to Bento and Bento (2004), science requires the exploration of cause and effect, showing the researcher the history of effects that leads to the final result (Kenny 2008). In fact, causal analysis reasoning is extremely useful in service life prediction models. In this study, as a starting point, it is only known that the aim of the study is to model the degradation and service life of façade claddings, and therefore this will be the dependent variable of the regression model. Otherwise, the existing information is very sparse. Empirical knowledge (along with the multiple linear regression performed in this chapter) shows that age contributes significantly to the explanation of the claddings' degradation. Multiple linear regression allows identifying the variables with more relevance in the degradation of the claddings under analysis. It seems then relevant to analyse the causal relationships between variables, obtaining a general idea of how different factors (environmental agents, exposure conditions, quality of materials, design quality, maintenance, among other effects) contribute to the coatings' degradation. To evaluate the causal effects between variables, two different approaches (based on a hierarchical multiple regression analysis): (i) test to moderation effects; (ii) and a path analysis are used in this study.

3.2.3.1 Test to Moderation Effects

Moderation effects (also referred as interaction effects) influence the relationship between the dependent variable (x) and the independent variable (y) present in the linear regression model (Aiken and West 1991). A moderator is a variable that changes the strength, the magnitude or the direction between an independent variable and the dependent variable (outcome) (James and Brett 1984; Baron and Kenny 1986; Holmbeck 1997). Frazier et al. (2004) refer that a moderator effect is nothing more than an interaction between independent variables, wherein the effect of one variable depends on the magnitude of another. Moreover, various authors (Preacher et al. 2007; Fairchild and McQuillin 2010) indicate that these effects occur when the relationship between two variables depends on a third; in other words, the moderating variable (W) interacts with the independent variable (X) in the prediction of the dependent variable (Y). This effect can be described using a hierarchical regression analysis, as described in Eq. (3.22).

$$Y = B_0 + B_1X + B_2W + B_3XW + \varepsilon \quad (3.22)$$

To test the moderation effects, it is necessary to investigate the interaction between variable X and variable W. Once variable X is statistically significant in the explanation of the variability of the variable Y, then it is necessary to perform a new regression analysis with a new independent variable XW, taking into account the interaction effects between the X and W variables (product between the two variables). If XW is statistically significant in the explanation of the variability of the dependent variable (Y), it is considered that there is a significant interaction between variable X and variable W and that the relationship between X and Y is moderated by W. These theoretical concepts are clarified in the next section with the illustration of the determination of the moderation effects on service life prediction models applied to façade claddings. However, to evaluate the moderation effects between variables, it is not enough to use a purely statistical view. It does not seem reasonable to analyse all possibilities, especially when dealing with an extensive range of variables. It is important to analyse the context of the study, only examining the causal relationships with physical sense.

Natural Stone Claddings

In this chapter (Sect. 3.2.2.1), it was found that the most relevant variables in the explanation of the degradation severity of natural stone claddings are age, distance from the sea, size of the stone plates and exposure to damp. In the analysis of the moderation effects, the following hypotheses are explored: (i) exposure to damp and wind-rain action (DWR); (ii) distance from the sea and wind-rain action (SWR); (iii) area of the stone plates and thickness of the stone plates (AST); (iv) area of the stone plates and type of stone (ASTS); and (v) exposure to damp and height of the building (DH).

To test the moderation effects, a new regression analysis is thus performed, evaluating the statistical significance of the five new variables resulting from the product between the original variables (considered explanatory in the severity of degradation of stone claddings) and a new variable. This analysis reveals as explanatory variables age, distance from the sea, area of the stone plates and a new variable, DWR, which includes the moderation effects between exposure to damp and wind-rain action. Table 3.25 presents the summary of the model obtained including the moderation effects. This model leads to a coefficient of determination (R^2) equal to that obtained for the original model. Table 3.26 presents the regression coefficients of the model, which allows obtaining a new mathematical equation that reflects the degradation severity of stone claddings according to the explanatory variables considered (Eq. 3.23).

Table 3.25 Summary of the multiple linear regression analysis including moderation effects (stone claddings)

Model[e]	R	R^2	Adjusted R^2	Standard error of the estimate
1	0.799[a]	0.639	0.637	0.050
2	0.830[b]	0.689	0.686	0.046
3	0.837[c]	0.701	0.696	0.046
4	0.843[d]	0.711	0.706	0.045

[a]Predictors: (constant), age (years)
[b]Predictors: (constant), age (years), distance from the sea
[c]Predictors: (constant), age (years), distance from the sea, area of the stone plates
[d]Predictors: (constant), age (years), distance from the sea, area of the stone plates, DWR
[e]Dependent variables: degradation severity (%)

Table 3.26 Regression coefficients of multiple linear regression including moderation effects (stone claddings)

Model		Non-standardized coefficients		Standardized coefficients	t	Sig.	95 % confidence interval for B	
		B	Standard error	Beta			Lower bound	Upper bound
4	(Constant)	0.772	0.119		6.475	0.000	0.537	1.007
	Age (years)	0.003	0.000	0.879	18.169	0.000	0.003	0.003
	Distance from the sea	−0.431	0.097	−0.183	−4.433	0.000	−0.622	−0.239
	Area of the stone plates	−0.195	0.069	−0.115	−2.808	0.005	−0.331	−0.058
	Exposure to damp	−0.175	0.064	−0.119	−2.725	0.007	−0.302	0.048

$$S_w = 0.003 \cdot A - 0.431 \cdot S - 0.195 \cdot \mathrm{AS} - 0.175 \cdot (D \times WR) + 0.772 \quad (3.23)$$

where S_w represents degradation severity, A age of the stone claddings, S distance from the sea, AS area of the stone plates, D exposure to damp, and WR wind rain action. To apply the mathematical equation it is necessary to replace the variables by their numerical value, presented in Table 3.9.

Rendered Façades

The multiple regression analysis performed before reveals that the predictors of degradation severity of rendered façades were: age, orientation; existence of balconies copings; render type; exposure to damp; and façades protection level. To evaluate eventual moderation effects, the following relationships are analysed: (i) exposure to damp and distance from the sea (DS); (ii) façades protection level and height of the building (PH); (iii) façades protection level and building shape

(PV); (iv) balconies copings and design level (BDL); (v) orientation and height of the building (OH); (vi) orientation and distance from the sea (OS). Six new variables are then created, which result from the product between a statistically explanatory variable of the degradation severity and one of the variables excluded from the global model.

A new regression analysis is performed in order to analyse the presence of moderation effects. Regarding the degradation severity of rendered façades, this analysis reveals that the variables with significant explanatory power are age and three new variables, OS (which contemplates the interaction effects between orientation and distance from the sea), BDL (product between the variables balconies copings and design level, exposure to damp, render type) and PH (a combination of façades protection level and height of the building). Therefore, the variables orientation, balconies copings and façades protection level are replaced by new variables, which take into account the moderation effects between these variables and three other variables initially excluded.

From the analysis of the summary of the model (Table 3.27), it is found that the coefficient of determination obtained by this model is equal to 0.904. This model thus provides a reduction of its explanatory power of approximately 0.1 % relative to the initial model (defined in Sect. 3.2.2.2). This result suggests that the multiple linear regression model without interaction effects may be more appropriate. It also confirms the conclusions of Lubinski and Humphreys (1990) according to whom the identification of moderator variables is not always easy and, often, the interaction effects are not significant, because the best regression model, rather than being additive or multiplicative, is not linear. Table 3.28 shows the regression coefficients of the model.

Equation (3.24) allows estimating the degradation severity of rendered façades, taking into account the interaction effects between explanatory variables.

Table 3.27 Summary of the multiple linear regression analysis including moderation effects (rendered façades)

Model[g]	R	R^2	Adjusted R^2	Standard error of the estimate
1	0.936[a]	0.876	0.875	0.077
2	0.941[b]	0.886	0.884	0.074
3	0.944[c]	0.892	0.888	0.072
4	0.946[d]	0.896	0.891	0.071
5	0.949[e]	0.900	0.895	0.070
6	0.951[f]	0.904	0.897	0.069

[a]Predictors: (constant), age (years)
[b]Predictors: (constant), age (years), OS
[c]Predictors: (constant), age (years), OS, BDL
[d]Predictors: (constant), age (years), OS, BDL, exposure to damp
e. Predictors: (constant), age (years), OS, BDL, exposure to damp, render type
[f]Predictors: (constant), age (years), OS, BDL, exposure to damp, render type, PH
[g]Dependent variables: degradation severity (%)

Table 3.28 Regression coefficients of multiple linear regression including moderation effects (rendered façades)

Model		Non-standardized coefficients		Standardized coefficients	t	Sig.	95 % confidence interval for B	
		B	Standard error	Beta			Lower bound	Upper bound
6	(Constant)	0.662	0.145		4.569	0.000	0.374	0.950
	Age (years)	0.013	0.001	0.865	20.803	0.000	0.012	0.014
	OS	−0.138	0.070	−0.068	−1.981	0.051	−0.277	0.000
	BDL	−0.059	0.032	−0.063	−1.877	0.064	−0.122	0.003
	Exposure to damp	−0.086	0.054	−0.057	−1.606	0.112	−0.193	0.020
	Render type	−0.169	0.066	−0.104	−2.539	0.013	−0.301	−0.037
	PH	−0.165	0.091	−0.065	−1.812	0.073	−0.345	0.016

$$S_w = 0.013 \cdot A - 0.138 \cdot (O \times S) - 0.059 \cdot (B \times DL) - 0.086 \cdot D - 0.169 \cdot R - 0.165 \cdot (P \times H) + 0.662 \tag{3.24}$$

where S_w represents degradation severity, *A* age of the rendered façades, *O* façades orientation, *S* distance from the sea, *B* presence of balconies copings, DL design level, *D* exposure to damp, *R* render type, *P* façade protection level and *H* height of the building. To apply the mathematical equation it is necessary to replace the variables by their numerical value, presented in Table 3.13.

Painted Surfaces

Concerning painted surfaces, the multiple linear regression model from Sect. 3.2.2.3 includes as explanatory variables age of the painted surfaces, orientation, colour of the paint and distance from the sea. In the analysis of the moderation effects, the following hypotheses are analysed: (i) orientation and exposure to damp (OD); (ii) orientation and wind-rain action (OWR); (iii) orientation and height of the building (OH); (iv) colour of the paint and type of product (CTP); (v) colour of the paint and texture (CT); (vi) distance from the sea and exposure to damp (SD); (vii) distance from the sea and height of the building (SH); and (viii) distance from the sea and wind-rain action (SWR).

A new regression analysis is then performed to test the existence of moderation effects. This new model includes as explanatory variables age and two new variables, OD and CTP, which reflect the interaction effects between the variables orientation and exposure to damp and colour of the paint and type of product, respectively. Table 3.29 presents the summary of the new model, with the inclusion

Table 3.29 Summary of the multiple linear regression analysis including moderation effects (painted surfaces)

Model[d]	R	R^2	Adjusted R^2	Standard error of the estimate
1	0.905[a]	0.820	0.819	0.073
2	0.911[b]	0.831	0.829	0.071
3	0.916[d]	0.839	0.837	0.069

[a]Predictors: (constant), age (years)
[b]Predictors: (constant), age (years), OD
[c]Predictors: (constant), age (years), OD, CTP
[d]Dependent variables: degradation severity (%)

Table 3.30 Regression coefficients of multiple linear regression including moderation effects (painted surfaces)

Model		Non-standardized coefficients		Standardized coefficients	t	Sig.	95 % confidence interval for B	
		B	Standard error	Beta			Lower bound	Upper bound
3	(Constant)	0.466	0.114		4.074	0.000	0.241	0.692
	Age (years)	0.035	0.001	0.940	32.685	0.000	0.033	0.037
	OD	−0.435	0.109	−0.115	−3.999	0.000	−0.649	−0.221
	CTP	−0.141	0.042	−0.092	−3.362	0.001	−0.224	−0.059

of moderation effects, obtaining a coefficient of determination (R^2) of 0.839. Table 3.30 shows the regression coefficients for this model and Eq. (3.25) allows obtaining the degradation severity of painted surfaces according to the new variables included in the model.

$$S_w = 0.035 \cdot A - 0.435 \cdot (O \times D) - 0.141 \cdot (C \times \mathrm{TP}) + 0.466 \tag{3.25}$$

where S_w represents degradation severity, A age of the painted surfaces, O façades orientation, D exposure to damp, C colour of the painted surfaces and TP type of product. To apply the mathematical equation it is necessary to replace the variables by their numerical value, presented in Table 3.17.

Ceramic Tiling Systems

Regarding ceramic tiling systems, the linear multiple regression model obtained in Sect. 3.2.2.4 includes, as significant variables in the explanation of the degradation of ceramic claddings, age, exposure to damp and size of the ceramic tiles. In the analysis of the existence of the moderation effects, the following hypotheses are explored: (i) exposure to damp and wind-rain action (DWR); (ii) exposure to damp and distance from the sea (DS); (iii) exposure to damp and orientation (OD); (iv) exposure to damp and height of the building (DH); (v) size of tiles and existence of peripheral joints (TPJ); (vi) size of tiles and type of surface (TSu); (vii) size of tiles and type of substrate (TSB); and (viii) size of tiles and existence of peripheral protection (TPP).

In the new regression analysis performed, to evaluate the moderation effects, age and two new variables, OD and TSu, which reflect the interaction effects between the variables orientation and exposure to damp and dimension of tiles and type of substrate, respectively, were included as explanatory variables. The summary of the multiple linear regression analysis including the moderation effects is presented in Table 3.31.

This new model leads to a coefficient of determination (R^2) of 0.703, higher than the value obtained without the inclusion of the moderation effects. The regression coefficients are shown in Table 3.32 and Eq. (3.26) allows obtaining the severity of degradation of ceramic tiling systems according to the new variables included in the model.

$$S_w = 0.006 \cdot A - 0.343 \cdot (D \times O) - 0.373 \cdot (T \times Su) + 0.642 \tag{3.26}$$

where S_w represents degradation severity, A age of the ceramic claddings, D exposure to damp, O façades orientation, T tiles size and Su type of substrate. For

Table 3.31 Summary of the multiple linear regression analysis including moderation effects (ceramic tiling systems)

Model[d]	R	R^2	Adjusted R^2	Standard error of the estimate
1	0.812[a]	0.659	0.657	0.079
2	0.831[b]	0.690	0.687	0.076
3	0.839[c]	0.703	0.699	0.074

[a]Predictors: (constant), age (years)
[b]Predictors: (constant), age (years), OD
[c]Predictors: (constant), age (years), OD, TSu
[d]Dependent variables: degradation severity (%)

Table 3.32 Regression coefficients of multiple linear regression including moderation effects (ceramic tiling systems)

Model		Non-standardized coefficients		Standardized coefficients	t	Sig.	95 % confidence interval for B	
		B	Standard error	Beta			Lower bound	Upper bound
3	(Constant)	0.642	0.142		4.520	0.000	0.362	0.923
	Age (years)	0.006	0.000	0.801	19.973	0.000	0.005	0.007
	OD	−0.343	0.082	−0.166	−4.165	0.000	−0.506	−0.181
	TSu	−0.373	0.126	−0.118	−2.967	0.003	−0.620	−0.125

the application of the mathematical equation it is necessary to replace the variables by their numerical value, presented in Table 3.21.

3.2.3.2 Path Analysis

Path analysis is one of the most powerful methods used in the explanation of causality (Bento and Bento 2004). This methodology is a generalization of multiple linear regression, which allows defining the regression models based on causal interpretations (Cohen et al. 1993). The intensity and direction of the causal relationships between variables are given by the coefficients' trajectory or path coefficients (Chesterton et al. 1989), which are given by the standardized regression coefficients. Usually, the application of this statistical technique requires that some assumptions are defined a priori, based on the experience in the study area (Woods et al. 2003). Everitt and Dunn (1991) refer that "however convincing, respectable and reasonable a path diagram… may appear, any causal inferences extracted are rarely more than a form of statistical fantasy". This statement reinforces what has been mentioned throughout this chapter, i.e. statistics does not replace logical reasoning. The aim of this book is to define service life prediction models, showing their applicability to the façades claddings. Therefore, all the conclusions obtained in this study, regardless of their statistical significance, should be coherent with the reality they intend to model.

Path analysis can be seen as a recursive model, i.e. the causality effects can only be unidirectional with no feedback loops (Musil et al. 1998). In these models, a variable can have a direct and indirect effect on other variables (Curtis et al. 1988; Shipley 2000). Furthermore, a variable may act as a dependent variable in one direction and as an independent variable in another (Rougoor et al. 1997; Woods et al. 2003).

In the multiple linear regression analysis, once the predictors for the degradation severity of the façade claddings are defined, a variant of the path analysis is implemented. The purpose of this analysis is to decompose the association between variables into different effects, direct and indirect, that can be observed in a set of cause–effect relationships (Maroco 2007). Consequently, new regression models are established, each with one of the predictors as the dependent variable. Using the stepwise technique, the predictors of the new regression models are selected enabling the estimation of the corresponding regression coefficients. For each type of cladding analysed, a path diagram is defined, which schematically shows the effect of the different factors that even though not contributing directly to the explanation of the degradation severity, they do so indirectly through their influence on the regression parameters considered representative to the explanation of the degradation severity. In the diagram, the regression coefficients of each variable are given by the arrows portraying their relationship.

Natural Stone Claddings

Knowing that the explanatory variables of the variability of the degradation severity of stone claddings are age, distance from the sea, area of the stone plates and exposure to damp, three new regression analyses are done to evaluate the causal relationships between the explanatory variables and other claddings characteristics (previously excluded of the multiple linear regression model). The three new models proposed are presented in Eqs. (3.27)–(3.29).

$$D = 1.361 \cdot \mathrm{LC} + 0.112 \cdot \mathrm{TS} + 0.335 \cdot O + 0.148 \cdot H - 0.954 \tag{3.27}$$

$$S = 0.810 \cdot \mathrm{LC} + 0.031 \cdot \mathrm{TS} + 0.168 \tag{3.28}$$

$$\mathrm{AS} = 1.377 \cdot \mathrm{TF} - 0.359 \tag{3.29}$$

where S_w represents degradation severity, D exposure to damp, LC location of the cladding, TS type of stone, O façades orientation, H height of the building, S distance from the sea, AS area of stone plates, and TF type of finishing.

Figure 3.16 shows the path diagram obtained for stone claddings, indicating nonstandard regression coefficients; the direction of the arrows establishes the direction of the cause–effect relationship between variables.

Equation (3.30) allows estimating the degradation severity of the stone claddings according to the path analysis model; in this equation, the explanatory variables in the initial multiple linear regression model are replaced by Eqs. (3.27)–(3.29).

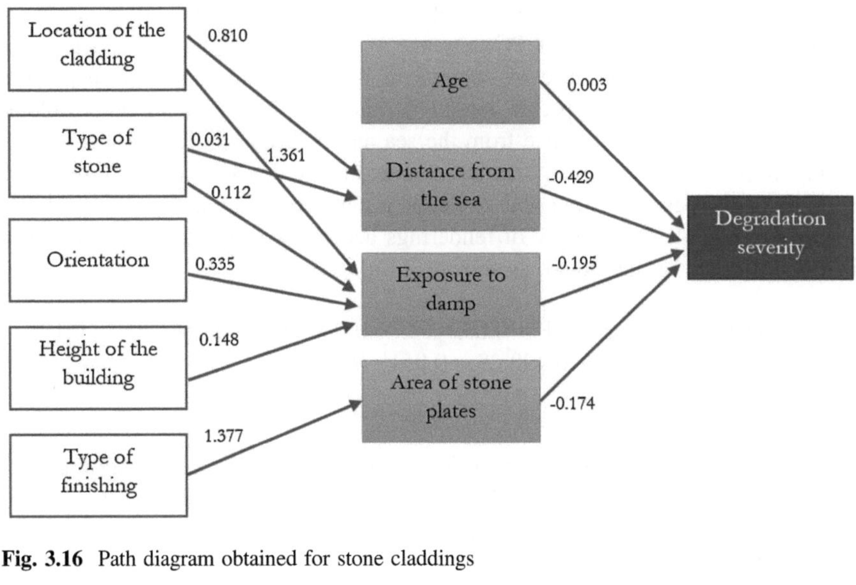

Fig. 3.16 Path diagram obtained for stone claddings

$$S_w = 0.003 \cdot A - 0.584 \cdot \mathrm{LC} - 0.033 \cdot \mathrm{TS} - 0.269 \cdot \mathrm{TF} - 0.058 \cdot O - 0.026 \cdot H + 0.936 \tag{3.30}$$

where S_w represents degradation severity, A age of the stone claddings, LC location of the cladding, TS type of stone, O façades orientation, H height of the building and TF type of finishing. For the application of the mathematical equation it is necessary to replace the variables by their numerical value, presented in Table 3.9.

Rendered Façades

Concerning rendered façades, it was found that the statistically significant variables are age, façade orientation, existence of balcony copings, render type, exposure to damp and façades protection level. Thus, five new models have been defined (Eqs. 3.31–3.35).

$$O = 5.037 \cdot \mathrm{GP} + 0.236 \cdot \mathrm{PC} + 0.207 \cdot \mathrm{EP} - 4.448 \tag{3.31}$$

$$B = 0.755 \cdot \mathrm{BV} + 0.370 \cdot \mathrm{DL} - 0.071 \tag{3.32}$$

$$R = 1.266 \cdot H + 0.473 \cdot \mathrm{DL} - 0.226 \cdot C - 0.433 \tag{3.33}$$

$$D = 3.739 \cdot S - 7.914 \cdot \mathrm{GP} - 0.198 \cdot \mathrm{DL} - 0.410 \cdot H + 5.843 \tag{3.34}$$

$$P = 0.545 \cdot S - 1.033 \cdot H + 1.547 \tag{3.35}$$

where S_w represents degradation severity, O façade orientation, GP presence of ground floor protection (socle), PC presence of platbands copings, EP existence of eaves' protection, B presence of copings in balconies, BV building shape, DL design level, R render type, H height of the building, C renderings colour, D exposure to damp, S distance from the sea and P façade protection level.

Figure 3.17 shows the path diagram obtained for rendered façades. Through the analysis of the paths, it is possible to define a mathematical expression in order to assess the degradation severity of renderings according to the different characteristics considered in the model, as shown in Eq. (3.36).

$$S_w = 0.013 \cdot A + 0.026 \cdot \mathrm{GP} - 0.039 \cdot \mathrm{PC} - 0.034 \cdot \mathrm{EP} - 0.066 \cdot \mathrm{BV} - 0.106 \cdot \mathrm{DL} + 0.045 \cdot C - 0.490 \cdot S - 0.051 \cdot H + 0.716 \tag{3.36}$$

where S_w represents degradation severity, A age of rendered façades, GP presence of ground floor protection (socle), PC presence of platbands copings, EP existence of eaves' protection, BV building shape, DL design level, C renderings colour, S distance from the sea and H height of the building. For the application of the

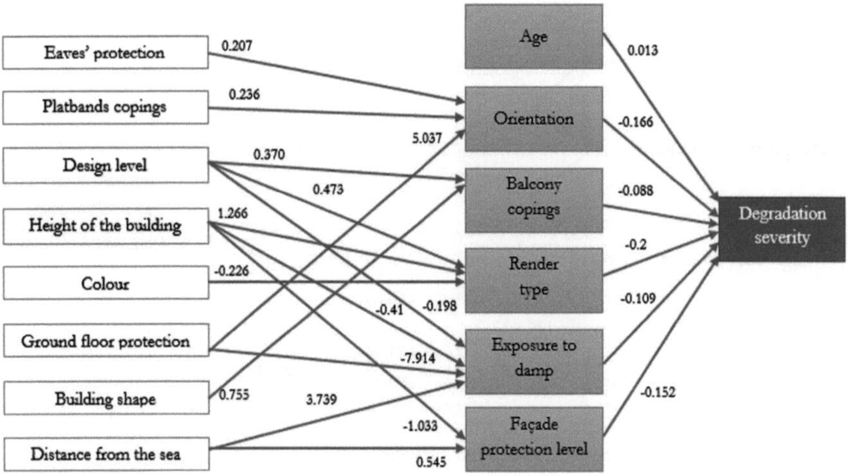

Fig. 3.17 Path diagram obtained for rendered façades

mathematical equation it is necessary to replace the variables by their numerical value, presented in Table 3.13.

Painted Surfaces

For painted surfaces, the global model includes, as explanatory variables age, façade orientation, colour of the paint and distance from the sea. Equations (3.37)–(3.39) represent the three new models defined in order to evaluate cause–effect relationship with other variables excluded from the initial model.

$$C = 2.365 \cdot H - 1.402 \tag{3.37}$$

$$O = 0.288 \cdot \mathrm{WR} + 0.697 \tag{3.38}$$

$$S = 0.758 \cdot D - 0.120 \cdot H + 0.031 \cdot \mathrm{BV} + 0.326 \tag{3.39}$$

where S_w represents degradation severity, A age of painted surfaces, C paint colour, H height of the building, O façades orientation, WR wind-rain action, S distance from the sea, D exposure to damp and BV building shape.

The path diagram obtained for painted surfaces is shown in Fig. 3.18, indicating the non-standardized regression coefficients; the direction of the arrows shows the direction of the cause–effect relationship between variables. The degradation severity of painted surfaces using the path analysis can be estimated as presented in Eq. (3.40).

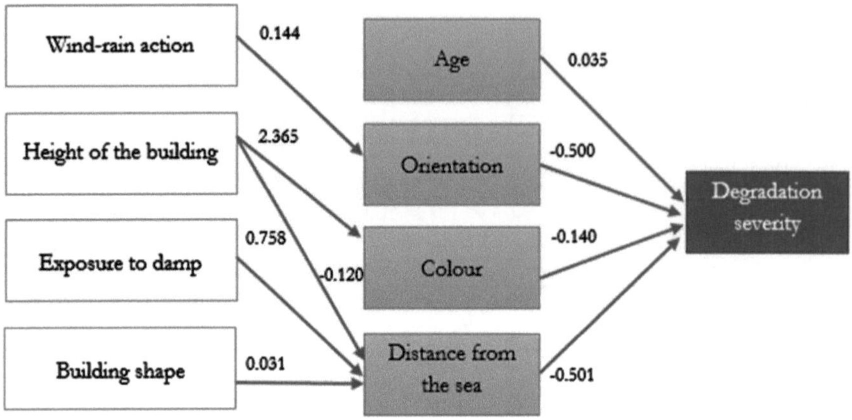

Fig. 3.18 Path diagram obtained for painted surfaces

$$S_w = 0.035 \cdot A - 0.144 \cdot \mathrm{WR} - 0.271 \cdot H - 0.380 \cdot D - 0.016 \cdot \mathrm{BV} + 0.713 \tag{3.40}$$

where S_w represents degradation severity, C paint colour, H height of the building, O façade orientation, WR wind-rain action, S distance from the sea, D exposure to damp, and BV building shape.

Ceramic Tiling Systems

In the path analysis performed for ceramic tiling systems, two new regression models are defined that assess the causal relationship between the independent variables (age, exposure to damp and tiles' size) and the other characteristics of ceramic tiles—Eqs. (3.41) and (3.42).

$$D = 1.105 \cdot \mathrm{WR} + 0.667 \cdot S + 0.786 \cdot H - 1.562 \tag{3.41}$$

$$T = 1.285 \cdot \mathrm{PP} - 0.067 \cdot \mathrm{TS} - 0.57 \cdot \mathrm{Su} + 0.338 \tag{3.42}$$

where D represents exposure to damp, WR wind-rain action, S distance from the sea, H height of the building, T size of tiles, PP peripheral protection, TS type of surface, and Su type of substrate.

Figure 3.19 presents the path diagram obtained for ceramic tiling systems. Applying this model, the degradation severity of ceramic claddings can be estimated as presented in Eq. (3.43).

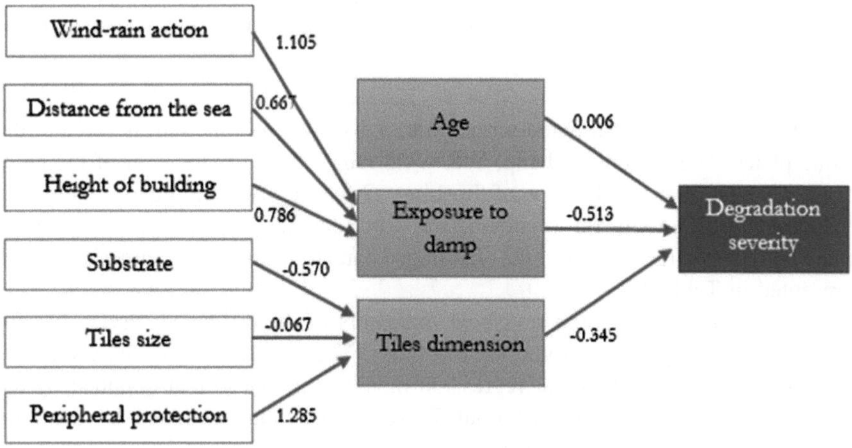

Fig. 3.19 Path diagram obtained for ceramic tiling systems

$$S_w = 0.006 \cdot A - 0.567 \cdot \mathrm{WR} - 0.342 \cdot S - 0.403 \cdot H - 0.443 \cdot \mathrm{PP} + 0.023 \cdot \mathrm{TS} + 0.197 \cdot \mathrm{Su} + 1.463 \tag{3.43}$$

where S_w represents degradation severity, A age of ceramic claddings, WR wind-rain action, S distance from the sea, H height of the building, PP peripheral protection, TS type of surface, and Su type of substrate. For the application of the mathematical equation it is necessary to replace the variables by their numerical value, presented in Table 3.21.

3.2.4 Results and Discussion

3.2.4.1 Natural Stone Claddings

From the application of multiple linear regression analysis and the analysis of the causal relationships between variables, three models were obtained for stone claddings according to the independent variables considered. Equations (3.44) to (3.46) allow obtaining the estimated service life of stone claddings based on multiple linear regression, on the model contemplating the moderation effects and the path analysis model, respectively, assuming that a stone coating reaches the end of its service life when it has a degradation severity of 20 %.

$$\mathrm{ESL} = (0.2 + 0.429 \cdot S + 0.195 \cdot \mathrm{AS} + 0.174 \cdot D - 0.772)/0.003 \tag{3.44}$$

$$\mathrm{ESL} = (0.2 + 0.431 \cdot S + 0.195 \cdot \mathrm{AS} + 0.175 \cdot (D \times WR) - 0.772)/0.003 \tag{3.45}$$

$$\text{ESL} = (0.2 + 0.584 \cdot \text{LC} + 0.033 \cdot \text{TS} + 0.269 \cdot \text{TF} + 0.058 \cdot O + 0.026 \cdot H - 0.936)/0.003 \quad (3.46)$$

where ESL represents estimated service life, *S* distance from the sea, AS area of the stone plates, *D* exposure to damp, WR wind-rain action, LC location of the cladding, TS type of stone, TF type of finishing, *O* façade orientation and *H* height of the building. To apply this mathematical formulation, the explanatory variables must be replaced (according to the claddings characteristics) by its numerical value, presented in Table 3.9.

Table 3.33 presents some statistical indicators of the estimated service life for the three models proposed for stone claddings. The average estimated service life obtained by the multiple linear regression model is similar to that obtained by the model with interaction effects, about 77 years; the value obtained by the path analysis is slightly lower, about 75 years.

The results show that, in otherwise similar conditions: (i) claddings with low exposure to damp have higher estimated service lives; (ii) claddings with medium-size stone plates present an average ESL higher than those with large-size stone plates; (iii) claddings in coastal areas (at less than 5 km from the sea) have lower ESL than those located at more than 5 km. The maximum ESL obtained is equal to 86.6 years and is obtained for a case study with the most favourable conditions, i.e. located at more than 5 km from the sea, with low exposure to damp and medium-size stone plates; the case study with the lowest ESL (62.2 years) corresponds to a cladding located in a coastal area, with high exposure to damp and large-size stone plates. In the case of the model that includes the interaction effects (between variables exposure to damp and wind-rain action), it appears that the results are similar to those obtained by the previous model; the case with the highest ESL (86.6 years) corresponds to claddings located at more than 5 km from the sea, with low exposure to damp, moderate exposure to wind-rain action and medium-size stone plates; conversely, the case with the lowest ESL (62.2 years) corresponds to claddings located at less than 5 km from the sea, with high exposure

Table 3.33 Statistical indicators of the estimated service life for the three models proposed for stone claddings

Statistical indicator	Multiple linear regression	Moderation effects	Path analysis
Average ESL (years)	77.213	77.222	74.968
Median ESL (years)	79.675	79.742	75.901
Maximum ESL (years)	86.599	86.614	82.495
Minimum ESL (years)	62.219	62.169	65.747
Amplitude of the ESL (years)	24.380	24.445	16.748
Standard deviation of ESL (years)	7.650	7.646	4.310
95 % confidence interval	±1.052	±1.052	±0.593

to damp and severe exposure to wind-rain action, with large-size stone plates. For the model obtained by path analysis, the case study with the lowest ESL (65.7 years) corresponds to an integral stone cladding located in a tall building, facing West, in marble with rough finishing; the maximum ESL (82.5 years) corresponds to a wall stone cladding, in a current building (less than 5 floors), facing South, in granite with smooth finishing. The results obtained are consistent and confirm the empirical perception regarding the degradation of stone claddings.

3.2.4.2 Rendered Façades

Concerning rendered façades, Eqs. (3.47)–(3.49) allow obtaining the estimated service life of rendered façades based on multiple linear regression, on the model with moderation effects and the path analysis model, respectively, assuming a maximum acceptable degradation level of 20 %.

$$\mathrm{ESL} = (0.20 + 0.166 \cdot O + 0.088 \cdot B + 0.2 \cdot R + 0.109 \cdot D + 0.152 \cdot P - 0.757)/0.013 \tag{3.47}$$

$$\begin{aligned}\mathrm{ESL} = {} & (0.2 + 0.138 \cdot (O \times S) + 0.059 \cdot (B \times DL) + 0.086 \cdot D + 0.169 \cdot R \\ & + 0.165 \cdot (P \times H) - 0.662)/0.013\end{aligned} \tag{3.48}$$

$$\begin{aligned}\mathrm{ESL} = {} & (0.2 - 0.026 \cdot GF + 0.039 \cdot \mathrm{PC} + 0.034 \cdot \mathrm{EP} + 0.066 \cdot \mathrm{BV} + 0.106 \cdot \mathrm{DL} \\ & -0.045 \cdot C + 0.490 \cdot S + 0.051 \cdot H - 0.716/0.013\end{aligned} \tag{3.49}$$

where ESL represents estimated service life, *O* façade orientation, *B* presence of copings in balconies, *R* render type, *D* exposure to damp, *P* façade protection level, DL design level, *H* height of the building, GP presence of ground floor protection (socle), PC presence of platbands copings, EP existence of eaves' protection, BV building shape, *C* renderings colour, *S* distance from the sea, and *H* height of the building. For the application of the mathematical equation it is necessary to replace the variables by their numerical value, presented in Table 3.13.

Various statistical indicators of the estimated service life for the three models proposed for rendered façades are shown in Table 3.34. It was found that the average estimated service life obtained is practically equal to that obtained by simple linear regression. Furthermore, the values obtained are in accordance with the studies performed by Gaspar (2009) and Shohet and Paciuk (2004), which obtained an estimated service life of approximately 15 years for rendered façades.

Due to the high number of variables included in the models, the analysis of the results is not entirely clear. However, it is possible to identify a degradation trend according to the characteristics of rendered façades. In the proposed models, the minimum ESL is obtained for a lime–cement rendering, located in a building facing

Table 3.34 Statistical indicators of the estimated service life for the three models proposed for rendered façades

Statistical indicator	Multiple linear regression	Moderation effects	Path analysis
Average ESL (years)	15.715	15.560	15.736
Median ESL (years)	15.605	15.847	15.973
Maximum ESL (years)	22.225	23.199	19.265
Minimum ESL (years)	7.658	7.019	12.181
Amplitude of the ESL (years)	14.568	16.181	7.084
Standard deviation of ESL (years)	3.021	3.004	1.547
95 % confidence interval	±0.595	±0.592	±0.305

West, lacking copings in balconies, with current conditions of exposure to damp, in a façade without protection. This conclusion seems coherent, since the façades with lowest ESL correspond to the less durable type of render exposed to adverse environmental agents. In the multiple linear regression model, the maximum value of ESL corresponds to a current cement rendering, facing Southwest, with favourable conditions of exposure to damp and in a façade protected from atmospheric agents. In the model with interaction effects, the maximum ESL is obtained for renderings with crushed marble, in a medium-height building with a high design level, lacking copings in balconies, facing East, located at more than 3.5 km from the sea, with favourable exposure conditions to damp and located in a façade with protection. Regarding the model obtained through path analysis, the maximum estimated service life is obtained for renderings with light colours, located in a tall building, with irregular shape, without eaves' protection but with platbands copings, with ground floor protection, with a current design level and located at more than 5 km from the sea. The results obtained seem realistic from a physical point of view, since the maximum values of ESL corresponds to façades with a favourable combination of conditions, i.e. for both models, the most durable façades are protected from atmospheric agents and favourably exposed to the effect of damp and at a greater distance from marine environments.

3.2.4.3 Painted Surfaces

The multiple linear regression models, with interaction effects and using path analysis, shown in Eqs. (3.50)–(3.52), allow estimating the service life of painted surfaces, depending on variables included in each of the models.

$$\mathrm{ESL} = (0.2 + 0.500 \cdot O + 0.140 \cdot C + 0.501 \cdot S - 1.029)/0.035 \tag{3.50}$$

Table 3.35 Statistical indicators of the estimated service life for the three models proposed for painted surfaces

Statistical indicator	Multiple linear regression	Moderation effects	Path analysis
Average ESL (years)	8.487	8.450	8.503
Median ESL (years)	8.655	8.549	8.287
Maximum ESL (years)	9.389	9.405	8.975
Minimum ESL (years)	6.009	6.028	8.009
Amplitude of the ESL (years)	3.380	3.377	0.966
Standard deviation of ESL (years)	0.701	0.693	0.341
95 % confidence interval	±0.093	±0.092	±0.045

$$\text{ESL} = (0.2 + 0.435 \cdot (D \times O) + 0.141 \cdot (C \times TP) - 0.466)/0.035 \tag{3.51}$$

$$\text{ESL} = (0.2 + 0.144 \cdot \text{WR} + 0.271 \cdot H + 0.380 \cdot D + 0.016 \cdot \text{BV} - 0.713)/0.035 \tag{3.52}$$

where ESL represents estimated service life, O façade orientation, C paint colour, S distance from the sea, D exposure to damp, TP type of product, WR wind-rain action, H height of the building, D exposure to damp, and BV building shape. For the application of the mathematical equation it is necessary to replace the variables by their numerical value, presented in Table 3.17.

Table 3.35 presents various statistical indicators of the estimated service life obtained by the proposed models. The average estimated service life with these models is approximately 8.5 years.

The results show that in otherwise similar conditions: (i) coatings oriented East and North have average estimated service life higher than those oriented South and West; (ii) white-painted surfaces and with light colours have an average ESL higher than those with dark colours; (iii) painted surfaces located in coastal areas (at less than 5 km from the sea) have an average ESL lower than those at more than 5 km from the sea. The maximum ESL is 9.4 years and it is obtained for the case study that has the most favourable conditions, oriented East, at more than 5 km from the sea and painted white. On the other hand, the minimum ESL value is equal to 6 years and is obtained in case studies oriented South, at more than 5 km from the sea and in dark colours (none of the case studies analysed corresponds to a combination of all unfavourable conditions). The model with moderation effects leads to a maximum ESL of 9.4 years (corresponding to a white and textured painted surface, located in a façade oriented East, with low exposure to damp) and a minimum ESL of 6 years (corresponding to an elastic membrane with dark colours, facing South with low exposure to damp). The model obtained by path analysis allows obtaining a maximum ESL of 9 years (low exposure to wind-rain action and low exposure to damp, in a low-rise building with compact shape) and a minimum ESL of 8 years (with severe exposure to wind-rain action and high exposure to damp, in a high building with irregular shape). It is considered that the values obtained are consistent and adequate in the description of the loss of performance of painted surfaces over time.

3.2.4.4 Ceramic Tiling Systems

Equations (3.53)–(3.55) allow obtaining the estimated service life of ceramic tiling systems based on multiple linear regression, on the model with moderation effects and the path analysis model, respectively, assuming a maximum acceptable degradation level of 20 %.

$$\mathrm{ESL} = (0.2 + 0.513 \cdot D + 0.345 \cdot T - 0.779)/0.006 \tag{3.53}$$

$$\mathrm{ESL} = (0.2 + 0.343 \cdot (D \times S) + 0.373 \cdot (T \times Su) - 0.642)/0.006 \tag{3.54}$$

$$\begin{aligned}\mathrm{ESL} = {} & (0.2 + 0.567 \cdot \mathrm{WR} + 0.342 \cdot S + 0.403 \cdot H + 0.443 \cdot \mathrm{PP} - 0.023 \cdot \mathrm{TS} \\ & - 0.197 \cdot \mathrm{Su} - 1.463/0.006\end{aligned} \tag{3.55}$$

where ESL represents estimated service life, D exposure to damp, T tiles size, Su type of substrate, WR wind-rain action, S distance from the sea, H height of the building, PP peripheral protection, and TS type of surface. For the application of the mathematical equation it is necessary to replace the variables by their numerical value, presented in Table 3.21.

Table 3.36 shows the statistical indicators of the estimated service life for the three models proposed for ceramic tiling systems. The average estimated service life obtained is approximately 46 years. The results obtained lead to the following conclusions: (i) ceramic tiling systems with medium-size tiles have higher estimated service lives; (ii) ceramic claddings with high exposure to damp present lower

Table 3.36 Statistical indicators of the estimated service life for the three models proposed for ceramic tiling systems

Statistical indicator	Multiple linear regression	Moderation effects	Path analysis
Average ESL (years)	46.023	46.480	45.977
Median ESL (years)	43.900	45.478	46.321
Maximum ESL (years)	50.962	55.456	51.317
Minimum ESL (years)	35.209	36.615	41.113
Amplitude of the ESL (years)	15.753	20.841	10.204
Standard deviation of ESL (years)	4.524	4.977	2.642
95 % confidence interval	±0.635	±0.699	±0.371

estimated service lives. The maximum ESL (around 51 years) is obtained for ceramic claddings that have the most favourable conditions, with low exposure levels to damp and with ceramic tiles whose size is less than 20 cm. The minimum ESL (35.2 years) is obtained for the opposite situation. In the model including moderation effects, the case study with the highest estimated service life (55.5 years) corresponds to a ceramic cladding oriented East, with low exposure levels to damp, with a size of ceramic tiles lower than 20 cm, adhered to a concrete substrate. The case study with the lowest ESL (34.9 years) corresponds to a cladding oriented South, highly exposed to damp, with ceramic tiles larger than 20 cm, adhered to a concrete substrate. In the model based on a path analysis, the maximum value obtained (51.3 years) corresponds to a ceramic cladding with low exposure to wind-rain action, located at more than 5 km from the sea, in a tall building without peripheral protection, with glazed tiles, on a concrete substrate. Conversely, the case with the lowest ESL (around 39.3 years) corresponds to a ceramic cladding with severe exposure conditions to wind-rain action, located in a coastal area, in a low-rise building, with peripheral protection, with unglazed tiles, on a masonry substrate. It is important to notice that these results are obtained based on a combination of several variables. The maximum and minimum values of ESL correspond to the most favourable and unfavourable case studies, respectively, present in the sample analysed. Based on this sample, it is considered that the values obtained are consistent and suitable in the definition of the degradation process of ceramic tiling systems.

3.3 Multiple Nonlinear Regression Analysis

3.3.1 Theoretical Foundations

Single and multiple linear functions, as well as polynomials, are the functions most commonly applied to the service life prediction of building elements [see, for example, the typical degradation patterns proposed by Shohet et al. (1999)]. However, the degradation phenomena, whether physical, chemical or biological, are in most cases best represented by other expressions. These expressions can be grouped into two distinct sets: intrinsically linear functions, which can be linearized by transformation; and intrinsically non-linear models (Draper and Smith 1998). In multiple linear regression, it is possible to transform the independent variables in, e.g. x^2, x^3, e^x, $\ln(x)$ (linearization of the independent variables by transformation).

Multiple nonlinear regression is an extension of multiple linear regression and uses a non-linear function in the approximation between the model and the data. In most cases, non-linear models have shown to be effective in modelling the degradation phenomena (Gallant 1975; Motulsky and Ransnas 1987). In these models, the arithmetic model that best fits the data set is unknown and the researcher does not know with certainty the variables that should be included in the model and must therefore perform a sensitivity analysis (Silva et al. 2016). Furthermore, the model can still be represented by interaction relationships or a combination of nonlinear relationships between the various independent variables. The incorrect selection of the regression model can lead to the presence of type I or type II errors, or both.

Analysing the specific case of this study, the arithmetic model that reflects the degradation of the façade claddings is unknown. There is only an empirical knowledge of how certain variables affect the degradation of the façades. Because of this difficulty, in the next sections, different nonlinear multiple regression models are proposed, in order to evaluate which is the most appropriate model to describe the degradation of façades claddings.

3.3.1.1 Polynomial Regression Models

The multiple polynomial regression can be seen as an extension of the multiple linear regression. In fact, a first degree polynomial is nothing but a straight line. Therefore, all assumptions concerning the adjustment and statistical inference of the models are the same. Polynomial models are very useful and widely used in approximating nonlinear equations. Montgomery (1984) refers that, generally, it is preferable to use the lowest order polynomial that conveniently describes the phenomena under analysis. The author argues that the decision on the order of polynomial to use is not always easy; as the degree of the polynomial function increases, so does the complexity of the model and the risk of overfitting to the data, compromising the usefulness and generalization capability of the model. In fact, it is possible to adjust an $n - 1$ order degree polynomial to n data points, but this model is practically useless, since it is unable to extrapolate and predict the behaviour of new case studies.

In the nonlinear simple regression analysis (in Sect. 3.1 of this chapter), it was found that second- and third-order polynomial functions are sufficiently accurate in the modelling of claddings' degradation over time. Furthermore, the application of higher order polynomial functions lead to degradation curves with no physical meaning. Therefore, in this study, it does not seem appropriate to consider the application of polynomial models higher than a third-order degree.

According to various authors (Yan and Su 2009; Azizi et al. 2012) and despite the term concerning the interaction between explanatory variables, the generic model of polynomial multiple regression with variables i and degree k can be defined by Eq. (3.56).

$$y = B_0 + B_1x_1 + B_2x_2 + \cdots + B_ix_i + B_{12}x_1^2 + B_{22}x_2^2 + \cdots + B_{i2}x_i^2 + \cdots + B_{ik}x_i^k + \varepsilon \quad (3.56)$$

where y represents the dependent variable, B_i the regression coefficients, x_i the independent variables, and ε the model random errors.

3.3.1.2 Gompertz Curve

The principles of the modelling of Gompertz curve were shown for the first time in 1825, in the article "On the nature of the function expressive of the law of human mortality and on the new mode of determining the value of life contingencies", from the author Benjamin Gompertz, demonstrating that the human mortality rates over time can be described by nonlinear functions. Currently, Gompertz curves have been successfully applied in several areas of knowledge, including the durability of building elements, namely in the service life prediction of painted and rendered façades (Garrido et al. 2012; Paulo et al. 2011). The Gompertz model is generally represented by Eq. (3.57) (Gompertz 1825).

$$y = A\mathrm{e}^{-B_0 \mathrm{e}^{-B_1 t}} \tag{3.57}$$

where y represents the dependent variable, A the asymptotic weight, B the integration constants, e the natural logarithm base and t the age of the element under analysis.

The generic equation of Gompertz curve only considers age as the independent variable and can be used to describe the degradation evolution of claddings over time. However, this study aims to investigate which characteristics of the claddings affect their degradation process as well. Using the Gompertz curve, there are two possible ways to evaluate the relationship between the characteristics of the claddings and their degradation: (i) define a distinct Gompertz curve for each claddings' characteristic; (ii) use the Gompertz curve with more than one independent variable. Therefore, in this study it is proposed to use of a modified Gompertz curve model (Eq. 3.58) that includes, in addition to age, the explanatory variables.

$$y = A\mathrm{e}^{-B_0 \mathrm{e}^{-B_1 x_1 - B_2 x_2 - \cdots - B_i x_i}} \tag{3.58}$$

where y represents the dependent variable, A the asymptotic weight, B the integration constants, e the natural logarithm base and x the independent variables.

3.3.1.3 Von Bertalanffy Curve

Von Bertalanffy proposed in 1938 a theory that allowed quantifying the growth of living organisms, relating their weight or length with their age (this model has been widely applied in aquaculture). The author based his model on the assumption that the growth is represented by the difference between anabolic and catabolic processes in the tissue (Lipinski and Roeleveld 1990). Equation (3.59) indicates the generic equation proposed by Von Bertalanffy (1938, 1957).

$$y = A\left(1 - B_0 \mathrm{e}^{-B_1 t}\right)^3 \quad (3.59)$$

where y represents the dependent variable, A the asymptotic weight, B the integration constant, e the natural logarithm base and t the age of the element under analysis.

With the necessary adaptations, it is considered that the sigmoidal curve proposed by Von Bertalanffy can be useful in the definition of a mathematical model (Eq. 3.60) that reflects the growth of degradation over time, depending on the characteristics of the elements under analysis.

$$y = A\left(1 - B_0 \mathrm{e}^{-B_1 x_1 - B_2 x_2 - \cdots - B_i x_i}\right)^3 \quad (3.60)$$

where y represents the dependent variable, A the asymptotic weight, B the integration constants, e the natural logarithm base and x the independent variables.

3.3.1.4 Richards Curve

The sigmoidal function proposed by Richards (1959) gave a very important contribution to the new paradigm of growth curves. The parameters included in the Richards curve have a specific biological meaning, considering the size, rate and shape of the growth process (Brisbin et al. 1987). The Richards curve is expressed by a four-parameter function with a variable inflection point (Eq. 3.61) (Goonewardene et al. 2003). Brown et al. (1976) refer that the use of this curve can be difficult because of the risk of non-convergence of iterative solution, due to the high correlation between the parameters B and m.

$$y = A\left(1 - B_0 \mathrm{e}^{-B_1 t}\right)^m \quad (3.61)$$

where y represents the dependent variable, A the asymptotic weight, B the integration constant, e the natural logarithm base, t the age of the element under analysis and m the parameter that shapes the curve.

According to Ratkowsky (1983), among the various models relating to sigmoidal curves, the Richards curve exhibits the more significant intrinsic curvature, differing from the linear models. Applying the same approach used in previous formulations, the application of the Richards's model is proposed to model the degradation of claddings over time, according to the variables statistically significant. For this purpose, a modified Richards's function, seen in Eq. (3.62), is proposed.

$$y = A\left(1 - B_0 \mathrm{e}^{-B_1 x_1 - B_2 x_2 - \cdots - B_i x_i}\right)^m \quad (3.62)$$

where y represents the dependent variable, A the asymptotic weight, B the integration constant, e the natural logarithm base, x the independent variables and m the parameter that shapes the curve.

3.3.1.5 Morgan–Mercer–Flodin (MMF) Curve

The model proposed by Morgan–Mercer–Flodin in 1975 intended to express the nutrition of higher organisms (animals and human), based on two existing models (those of Michaelis–Menten and Hill) (Morgan et al. 1975; Ratkowsky 1983). The model of Morgan–Mercer–Flodin can be described by Eq. (3.63).

$$y = \frac{B_1 B_2 + A x^{B_3}}{B_2 + x^{B_3}} \tag{3.63}$$

where y represents the dependent variable, A the asymptotic weight, B the integration constant (B_3 can be seen as the parameter that shapes the curve or the inflection point parameter), e the natural logarithm base and x the independent variable.

As the Richards model, the Morgan–Mercer–Flodin model also has four parameters (Pedersen and Skovsgaard 2009), which gives greater flexibility to the model but leads to a greater risk of non-convergence. The application of this model to the degradation of façade's claddings can be accomplished through the use of a modified Morgan–Mercer–Flodin curve, as shown in Eq. (3.64).

$$y = \frac{B_1 B_2 + A\left(x_1^{B_3} + x_2^{B_4} + \cdots + x_i^{B_{i+2}}\right)}{B_2 + x_1^{B_3} + x_2^{B_4} + \cdots + x_i^{B_{i+2}}} \tag{3.64}$$

where y represents the dependent variable, A the asymptotic weight, B the integration constant (B_3 can be seen as the parameter that shapes the curve or the inflection point parameter), e the natural logarithm base and x the independent variables.

3.3.1.6 Weibull Curve

The Weibull distribution results from the research of Waloddi Weibull relative to material's fatigue that was published in 1951 (Weibull 1951; Moss 2005). The sigmoidal Weibull curve (intrinsically related to the Weibull distribution) can be expressed by Eq. (3.65) (Seber and Wild 1989; Karadavut 2009). The increasing application of the Weibull curve is mainly due to its high flexibility and the fact that it can be used even in small samples (Dodson and Nolan 1999). In fact, these curves have been employed to model different physical phenomena, particularly in reliability models (Dhillon 1982; Xie and Lai 1996).

$$y = A - B_0 e^{-B_1 x^{B_2}} \tag{3.65}$$

where y represents the dependent variable, A the asymptotic weight, B the integration constant, e the natural logarithm base and x the independent variable.

In order to apply the Weibull curve to model the degradation of claddings, the use of a modified curve is suggested, as shown in Eq. (3.66).

$$y = A - B_0 e^{-B_1 \left(x_1^{B_2} + x_2^{B_3} + \cdots + x_i^{B_i+1}\right)} \tag{3.66}$$

where y represents the dependent variable, A the asymptotic weight, B the integration constant, e the natural logarithm base and x the independent variables.

3.3.1.7 Brody's Model

The model proposed by Brody (1945) has been successfully applied to characterize animal's growth rate. According to Kersey and Brink (1985), this function, despite having three parameters, is easier to calculate than other nonlinear functions of this type. Equation (3.67) presents the mathematical formulation of the Brody's model, where the similarities with the Richards curve (Eq. 3.61) are evident.

$$y = A\left(1 - B_0 e^{-B_1 t}\right) \tag{3.67}$$

where y represents the dependent variable, A the asymptotic weight, B the integration constant, e the natural logarithm base and t the age of the element under analysis.

When applied to the modelling of the degradation of façade's claddings, Brody's model can be described as indicated in Eq. (3.68).

$$y = A\left(1 - B_0 e^{-B_1 x_1 - B_2 x_2 - \cdots - B_i x_i}\right) \tag{3.68}$$

where y represents the dependent variable, A the asymptotic weight, B the integration constant, e the natural logarithm base and x the independent variables.

3.3.1.8 Exponential Model

As a special case of the Weibull distribution (Gupta and Kundu 2007), the exponential distribution is also suitable in the definition of reliability models. In science, there are various problems whose experimental observations are best represented by exponential functions (Gardner et al. 1959). This type of functions are extremely useful in the modelling of data on life cycles and in the description of survival functions; for example, Cox and Snell (1981) used an exponential model to estimate the survival times of adult patients with leukemia. These models have also been applied to the analysis of materials and equipment's lifecycle. Various authors (Marshall and Olkin 1995; Roy 2002) state that the exponential function is characterized by the "lack of memory property". Therefore, past events do not influence the probability of future events. An exponential model with more than one explanatory variable can be described by Eq. (3.69).

$$y = Ae^{B_1x_1 + B_2x_2 + \cdots + B_ix_i} \tag{3.69}$$

where y represents the dependent variable, A the asymptotic weight, B the integration constant, e the natural logarithm base and x the independent variables.

3.3.1.9 Potential Model

The model proposed by Wright (1936) was pioneer in the application of potential models. There are several studies that address various applications of this type of models, emphasizing the major advantage of its mathematical simplicity (Globerson and Gold 1997; Blancett 2002; Vits and Gelders 2002). These models have been scrutinized over the years and various authors (Yelle 1979; Vigil and Sarper 1994) state that one of the main disadvantages of potential curves is their deficient adjustment ability and their low predictive power, when the sample used presents high variability. The potential curves have also some limitations when compared to exponential models. The nature of exponential curves allows obtaining more information regarding the modelled phenomenon, generating more accurate predictions than those provided by the potential curves (Nembhard and Uzumeri 2000). To overcome these limitations, Knecht (1974) proposed a model that uses a combination of exponential and potential factors. Nevertheless, potential models have been successfully applied by Paulo et al. (2011) and Garrido et al. (2012) in the service life prediction of painted surfaces in old buildings. In this book, it is also tested the applicability of potential model to the service life prediction of façade claddings, using the formulation shown in Eq. (3.70).

$$y = A\left(x_1^{B_1} + x_2^{B_2} + \cdots + x_i^{B_i}\right) \tag{3.70}$$

where y represents the dependent variable, A the asymptotic weight, B the integration constant, e the natural logarithm base and x the independent variables.

3.3.2 Application of Multiple Linear Regression to Façade Claddings

The nonlinear regression models described in the previous section are composed of a dependent variable, one or more independent variables and estimation parameters (or regression coefficients). Various authors (Perl 1960; Worsley and Lax 1962;

Lipton and McGilchrist 1964) have focused on different methodologies to calculate these parameters; however, Foss (1970) argues that the main disadvantage of most of the proposed methods is that they are "by-eye" methods.

Nonlinear regression models are inherently more difficult to solve than linear models, since more time is needed in the model's definition phase, a larger number of parameters are required, and the mathematical formulas are more complex to such an extent that numerical solutions can be more difficult to achieve. Generally, in nonlinear models, the estimation of the regression coefficients is performed by iterative processes. In this Book, SPSS software and the method of Levenberg–Marquardt are used in the adjustment of nonlinear models (and the determination of the regression coefficients) (Levenberg 1944; Marquardt 1963), thus resulting in a specific degradation curve with specific parameters for each sample. The Levenberg–Marquardt method has proved to be an effective tool to solve nonlinear least squares problems (Griva et al. 2009; Davis 1993; Mondragon 2003).

The definition of the models is also an iterative process; the generic expression of the models is known, but not the explanatory variables to be included in the models, making it necessary to carry out a sensitivity analysis. In this analysis, it is necessary to insert one by one a new variable in the model, checking whether the new variable produces an increase in the model's statistical significance. The final model should combine statistical accuracy and simplicity. The significance of the various models proposed is analysed according to their coefficient of determination (R^2), the mean square error and the quadratic sum-squares errors. By default, SPSS considers as convergence criterion a value of 10^{-8}; this criterion is the reduction of the quadratic sum of the errors between consecutive iterations.

3.3.2.1 Natural Stone Claddings

The nonlinear regression models applied to stone claddings are presented in Eqs. (3.71)–(3.79). All models include as explanatory variables age, distance from the sea, exposure to damp and the area of stone plates; additionally, almost all models include the variable type of stone (except polynomial model) and some models include the variables orientation and the exposure to wind-rain action.

$$\underline{\text{Polynomial model:}}\, S_w = -4.213 + 4.033^{-05} A^2 - 0.08\, S^2 - 0.083\, \mathrm{AS}^2 - 0.075\, D^2 + 4.659\, \mathrm{WR}^2 - 0.069\, O^2 \tag{3.71}$$

$$\underline{\text{Gompertz curve:}}\, S_w = \mathrm{e}^{-0.097\mathrm{e}^{-0.017 A + 1.44 S + 0.504 D + 1.058\, \mathrm{AS} + 0.553\, O + 0.447\, \mathrm{TS}}} \tag{3.72}$$

$$\underline{\text{von Bertalanffy curve:}}\, S_w = \left(1 - \mathrm{e}^{-0.05\mathrm{e}^{-0.011 A + 1.145 S + 0.359 D + 0.740 AS + 0.299 TS + 0.366 O}}\right)^3 \tag{3.73}$$

$$\underline{\text{Richards curve:}}\ S_w = \left(1 - e^{-1.896^{-04} e^{-0.018 I + 1.12 S + 0.938 D + 0.884 \text{AS}}}\right)^{1606.351} \tag{3.74}$$

$$\underline{\text{MMF curve:}}\ S_w = \frac{-76.557\left(A^{1.765} + S^{-74.741} + D^{-32.849} + \text{AS}^{-49.303} + \text{TS}^{-53.132}\right)}{1594.935 + A^{1.765} + S^{-74.741} + D^{-32.849} + \text{AS}^{-49.303} + \text{AS}^{-53.132}} \tag{3.75}$$

$$\underline{\text{Weibull curve:}}\ S_w = 588.913 - 588.911 e^{-1.66^{-09}\left(A^{2.863} + S^{-189.674} + D^{-107.606} + \text{AS}^{-164.604} + \text{TS}^{-216.957}\right)} \tag{3.76}$$

$$\underline{\text{Brody's model:}}\ S_w = 1064.756\left(1 - 0.999 e^{-2.595^{-06} A + 4.148^{-04} S + 1.213^{-04} D + 2.06^{-04} \text{AS} + 3.013^{-05} \text{TS}}\right) \tag{3.77}$$

$$\underline{\text{Exponential model:}}\ S_w = 7.478 e^{0.035 A - 1.501 S - 1.756 D - 1.777 \text{AS} - 1.062 \text{TS}} \tag{3.78}$$

$$\underline{\text{Potential model:}}\ S_w = 1.907^{-06}\left(A^{2.718} + S^{-167.949} + D^{-102.858} + \text{AS}^{-152.310} + \text{TS}^{22.006}\right) \tag{3.79}$$

where S_w represents degradation severity, A age of the stone claddings, S distance from the sea, AS area of the stone plates, D exposure to damp, WR wind-rain action, O façade orientation, and TS type of stone. For the application of the mathematical equation it is necessary to replace the variables by their numerical value, presented in Table 3.9.

Table 3.37 shows the statistical indicators used to evaluate the predictive ability of the models proposed applied to natural stone claddings. Table 3.38 presents

Table 3.37 Statistical indicators of predictive ability of the models proposed applied to natural stone claddings

Model	R^2	Quadratic sum-squares error	Root mean squared error	Maximum squared error
Polynomial model	0.785	0.298	0.038	0.174
Gompertz curve	0.813	0.259	0.036	0.204
von Bertalanffy curve	0.793	0.288	0.038	0.160
Richards curve	0.802	0.275	0.037	0.156
Morgan–Mercer–Flodin curve	0.718	0.392	0.044	0.189
Weibull curve	0.804	0.272	0.037	0.194
Brody's model	0.715	0.396	0.044	0.194
Exponential model	0.831	0.235	0.034	0.173
Potential model	0.796	0.283	0.037	0.184

Table 3.38 Statistical indicators of the estimated service life for the nonlinear regression models proposed for natural stone claddings

Statistical indicator	Polynomial model	Gompertz curve	von Bertalanffy curve	Richards curve	MMF curve	Weibull curve	Brody's model	Exponential model	Potential model
Average ESL (years)	69.110	72.775	73.585	69.380	76.219	68.563	79.370	73.193	68.008
Median ESL (years)	69.524	72.107	73.347	70.408	77.893	69.034	80.196	71.694	67.707
Maximum ESL (years)	75.609	88.595	90.464	75.318	83.124	71.390	92.286	87.387	70.345
Minimum ESL (years)	62.216	57.253	56.593	60.014	64.970	63.707	62.185	60.143	64.026
Amplitude of the ESL (years)	13.393	31.342	33.871	15.304	18.154	7.682	30.101	27.243	6.320
Standard deviation of ESL (years)	2.873	7.489	8.114	4.642	5.610	1.980	8.358	7.422	1.828
95 % confidence interval	±0.395	±1.030	±1.116	±0.639	±0.772	±0.272	±1.150	±1.021	±0.252

some statistical indicators regarding the estimated service life of stone claddings according to the nonlinear multiple regression models proposed. Generally, nonlinear multiple regression models lead to results closer to the reality than linear models; however, these models are significantly more complex and the determination of models' parameters is more difficult, requiring an iterative process. The Gompertz curve, the Richards and the exponential model are those with a greater correlation between the model and the dataset (leading to smaller estimation errors). The Richards curve includes the same explanatory variables adopted in multiple linear regression; this nonlinear model presents a statistical significance 9.1 % higher than the linear model, with an average ESL lower than the average ESL obtained by multiple linear regression model. The exponential model is the model with the highest coefficient of determination, revealing that the five variables included in the model (age, distance from the sea, exposure to damp, area of the stone plates and type of stone) are able to explain 83.1 % of the variability of the degradation severity of stone claddings. The average ESL predicted by the exponential model are similar to the values obtained by the multiple linear regression model. The results obtained seem realistic and in accordance with the empirical knowledge regarding the durability of stone claddings, and are also in accordance with similar studies existing in the literature (Silva et al. 2012).

3.3.2.2 Rendered Façades

Concerning rendered façades, the nonlinear models proposed included as explanatory variables age of the façade, its orientation, render type, exposure to damp and façade's protection level. A sixth variable is also added: in some models, distance from the sea and, in others, existence of copings in balconies and, in the potential model, height of the building. The adoption of different variables is explained by the fact that the models are defined by an iteration process, aiming to obtain the highest predictive capability. Equations (3.80)–(3.88) present the mathematical formulation for the multiple nonlinear models obtained.

$$\underline{\text{Polynomial model:}}\, S_w = 0.379 + 0.013\,A - 0.08\,O^2 - 0.088\,R^2 - 0.055\,D^2 - 0.04\,B^2 - 0.072\,P^2 \tag{3.80}$$

$$\underline{\text{Gompertz curve:}}\, S_w = 1.01^{-14} \cdot \mathrm{e}^{31.959\mathrm{e}^{0.002\,A - 0.012\,R - 0.009\,P - 0.037\,O - 0.04\,D + 0.027\,S}} \tag{3.81}$$

$$\underline{\text{von Bertalanffy curve:}}\, S_w = 8.77^{-08}\left(1 - \mathrm{e}^{10.701\mathrm{e}^{0.002\,A - 0.012\,R - 0.009\,P - 0.037\,O - 0.04\,D + 0.027\,S}}\right)^3 \tag{3.82}$$

$$\underline{\text{Richards curve:}}\ S_w = 1.61^{-08}\left(1 - e^{2.606e^{0.001A - 0.011R - 0.008P - 0.031O - 0.033D + 0.021S}}\right)^{14.754} \tag{3.83}$$

$$\underline{\text{MMF curve:}}\ S_w = \frac{5.69^{+08} + 8.65^{+06}\left(A^{1.026} + R^{-16.561} + O^{-9.728} + D^{-4.677} + P^{-12.281} + B^{-13.213}\right)}{7.3^{+08} + A^{1.026} + R^{-16.561} + O^{-9.728} + D^{-4.677} + P^{-12.281} + B^{-13.213}} \tag{3.84}$$

$$\underline{\text{Weibull curve:}}\ S_w = 1904.388 - 1904.466e^{-6E-06\left(A^{1.027} + R^{-16.562} + O^{-9.728} + D^{-4.677} + P^{-12.281} + B^{-13.214}\right)} \tag{3.85}$$

$$\underline{\text{Brody's model:}}\ S_w = 1428.246\left(1 - 0.999e^{-9.19E-06A + 1.40E-04R + 1.07E-04P + 1.16E-04O + 7.61E-05D + 6.14E-05B}\right) \tag{3.86}$$

$$\underline{\text{Exponential model:}}\ S_w = 1.674e^{0.048A - 0.415R - 0.426P - 1.006O - 1.08H - 0.077B} \tag{3.87}$$

$$\underline{\text{Potential model:}}\ S_w = 0.004\left(A^{1.318} + R^{-26.833} + O^{-9.549} + P^{-23.809} + D^{-7.299} + H^{-68.465}\right) \tag{3.88}$$

where S_w is degradation severity, A age the of rendered façade, O façade orientation, B level of protection of the balconies, R render type, D exposure to damp, P façade's protection level, S distance from the sea, and H height of the building. For the application of the mathematical equation it is necessary to replace the variables by their numerical value, presented in Table 3.17.

Table 3.39 presents the statistical indicators that allow analysing the explanatory power of the different proposed models. The formulation of the Brody and the

Table 3.39 Statistical indicators of predictive ability of the models proposed applied to rendered façades

Model	R^2	Quadratic sum-squares error	Root mean squared error	Maximum squared error
Polynomial model	0.904	0.441	0.067	0.213
Gompertz curve	0.872	0.590	0.077	0.296
von Bertalanffy curve	0.872	0.590	0.077	0.296
Richards curve	0.872	0.587	0.077	0.295
Morgan–Mercer–Flodin curve	0.902	0.448	0.067	0.216
Weibull curve	0.902	0.448	0.067	0.216
Brody's model	0.905	0.438	0.067	0.213
Exponential model	0.875	0.576	0.076	0.291
Potential model	0.900	0.458	0.068	0.209

exponential model is similar to the Gompertz, Richards and von Bertalanffy curves. However, using the sample analysed, the Morgan–Mercer–Flodin and the Brody's model are the most effective. In addition to these models, the polynomial model, the Weibull curve and the potential model have very satisfactory results, with a high coefficient of determination and, consequently, with minor estimation errors. The polynomial model has some advantages: (i) it is the one whose application is simpler; (ii) it needs fewer iterations for the estimation of parameters of mathematical equation; (iii) the application of the mathematical equation in the estimating of the ESL of each case study is easier.

The statistical indicators of the estimated service life for the nonlinear regression models proposed for rendered façades are presented in Table 3.40. The average ESL by the polynomial, the Morgan–Mercer–Flodin, the Weibull and the Brody models is similar to the value obtained by multiple linear regression analysis. The studies of Gaspar (2009) and Shohet and Paciuk (2004) lead to similar results, with an average estimated service life of 15 years. The other models lead to higher values, which seem unrealistic compared to a maximum level of 20 % degradation and for façades not subjected to any maintenance action.

3.3.2.3 Painted Surfaces

In painted surfaces, all the multiple nonlinear regression models include five explanatory variables: age, façade orientation, colour of the coating, distance from the sea and exposure to wind-rain action. The polynomial model includes the type of finishing as an additional variable. Equations (3.89)–(3.97) show the various models of multiple nonlinear regression models proposed for painted surfaces.

$$\underline{\text{Polynomial model:}}\ S_w = 1.567 + 0.002\,A^2 - 0.166\,O^2 - 0.022\,C^2 - 0.177\,S^2 - 0.959\,\text{TF}^2 - 0.248\,WR^2 \tag{3.89}$$

$$\underline{\text{Gompertz curve:}}\ S_w = 1.598\text{e}^{-0.151\text{e}^{-0.108\,A + 1.156\,O + 0.077\,C + 0.669\,S + 1.788\,\text{WR}}} \tag{3.90}$$

$$\underline{\text{von Bertalanffy curve:}}\ S_w = 5.302\left(1 - 0.275\text{e}^{-0.036\,A + 0.379\,O + 0.03\,C + 0.23\,S + 0.597\,\text{WR}}\right)^3 \tag{3.91}$$

$$\underline{\text{Richards curve:}}\ S_w = \left(1 + 0.007\text{e}^{-0.201\,A + 2.201\,O + 0.11\,C + 1.201\,S + 3.399\,\text{WR}}\right)^{-2.343} \tag{3.92}$$

$$\underline{\text{MMF curve:}}\ S_w = \frac{-361.049 + 26.825\left(A^{1.927} + O^{-33.364} + C^{-3.186} + S^{-68.031} + \text{WR}^{-75.914}\right)}{10{,}029.150 + A^{1.927} + O^{-33.364} + C^{-3.186} + S^{-68.031} + \text{WR}^{-75.914}} \tag{3.93}$$

Table 3.40 Statistical indicators of the estimated service life for the nonlinear regression models proposed for rendered façades

Statistical indicator	Polynomial model	Gompertz curve	von Bertalanffy curve	Richards curve	MMF curve	Weibull curve	Brody's model	Exponential model	Potential model
Average ESL (years)	15.823	21.960	21.980	21.970	15.778	16.101	15.960	21.960	17.495
Median ESL (years)	15.814	23.000	23.000	23.000	16.000	17.000	16.000	22.000	18.000
Maximum ESL (years)	22.507	29.000	29.000	29.000	21.000	21.000	22.000	29.000	21.000
Minimum ESL (years)	8.032	10.000	10.000	10.000	8.000	8.000	8.000	11.000	6.000
Amplitude of the ESL (years)	14.475	19.000	19.000	19.000	13.000	13.000	14.000	18.000	15.000
Standard deviation of ESL (years)	2.982	4.262	4.286	4.294	2.776	2.866	2.972	4.157	3.348
95 % confidence interval	±0.587	±0.839	±0.844	±0.846	±0.547	±0.565	±0.586	±0.819	±0.660

$$\text{Weibull curve: } S_w = 668.745 - 668.781\mathrm{e}^{-4.205^{-06}\left(A^{1.903} + O^{-32.68} + C^{-3.097} + S^{-66.185} + \mathrm{WR}^{-74.097}\right)} \quad (3.94)$$

$$\text{Brody's model: } S_w = 1694.514\left(1 - 0.999\mathrm{e}^{-2.081^{-05}A + 2.927^{-04}O + 8.177^{-05}C + 2.88^{-04}S + 4.638^{-05}\mathrm{WR}}\right) \quad (3.95)$$

$$\text{Exponential model: } S_w = 13.941\mathrm{e}^{0.185A - 1.178O - 0.173C + 0.632S - 5.436\mathrm{WR}} \quad (3.96)$$

$$\text{Potential model: } S_w = 0.002\left(A^{2.108} + O^{-36.38} + C^{-3.26} + S^{-47.581} + \mathrm{WR}^{-67.986}\right) \quad (3.97)$$

where S_w is degradation severity, A age of painted surfaces, O façade orientation, C paint colour, S distance from the sea, TP type of finishing, and WR wind-rain action. For the application of the mathematical equation it is necessary to replace the variables by their numerical value, presented in Table 3.17.

Table 3.41 shows the statistical indicators used to evaluate the explanatory power of the different nonlinear multiple regression models proposed. All the models have high predictive power, with coefficients of determination greater than 0.8, revealing a high correlation between the values observed during fieldwork and the values predicted by the models. The Gompertz, Richards, and von Bertalanffy curves are the models with greater explanatory power. The Brody and exponential models are the worst performing models.

Table 3.42 shows the statistical indicators of the estimated service life for the nonlinear regression models proposed for painted surfaces.

Table 3.41 Statistical indicators of predictive ability of the models proposed applied to painted surfaces

Model	R^2	Quadratic sum-squares error	Root mean squared error	Maximum squared error
Polynomial model	0.898	0.652	0.055	0.288
Gompertz curve	0.900	0.638	0.054	0.274
von Bertalanffy curve	0.900	0.640	0.054	0.280
Richards curve	0.900	0.639	0.054	0.267
Morgan–Mercer–Flodin curve	0.895	0.675	0.055	0.279
Weibull curve	0.895	0.674	0.055	0.279
Brody's model	0.839	1.047	0.069	0.294
Exponential model	0.872	0.819	0.061	0.333
Potential model	0.890	0.703	0.057	0.276

Table 3.42 Statistical indicators of the estimated service life for the nonlinear regression models proposed for painted surfaces

Statistical indicator	Polynomial model	Gompertz curve	von Bertalanffy curve	Richards curve	MMF curve	Weibull curve	Brody's model	Exponential model	Potential model
Average ESL (years)	9.724	9.659	9.620	9.712	9.558	9.553	8.454	10.236	9.746
Median ESL (years)	9.654	9.632	9.603	9.672	9.614	9.612	8.678	10.219	9.795
Maximum ESL (years)	10.772	10.697	10.655	10.761	10.131	10.136	9.381	11.417	10.032
Minimum ESL (years)	8.596	8.481	8.403	8.589	8.564	8.551	5.999	8.765	8.993
Amplitude of the ESL (years)	2.176	2.215	2.252	2.172	1.567	1.585	3.382	2.651	1.039
Standard deviation of ESL (years)	0.502	0.579	0.574	0.588	0.399	0.404	0.685	0.611	0.237
95 % confidence interval	±0.066	±0.076	±0.076	±0.078	±0.053	±0.053	±0.091	±0.081	±0.031

The average ESL values obtained by various methods are very similar, around 10 years. This result seems realistic in the description of the service life of painted surfaces without maintenance actions, and is consistent with a number of studies focusing the service life of this type of coating (Chai et al. 2014; Adalberth 1997; Keoleian et al. 2001; Hed 1999; Fay et al. 2000).

3.3.2.4 Ceramic Tiling Systems

All the multiple nonlinear regression models obtained for ceramic tiling systems include as explanatory variables age, tiles size and exposure to damp. The Weibull curve considers only these three variables. The remaining models also include the variable distance from the sea (with high statistical significance in the explanation of the variability of the degradation severity of ceramic claddings) and other variables with explanatory power. Equations (3.98)–(3.106) show the mathematical formulations of the nonlinear regression models applied to ceramic claddings.

$$\underline{\text{Polynomial model:}}\ S_w = 0.339 + 8.48^{-05} A^2 - 0.116\, T^2 - 0.201\, D^2 - 0.041\, S^2 + 0.018\, \mathrm{WR}^2 \tag{3.98}$$

$$\underline{\text{Gompertz curve:}}\ S_w = 3.537\mathrm{e}^{-0.282\mathrm{e}^{-0.015A + 0.933T + 0.760D + 1.057S + 0.327O}} \tag{3.99}$$

$$\underline{\text{von Bertalanffy curve:}}\ S_w = \left(1 - 0.023\mathrm{e}^{-0.02A + 1.080T + 1.128D + 0.92S + 0.729O}\right)^3 \tag{3.100}$$

$$\underline{\text{Richards curve:}}\ S_w = \left(1 - 1.149^{-05}\mathrm{e}^{-0.028A + 1.618T + 1.45D + 1.459S + 0.866O}\right)^{2494.918} \tag{3.101}$$

$$\underline{\text{MMF curve:}}\ S_w = \frac{-184.021 + 2.275\left(A^{1.442} + T^{-29.953} + D^{-68.359} + S^{-278.412} + O^{-48.198}\right)}{2139.776 + A^{1.442} + T^{-29.953} + D^{-68.359} + S^{-278.412} + O^{-48.198}} \tag{3.102}$$

$$\underline{\text{Weibull curve:}}\ S_w = 79.072 - 79.704\mathrm{e}^{-1.602^{-07}\left(A^{2.448} + T^{-146.487} + D^{-56.602}\right)} \tag{3.103}$$

$$\underline{\text{Brody's model:}}\ S_w = 1470.207\left(1 - 0.999\mathrm{e}^{-3.973^{-06}A + 2.651^{-04}T + 2.640^{-04}D + 2.908^{-04}S + 1.654^{-05}O}\right) \tag{3.104}$$

$$\underline{\text{Exponential model:}}\ S_w = 286.349\mathrm{e}^{0.041A - 2.557T - 1.920D - 4.631S - 0.289O} \tag{3.105}$$

Table 3.43 Statistical indicators of predictive ability of the models proposed applied to ceramic tiling systems

Model	R^2	Quadratic sum-squares error	Root mean squared error	Maximum squared error
Polynomial model	0.781	0.782	0.063	0.255
Gompertz curve	0.795	0.732	0.061	0.245
von Bertalanffy curve	0.777	0.805	0.064	0.244
Richards curve	0.789	0.756	0.062	0.241
Morgan–Mercer–Flodin curve	0.731	0.962	0.070	0.273
Weibull curve	0.787	0.761	0.062	0.254
Brody's model	0.707	1.047	0.073	0.275
Exponential model	0.785	0.772	0.063	0.260
Potential model	0.788	0.758	0.062	0.258

$$\underline{\text{Potential model:}}\ S_w = 1.102^{-05}\left(A^{2.479} + T^{-56.522} + D^{-147.819} + S^{-522.919} + O^{-107.846}\right) \tag{3.106}$$

where S_w is degradation severity, A age of ceramic claddings, D exposure to damp, T tiles size, O façade orientation, S distance from the sea, and WR wind-rain action. To apply the mathematical equation it is necessary to replace the variables by their numerical value, presented in Table 3.21.

Some statistical indicators to assess the explanatory power of the various models proposed are shown in Table 3.43. For ceramic tiling systems, the Gompertz curve is the multiple nonlinear regression model with greater explanatory power, followed by the Richards's model and the potential model. The Weibull model presents a coefficient of determination relatively high (above 0.78, revealing a good correlation between the dataset and the model), with the additional advantage of only having three independent variables, which implies that it is easier and faster to apply. Moreover, the Brody's curve is the model with the worse performance, even though leading to very acceptable results (with a coefficient of determination of 0.707).

Table 3.44 shows some statistical indicators regarding the ESL of ceramic tiling systems obtained by the various nonlinear regression models proposed. The average ESL obtained by various methods varies between 46 and 52 years. The average ESL obtained by the Gompertz, Weibull and potential models is close to 50 years, similar to the value obtained by Galbusera et al. (2014), despite the fact that studies on the service life prediction of ceramic claddings are still incipient. A study by BCIS (2001) proposes an estimated service life of 35 years with a range between 20 and 50 years. Therefore, the results obtained in this study fall within the range assumed in this publication.

Table 3.44 Statistical indicators of the estimated service life for the nonlinear regression models proposed for ceramic tiling systems

Statistical indicator	Polynomial model	Gompertz curve	von Bertalanffy curve	Richards curve	MMF curve	Weibull curve	Brody's model	Exponential model	Potential model
Average ESL (years)	48.497	50.028	48.043	48.786	46.358	49.721	46.308	52.081	49.829
Median ESL (years)	48.282	48.902	46.843	47.546	45.908	50.113	45.387	52.775	50.663
Maximum ESL (years)	51.594	57.705	55.372	56.132	51.569	51.939	55.879	60.818	52.225
Minimum ESL (years)	42.318	38.255	36.383	37.047	35.668	45.435	31.898	40.565	45.047
Amplitude of the ESL (years)	9.276	19.449	18.988	19.085	15.901	6.504	23.981	20.252	7.178
Standard deviation of ESL (years)	2.540	4.112	4.033	3.993	3.516	2.158	5.073	4.627	2.135
95 % confidence interval	±0.357	±0.577	±0.566	±0.560	±0.493	±0.303	±0.712	±0.649	±0.300

3.4 Conclusions

Service life prediction models aims to anticipate and understand the behaviour of a given real phenomenon, using for that purpose mathematical equations. However, these equations are not perfect in the description of the modelled reality. The statistical modelling of the service life of the façade's claddings allows: the discussion of theoretical hypotheses initially established; analysing the degradation patterns of claddings according to their characteristics; making quantitative comparisons regarding the estimated service lives of claddings according to the exposure conditions.

In this chapter, the application of the simple and multiple regression analysis to model the degradation phenomena of façade's claddings is discussed. Intuitively, it is known that models that include more than one variable are more accurate, with higher precision in the description of the degradation phenomena. However, the addition of variables to the model should be carefully analysed. The introduction of more variables leads to more accurate models, but can result in an excessive number of parameter estimation, and may ultimately lead to the model's overfitting. Therefore, the parsimony principle should be respected, i.e. if two distinct models are equally accurate to describe a given reality, the model that requires a lower number of arithmetic parameters should be selected.

None of the models proposed is perfect, i.e. they do not fit perfectly the dataset. In fact, a predictive model should not be over adjusted to the model, ensuring that it is able to conveniently describe the behaviour of other case studies, external to the data that were used in the model's definition. An effective service life prediction model should be easy to upgrade, i.e. when extended to a new universe of cases, it should maintain its predictive capability.

For the claddings analysed, it was found that the linear multiple regression analysis produces useful service life prediction models. Using this statistical tool, it is possible to identify the durability factors that best describe the severity degradation values of the façade claddings. In this chapter, the causal relationships between variables are also analysed, in order to understand how the durability variables are interconnected and influence each other. These methods allow identifying a relative hierarchy of durability factors that most directly affect the service life of the construction elements.

This chapter also proposes the application of multiple nonlinear regression models. In general, nonlinear models provide a better understanding of the phenomenon under analysis. However, as mentioned before, these models have some disadvantages: (i) they consider a higher number of parameters, whose estimation is done by iterative processes; (ii) the mathematical equations that define the models are more complex; (iii) the manipulation of mathematical equations is intricate, and the analytical or numerical solutions can be difficult to achieve and (iv) they require a higher consumption of time in the definition of the models. All these issues should be taken into account in the selection of the best service life prediction model.

References

Abdi H (2010) Least squares. Encyclopaedia of research design, Neil Salkind edition. Sage Publications, Thousand Oaks, 7p

Adalberth K (1997) Energy use during the life cycle of buildings: a method. Build Environ 32(4):317–320

Aguirre-Basurko E, Ibarra-Berastegi G, Madariaga I (2006) Regression and multilayer perceptron-based models to forecast hourly O_3 and NO_2 levels in the Bilbao area. Environ Model Softw 21(4):430–446

Aiken LS, West SG (1991) Multiple regression: testing and interpreting interactions, 1st edn. Sage Publications, Newbury Park

Altman DG (1991) Statistics in medical journals: development in the 1980s. Stat Med 10(12):1897–1913

Atkinson AC (1985) Plots, transformations and regression: an introduction to graphical methods of diagnostic regression analysis. Oxford statistical science series. Oxford University Press, Oxford

Azizi A, Amir Yazid BA, Ping LW, Mohammadzadeh M (2012) A hybrid model of ARIMA and multiple polynomial regression for uncertainties modeling of a serial production line. International Conference on Engineering andTechnology Management (ICETM), Kuala Lumpur, Malaysia, 62, 63–68. P-ISSN 2010-376X and E-ISSN 2010-3778

Barnett V, Lewis T (1994) Outliers in statistical data, 3rd edn. Wiley, New York

Baron RM, Kenny DA (1986) The moderator–mediator variable distinction in social psychological research: conceptual, strategic, and statistical considerations. J Pers Soc Psychol 51(6):1173–1182

Barrett BE, Gray JB (1992) Efficient computation of subset influence in regression. J Comput Graph Stat 1(4):271–286

Bayhan GM, Bayhan M (1998) Forecasting using autocorrelated errors and multicollinear predictor variables. Comput Ind Eng 34(2):413–421

BCIS Life expectancy of buildings components (2001) Surveyor's experiences of building in use—a practical guide. Building Cost Information Service, London

Beebe KR, Pell RJ, Seasholtz MB (1998) Chemometrics: a practical guide, 1st edn. Wiley, New York

Bell PM, Wang H (1997) Fuzzy linear regression models for assessing risks of cumulative trauma disorders. Fuzzy Sets Syst 92(3):317–340

Belsley D, Kuh E, Welsch R (1980) Regression diagnostics: identifying influential data and sources of collinearity, 1st edn. Wiley, New York

Ben-Gal I (2005) Outlier detection. In: Maimon O, Rockach L (eds) Data mining and knowledge discovery handbook: a complete guide for practitioners and researchers, Kluwer Academic Publishers, Berlin

Bento A, Bento R (2004) The use of causal analysis techniques in information systems research: a methodological note. J Inf Technol Manage XV(3–4):44–51

Berry WD, Feldman S (1985) Multiple regression in practice. Series on quantitative applications in the social sciences. Sage Publications, California

Blalock HM Jr (1964) Causal inferences in nonexperimental research. University of North Carolina Press, Chapel Hill

Blancett RS (2002) Learning from productivity learning curves. Res Technol Manage 30(3):53–58

Blocken B, Carmeliet J (2004) A review of wind-driven rain research in building science. J Wind Eng Ind Aerodyn 92(13):1079–1130

Bordalo R, de Brito J, Gaspar P, Silva A (2011) Service life prediction modelling of adhesive ceramic tiling systems. Build Res Inf 39(1):66–78

Bowerman BL, O'Connel RT, Koehler AB (2005) Forecasting time series and regression, 4th edn. Brooks/Cole Thomson Learning Inc., USA

Box GEP, Draper NR (1987) Empirical model-building and response surfaces. Wiley series in probability and statistics, 1st edn. Wiley, New York

Briggen PM, Blocken B, Schellen HL (2009) Wind-driven rain on the facade of a monumental tower: numerical simulation, full-scale validation and sensitivity analysis. Build Environ 44(8):1675–1690
Brisbin IL, Collins CT, White GC, McCallum DA (1987) A new paradigm for the analysis and interpretation of growth data: the shape of things to come. Auk 104(3):552–554
Brody S (1945) Bioenergetics and growth: with special reference to the energetic efficiency complex in domestic animals, 1st edn. Reinhold Publishers, New York, USA
Brown JE, Fitzhugh HA, Cartwright TC (1976) A comparison of nonlinear models for describing weight-age relationships in cattle. J Anim Sci 42(4):810–818
BS 5385 Part 2: Wall and floor tiling—code of Practice for the design and installation of external ceramic wall tiling and mosaics (1991)
Casasola R, Rincón JM, Romero M (2012) Glass-ceramic glazes for ceramic tiles: a review. J Mater Sci 47(2):553–582
Chai C, de Brito J, Gaspar P, Silva A (2014) Predicting the service life of exterior wall painting: techno-economic analysis of alternative maintenance strategies. J Constr Eng Manage 140(3):04013057
Chatterjee S, Hadi AS, Price B (2000) Regression analysis by example, 3rd edn. Wiley, New York
Chesterton RN, Pfeiffer DU, Morris RS, Tanner CM (1989) Environmental and behavioural factors affecting the prevalence of foot lameness in New Zealand dairy herds—a case-control study. N Z Vet J 37(4):135–142
Cohen PR, Carlson A, Ballasteros L, St. Amant R (1993) Automating path analysis for building causal models for data. Computer science technical report 93-38. Experimental Knowledge Systems Laboratory, Department of Computer Science, University of Massachusetts, USA
Cook RD (1977) Detection of influential observations in linear regression. Technometrics 19(365):15–18
Cook RD, Weisberg S (1982) Residuals and influence in regression, 1st edn. Chapman & Hall, London, UK
Cottrell A (2003) Regression analysis: basic concepts. Wake Forest University, Department of Economics, ECN 215: Econometric theory and methods, 4p
Cox DR, Snell EJ (1981) Applied statistics, 1st edn. Chapman and Hall Publishers, London
Curtis CR, Salman MD, Strickland D, Edmonston B, Erb HN (1988) Path analysis using logistic regression: interpretational and methodological issues. Acta Veterinaria Scandinavica 84 (Supplementum):469–472
D'Urso P, Santoro A (2006) Goodness of fit and variable selection in the fuzzy multiple linear regression. Fuzzy Sets Syst 157(19):2627–2647
Dallal GE (2007) The little handbook of statistical practice. Biostatistics Unit, Jean Mayer USDA Human Nutrition Research Center on Aging at Tufts University, Boston, USA
Davis P (1993) Levenberg-Marquardt methods and nonlinear estimation. Soc Ind Appl Math (SIAM) News 26(6):1–12
Davis LW, Moldenhauer R, Davis HD (2014) Terrorism, homeland security, and risk assessment via research proposal, 3rd edn. Xlibris, USA
Deschepper E, Thas O, Ottoy JP (2006) Regional residual plots for assessing the fit of linear regression models. Comput Stat Data Anal 50(8):1995–2013
Dette H (2002) A consistent test for heteroscedasticity in nonparametric regression based on the kernel method. J Stat Plan Infer 103(1–2):311–329
Dhillon BS (1982) Reliability engineering in systems design and operation, 1st edn. Van Nostrand Reinhold Editors, New York
Dodson B, Nolan D (1999) Reliability engineering handbook, 1st edn. Quality Publishing, New York
Draper NR, Smith H (1998) Applied regression analysis, 3rd edn. Wiley, USA
Durbin J, Watson GS (1971) Testing for serial correlations in least squares regression III. Biometrika 58(1):1–19
El-Basyouny K, Sayed T (2010) A method to account for outliers in the development of safety performance functions. Accid Anal Prev 42(4):1266–1272

Ethington C (2010) Multicollinearity. Course EDPR 7/8542 statistical methods applied to education (II). University of Memphis, Memphis
Evans VP (1999) Strategies for detecting outliers in regression analysis: an introductory primer. In: Thompson B (ed) Advances in social science methodology. Annual meeting of the Southwest Educational Research Association, vol 5, pp 213–233
Everitt BS, Dunn G (1991) Applied multivariate data analysis, 1st edn. Edward Arnold Publishers, London
Fairchild AJ, McQuillin SD (2010) Evaluating mediation and moderation effects in school psychology: a presentation of methods and review of current practice. J Sch Psychol 48(1):53–84
Farrar DE, Glauber RR (1967) Multicollinearity in regression analysis: the problem revisited. Rev Econ Stat 49(1):92–107
Fay R, Treloar G, Iyer-Raniga U (2000) Life-cycle energy analysis of buildings: a case study. Build Res Inf 28(1):31–41
Fisher RA (1925) Applications of "Student's" distribution. Metron 5(3):90–104
Fisher RA (1935) The logic of inductive inference. J Roy Stat Soc 98(1):39–82
Florentzou F, Brandt E, Wetzel C (1999) MEDIC—a method for predicting residual service life and refurbishment investment budgets. In: 8th international conference on the durability of building materials and components, Ottawa, Canada, pp 1280–1288
Foss SD (1970) A method of exponential curve fitting by numerical integration. Biometrics 26(4):815–821
Franke L, Schumann I, van Hees R, van der Klugt L, Naldini S, Binda L, Baronio G, van Balen K, Mateus J (1998) Damage atlas: classification and analyses of damage patterns found in brick masonry. European Commission Research, report No. 8, vol 2, Fraunhofer IRB Verlag, Brussels
Frazier PA, Tix AP, Barron KE (2004) Testing moderator and mediator effects in counseling psychology research. J Couns Psychol 51(1):115–134
Freedman DA (2010) In: Collier D, Sekhon JS, Stark PB (eds) Statistical models and causal inference: a dialogue with social sciences, 1st edn. Cambridge University Press, New York
Galbusera MM, de Brito J, Silva A (2014) Application of the factor method to the prediction of the service life of ceramic external wall cladding. Journal of Performance of Constructed Facilities. doi:10.1061/(ASCE)CF.1943-5509.0000588
Gallant AR (1975) Nonlinear regression. Am Stat 29(2):73–81
Gardner DG, Gardner JC, Lausii G, Meinke WW (1959) Method for the analysis of multicomponent exponential decay curves. J Chem Phys 31(4):978–986
Garrido MA, Paulo PV, Branco FA (2012) Service life prediction of façade paint coatings in old buildings. Constr Build Mater 29:394–402
Gaspar P (2009) Service life of constructions: development of a methodology to estimate the durability of construction elements, Application to renders in current buildings (in Portuguese). PhD Thesis in Engineering Sciences at Instituto Superior Técnico, Lisbon, Portugal
Gigerenzer G, Stefan K, Oliver V (2004) The null ritual: what you always wanted to know about significance testing but were afraid to ask. In: Kaplan D (ed) Handbook of quantitative methodology for the social sciences. Sage Publications, Thousand Oaks, pp 391–408
Globerson S, Gold D (1997) Statistical attributes of the power learning curve model. Int J Prod Res 35(3):699–711
Gompertz B (1825) On the nature of the function expressive of the law of human mortality and on the new mode of determining the value of life contingencies. Philos Trans R Soc A 115:513–580
Goonewardene LA, Wang Z, Okine E, Zuidhof MJ, Dunk E, Onderka D (2003) Comparative growth characteristics of emus (*Dromaius novaehollandiae*). J Appl Poult Res 12(1):27–31
Griva I, Nash SG, Sofer A (2009) Linear and nonlinear optimization, 2nd edn. Society for Industrial and Applied Mathematics, Philadelphia
Guan WL, Alum J, Liu ZJ, Yang T (1997) Performance of external tiled-wall systems under tropical weathering. J Perform Constr Facil 11(1):24–34
Gunst RF, Eubank RL (1983) Regression diagnostics and approximate inference procedures for penalized least squares estimators. Technical report no. SHU/DS/TR—181. Research

sponsored by the National Aeronautics Space Administration, Department of Statistics—NASA Contract, Southern Methodist University, Texas, USA
Gupta RD, Kundu D (2007) Generalized exponential distribution: existing results and some recent developments. J Stat Plan Infer 137(11):3537–3547
Hagquist C, Stenbeck M (1998) Goodness of fit in regression analysis—R2 and G2 reconsidered. Qual Quant 32(3):229–245
Hair JF, Black WC, Babin B, Anderson RE, Tatham RL (2007) Multivariate data analysis, 6th edn. Prentice-Hall Publishers, Englewood Cliffs
Hed G (1999) Service life planning of building components. In: 8th international conference on durability of building materials and components, Vancouver, Canada, pp 1543–1551
Higgins J (2006) The radical statistician: a beginner's guide to unleashing the power of applied statistics in the real world, 5th edn. Jim Higgins Publishing, USA
Ho DCW, Lo SM, Yiu CY (2005) A study on the causes of external finishes defects in Hong Kong. Struct Surv 23(5):386–402
Hoaglin DC, Welsch RE (1978) The hat matrix in regression and ANOVA. Am Stat 32(1):17–22
Hocking RR (2003) Methods and applications of linear models: regression and analysis of variance. Wiley series in probability and statistics, 2nd edn. Wiley, New York
Hodge VJ, Austin J (2004) A survey of outlier detection methodologies. Artif Intell Rev 22(2):85–126
Holmbeck GN (1997) Toward terminological, conceptual, and statistical clarity in the study of mediators and moderators: examples from the child-clinical and paediatric psychology literatures. J Consult Clin Psychol 65(4):599–610
Hsieh FY, Lavori PW (2000) Sample-size calculations for the Cox proportional regression model with nonbinary covariates. Control Clin Trials 21(6):552–560
Hubbard R, Bayarri MJ (2003) Confusion over measures of evidence (p's) versus errors (α's) in classical statistical testing (with comments). Am Stat 57(3):171–182
Jahufer A, Jianbao C (2009) Assessing global influential observations in modified ridge regression. Stat Probab Lett 79(4):513–518
James LR, Brett JM (1984) Mediators, moderators, and tests for mediation. J Appl Psychol 69 (2):307–321
Jarque CM, Bera AK (1980) Efficient tests for normality, homoscedasticity and serial independence of regression residuals. Econ Lett 6(3):255–259
Jensen DR (2010) Leverage and subset efficiencies in regression. Stat Methodol 7(5):541–551
Johnson RA, Wichern DW (2007) Applied multivariate statistical analysis, 6th edn. Prentice-Hall Inc., Englewood Cliff
Judge GC, Griffiths WE, Hill RC, Lutkepohl H, Lee T-C (1985) The theory and practice of econometrics, 2nd edn. Wiley, USA
Karadavut U (2009) Non-linear models for growth curves of triticale plants under irrigation conditions. Turk J Field Crops 14(2):105–110
Keith T (2015) Multiple regression and beyond. An introduction to multiple regression and structural equation modeling, 2nd edn. Routledge Publisher, New York
Kendall MG, Buckland WR (1982) A dictionary of statistical terms, 4th edn. Longman Group, London
Kenny DA (2008) Reflections on mediation. Organ Res Methods 11(2):353–358
Keoleian GA, Blanchard S, Reppe P (2001) Life-cycle energy, costs, and strategies for improving a single-family house. J Ind Ecol 4(2):135–156
Kersey RS, Brinks JS (1985) Genetic and environmental aspects of the growth curve parameters in beef cows. J Anim Sci 61(6):1431–1440
King G (1986) How not to lie with statistics: avoiding common mistakes in quantitative political science. Am J Polit Sci 30(3):666–687
Knecht G (1974) Costing, technological growth and generalized learning curves. Oper Res Q 25(3):487–491
Kus H, Nygren K, Norberg P (2004) In-use performance assessment of rendered autoclaved concrete walls by long-term moisture monitoring. Build Environ 39(6):677–687

Kutner MH, Nachtsheim C, Neter J, Li W (2005) Applied linear statistical methods, 5th edn. McGraw-Hill, New York

Larsen WA, McCleary SJ (1972) The use of partial residual plots in regression analysis. Technometrics 14:781–790

Leclerc G, Pireaux JJ (1995) The use of least squares for XPS peak parameters estimation. Part 3. Multicollinearity, ill-conditioning and constraint-induced bias. J Electron Spectrosc Relat Phenom 71(2):179–190

Leon AR, Zhu Y (2008) ANOVA extensions for mixed discrete and continuous data. Comput Stat Data Anal 52(4):2218–2227

Leung AWT, Tam CM, Liu DK (2001) Comparative study of artificial neural networks and multiple regression analysis for predicting hoisting times of tower cranes. Build Environ 36 (4):457–467

Levenberg K (1944) A method for the solution of certain non-linear problems in least squares. Q Appl Math II(2):164–168

Lin L (2007) Using SPSS for multiple regression. Course UDP 520—quantitative methods in urban design and planning. University of Washington, USA

Lipinski MR, Roeleveld MA (1990) Minor extension of the von Bertalanffy growth theory. Fish Res 9(4):367–371

Lipton S, McGilchrist CA (1964) The derivation of methods for fitting exponential regression curves. Biometrika 51(3–4):504–508

Liu H, Shah S, Jiang W (2004) On-line outlier detection and data cleaning. Comput Chem Eng 28(9):1635–1647

Lubinski D, Humphreys LG (1990) Assessing spurious “moderator effects”: illustrated substantively with the hypothesized (“synergistic”) relation between spatial and mathematical ability. Psychol Bull 107(3):385–393

MacDonald B (2002) A teaching note on Cook’s distance—a guideline. Res Lett Inf Math Sci 3:127–128

Maroco J (2007) Statistical analysis using SPSS. (in Portuguese), 3rd edn. Sílabo Editions, Lisbon

Marin-Galiano M, Kunert J (2006) Comparison of ANOVA with Tobit model for analyzing sensory data. Food Qual Prefer 17(3–4):209–218

Marquardt DW (1963) An algorithm for least-squares estimation of nonlinear parameters. J Soc Ind Appl Math 11(2):431–441

Marquardt DW (1970) Generalized inverses, ridge regression, biased linear estimation and non-linear estimation. Technometrics 12(3):591–612

Marshall AW, Olkin I (1995) Multivariate exponential and geometric distributions with limited memory. J Multivar Analy 53(1):110–125

Martin JW, Saunders SC, Floyd FL, Wineburg JP (1996) Methodologies for predicting the service lives of coating systems. Federation of Societies for Coatings Technology, Blue Bell, Pennsylvania, USA

Mason CH, Perreault WD (1991) Collinearity, power, and interpretation of multiple regression analysis. J Mark Res 28(3):268–280

McRoberts RE (2009) Diagnostic tools for nearest neighbors techniques when used with satellite imagery. Remote Sens Environ 113(3):489–499

Meloun M, Militký J (2001) Detection of single influential points in OLS regression model building. Anal Chim Acta 439(2):169–191

Mondragon PF (2003) A comparison of nonlinear regression codes. Master thesis of Science in Mathematics with Operations Research and Statistics Option, New Mexico Institute of Mining and Technology, New Mexico

Montgomery DC (1984) Design and analysis of experiments, 2nd edn. Wiley, New York

Montgomery DC, Ruger GC (2003) Applied statistics and probability for engineers, 3rd edn. Wiley, USA

Montgomery DC, Peck EC, Vining GG (2001) Introduction to linear regression analysis, 3rd edn. Wiley, USA

Moore DS (1997) Statistics, concepts and controversies, 4th edn. W.H. Freeman and Company, New York

Morad S, Shacham M, Brenner A (2007) Utilization of collinearity in regression modeling of activated sludge processes. Chem Eng Process 46(3):222–229

Morgan PH, Mercer LP, Flodin NW (1975) General model for nutritional responses of higher organisms. Natl Acad Sci USA 72(11):4327–4331

Moss TR (2005) The reliability data handbook, 1st edn. Professional Engineering Publishing Limited and Bury St. Edmunds, London

Motulsky HJ, Ransnas LA (1987) Fitting curves to data using nonlinear regression: a practical and nonmathematical review. FASEB J 1(5):365–374

Musil CM, Jones SL, Warner CD (1998) Structural equation modelling and its relationship to multiple regression and factor analysis. Res Nurs Health 21(3):271–281

Myers RH (1990) Classical and modern regression with applications, 2nd edn. Duxbury press, CA

Nembhard DA, Uzumeri MV (2000) An individual-based description of learning within an organization. IEEE Trans Eng Manage 47(3):370–378

Neter J, Kutner MH, Nachtsheim CJ, Wasserman W (1996) In: Irwin RD (ed) Applied linear statistical models, 4th edn. Times Mirror Higher Education Group, Inc., Chicago

Osborne J, Waters E (2002) Four assumptions of multiple regression that researchers should always test. Pract Assess Res Eval 8(2):1–5

Paulo PV, Branco FA, de Brito J (2011) Deterministic and stochastic prediction models for the BuildingsLife platform. In: XII DBMC, 9th international conference on durability of building materials and components, Porto, pp 1–9

Pedersen RØ, Skovsgaard JP (2009) Impact of bias in predicted height on tree volume estimation: a case-study of intrinsic nonlinearity. Ecol Model 220(20):2656–2664

Perl W (1960) A method for curve-fitting by exponential functions. Int J Appl Radiat Isot 8(4):211–222

Piaw CY (2006) Basic research statistics (Book 2). McGraw-Hill, Kuala Lumpur (in Malay)

Pires JCM, Martins FG, Sousa SIV, Alvim-Ferraz MCM, Pereira MC (2008) Selection and validation of parameters in multiple linear and principal component regressions. Environ Model Softw 23(1):50–55

Pitts BJF, Pitts JN (2000) Chemistry of the upper and lower atmosphere. Theory, experiments and applications. Academic Press, New York

Preacher KJ, Rucker DD, Hayes AF (2007) Addressing moderated mediation hypotheses: theory, methods, and prescriptions. Multivar Behav Res 42(1):185–227

Ratkowsky DA (1983) Nonlinear regression modeling: a unified practical approach, 1st edn. Marcel Dekker Editors, New York

Rauch AF (1997) EPOLLS: an empirical method for predicting surface displacements due to liquefaction-induced lateral spreading in earthquakes. PhD thesis in Civil Engineering, Faculty of the Virginia Polytechnic Institute and State University, Virginia, USA

Rice WR (1989) Analyzing tables of statistical tests. Evolution 43(1):223–225

Richards JF (1959) A flexible growth function for empirical use. Journal of Experimental Botany 10(2):290–301

Rougoor CW, Dijkhuizen AA, Huirne RBM, Mandersloot F, Schukken YH (1997) Relationships between technical, economic and environmental results on dairy farms: an explanatory study. Livestock Prod Sci 47(3):235–244

Roy D (2002) On bivariate lack of memory property and a new definition. Ann Inst Stat Math 54(2):404–410

Royall RM (1997) Statistical evidence: a likelihood paradigm. Chapman and Hall Editors, New York

Samal AR, Mohanty MK, Fifarek RH (2008) Backward elimination procedure for a predictive model of gold concentration. J Geochem Explor 97(2–3):69–82

Satapathy SC, Murthy JVR, Reddy PVGDP, Misra BB, Dash PK, Panda G (2009) Particle swarm optimized multiple linear regression linear model for data classification. Appl Soft Comput 9(2):470–476

Savin NE, White MB (1977) The Durbin-Watson test for serial correlation with extreme sample sizes or array regressors. Econometrica 15:1989–1996
Sawalha Z, Sayed T (2006) Traffic accidents modeling: some statistical issues. Can J Civ Eng 33(9):1115–1124
Seber GAF, Wild CL (1989) Nonlinear Regression. Wiley series in probability and mathematical statistics. Wiley, New York
Seinfeld JH, Pandis SN (1998) Atmospheric chemistry and physics. From air pollution to climate change. Wiley, New York
Serdahl E (1996) An introduction to graphical analysis of residual scores and outlier detection in bivariate least squares regression analysis. Annual meeting of the Southwest Educational Research Association, New Orleans, USA
Shipley B (2000) Cause and correlation in biology: a user's guide to path analysis, structural equations and causal inference. Cambridge University Press, Cambridge
Shohet IM, Paciuk M (2004) Service life prediction of exterior cladding components under standard conditions. Constr Manage Econ 22(10):1081–1090
Shohet I, Rosenfeld Y, Puterman M, Gilboa E (1999) Deterioration patterns for maintenance management—a methodological approach. In: 8th international conference on durability of building materials and components, Vancouver, Canada, pp 1666–1678
Shohet IM, Puterman M, Gilboa E (2002) Deterioration patterns of building cladding components for maintenance management. Constr Manage Econ 20(4):305–314
Silva A, de Brito J, Gaspar PL (2012) Application of the factor method to maintenance decision support for stone cladding. Autom Constr 22:165–174
Silva A, Gaspar PL, de Brito J (2016) Comparative analysis of service life prediction methods applied to rendered façades. Mater Struct pp 1–18. doi:10.1617/s11527-016-0832-6
Sirkin RM (2006) Statistics for the social sciences, 3rd edn. Sage Publications, California, USA
Spanos A, McGuirk A (2002) The problem of near-multicollinearity revisited: erratic vs. systematic volatility. J Econom 108(2):365–393
Staniswalis JG (2006) On fitting generalized non-linear models with varying coefficients. Comput Stat Data Anal 50(7):1818–1839
Stevens JP (2009) Applied multivariate statistics for the social sciences, 5th edn. Routledge Publisher, New York
Sykes AO (1992) An introduction to regression analysis. The inaugural case lecture. Chicago working paper in Law & Economics. The University of Chicago, The Law School, 1 Dec 1992, 33p
Tabachnick BG, Fidell LS (1996) Using multivariate statistics, 3rd edn. Collins College Publishers, New York
Taylor J (2009) Multiple linear regression. Course statistics 191: introduction to applied statistics. Department of Statistics, Stanford University, Stanford, California, 2 Feb 2009, 38p
Teo EAL, Chew MYL, Harikrishna N (2005) An assessment of factors affecting the service life of external paint finish on plastered facades. In: 10th international conference on durability of building materials and components, Lyon, TT4-152
Velilla S (1998) A note on the behaviour of residual plots in regression. Stat Probab Lett 37(3):269–278
Velleman P, Welsch R (1981) Efficient computing of regression diagnostics. Am Stat 35(4): 234–242
Vigil DP, Sarper H (1994) Estimating the effects of parameter variability on learning curve model predictions. Int J Prod Econ 34(2):187–200
Viitaniemi P, Haapio A (2008) How workmanship should be taken into account in service life planning. In: 11th DBMC international conference on durability of buildings materials and components, DBMC 2008, Istanbul, Turkey, T45
Vits J, Gelders L (2002) Performance improvement theory. Int J Prod Econ 77(3):285–298
Von Bertalanffy L (1938) A quantitative theory of organic growth. Hum Biol 10(2):181–213
Von Bertalanffy L (1957) Quantitative laws in metabolism and growth. Q Rev Biol 32(3):217–230
Wallentin B, Ågren A (2002) Test of heteroscedasticity in a regression model in the presence of measurement errors. Econ Lett 76(2):205–211

Weibull W (1951) A statistical distribution function of wide applicability. J Appl Mech 18(3):293–297

Westberg K, Norén J, Kus H (2001) On using available environment data in service life estimations. Build Res Inf 29(6):428–439

Wheeler DC, Hickson DA, Waller LA (2010) Assessing local model adequacy in Bayesian hierarchical models using the partitioned deviance information criterion. Comput Stat Data Anal 6(1):1657–1671

Williams R (2004) Analytic strategies: simultaneous, hierarchical, and stepwise regression. Course of Sociology 63992, University of Notre Dame, Indiana

Williams GJ, Baxter RA, He HX, Hawkins S, Gu L (2002) A Comparative study of rnn for outlier detection in data mining. In: IEEE international conference on data-mining (ICDM'02), Maebashi City, Japan, CSIRO technical report CMIS-02/102

Woods PSA, Wynne HJ, Ploeger HW, Leonard DK (2003) Path analysis of subsistence farmers' use of veterinary services in Zimbabwe. Prev Vet Med 61(4):339–358

Wooldridge JM (2009) Introductory econometrics: a modern approach, 4th edn. South-Western College Publishing, Cengage Learning, 867p

Worsley BH, Lax LC (1962) Selection of a numerical technique for analyzing experimental data of the decay type with special reference to the use of tracers in biological systems. Biochim Biophys Acta 59(1):1–24

Wright TP (1936) Factors affecting the cost of airplanes. J Aeronaut Sci 3(4):122–128

Xie M, Lai CD (1996) Reliability analysis using an additive Weibull model with bathtub-shaped failure rate function. Reliab Eng Syst Saf 52(1):87–93

Yaffee RA (2004) Regression analysis with SPSS. Statistics, Mapping and Social Science Group Academic Computing Services, Information Technology Services, New York University, New York, USA

Yan X, Su X (2009) Linear regression analysis: theory and computing. World Scientific Publishing Company Inc., Singapore

Yelle LE (1979) The learning curve: historical review and comprehensive survey. Decis Sci 10 (2):302–328

Yiu CY, Ho DCW, Lo SM (2007) Weathering effects on external wall tiling systems. Constr Build Mater 21(3):594–600

Zar J (1999) Biostatistical analysis, 4th edn. Prentice-Hall, Inc., Englewood Cliff

Zhu Q, Guan WL (1997) Hygrothermal load on composite-facade performance. In: Conference on building envelope systems & technology, Bath, UK, pp. 435–440

Ziliak ST, McCloskey D (2009) The cult of statistical significance. In: Proceedings of the joint statistical meetings 2009, Washington, DC, pp 2302–2316

Chapter 4
Stochastic Models

4.1 Logistic Regression

4.1.1 Theoretical Foundations

Statistical models allow establishing an empirical relationship between variables through the estimation of parameters whose statistical validity can be tested. When the variables to be modelled are quantitative (continuous or discrete), it is usual to use linear or non-linear regression techniques (single or multiple). However, when the aim of the study is to identify the relationship between a qualitative dependent variable and categorical independent variables (factors) or numerical (covariates), it is necessary to use a categorical regression analysis (Trexler and Travis 1993).

This regression analysis has different definitions depending on the characteristics of the dependent variable. When the dependent variable is dichotomous (with only two mutually exclusive events), categorical regression is called logistic regression, a statistics technique used in the definition of prediction models (Choi and Lee 2010). Logistic regression can be expanded to multinomial logistic regression that is used when the dependent variable is nominal polytomous. Unlike binary logistic models in which the dependent variable is a binary choice, usually characterized by the presence or absence of a given characteristic, in multinomial logistic regression there may be more than two choices (the dependent variable has more than two mutually exclusive features) (Wang 2005). This technique describes the relationships between a categorical target variable (dependent variable) and the explanatory variables (independent or predictor variables) in terms of the probability of a given event (Witten and Frank 2005).

A. Silva et al., *Methodologies for Service Life Prediction of Buildings*,
Green Energy and Technology, DOI 10.1007/978-3-319-33290-1_4

4.1.1.1 Description

Contrary to what happens in continuous variables such as age, which may take an arbitrary number of values, categorical variables are measured by a limited number of options. In categorical variables, the answer is all-or-nothing, which means it is composed of only two possible answers "yes" or "no". Usually, categorical variables are coded into dummy variables, assuming values 1 or 0, which represents the presence or absence of a given attribute.

Logistic regression models the probability of success, where the word "success" represents the occurrence of a given attribute in a variable; usually the attribute is more relevant for the problem under analysis. The objective is to analyse the probability of each case study to reach the end of its service life. In this situation, there is "success" when a case study has reached the end of its service life ($Y = 1$) and "failure" in the opposite situation ($Y = 0$). Using this method, it is possible to relate the target variable with other variables, such as age, environmental exposure characteristics and the quality of the materials used in coatings, among others. The dichotomous categorical variables, with two attributes, where one is "success" and the other is "failure" can be characterized by the Bernoulli distribution, which assumes the following assumption: let $P[Y = 1]$ be the probability of "success" and $P[Y = 0]$ the probability of "failure"; then $P[Y = 1] + P\ [Y = 0] = 1$. The probability of "success" is given by $\hat{\pi} = \pi(1)/[1 - \pi(1)]$ and the probability of "failure" by $(1 - \hat{\pi}) = \pi(0)/[1 - \pi(0)]$. In simple terms, the probability of something happening is given by the ratio between the number of times that event occurs in a population and the total number of events ($[\#\{Y = 1\}/n]$).

The function used in the logistic regression to estimate the probability of a given realization j ($j = 1, \ldots, n$) of the dependent variable being a "success", $P[Y_j = 1] = \hat{\pi}_j$, is the logistic function whose generic form, for more than one independent variable X_i ($i = 1, \ldots, p$), is given by Eq. (4.1) (Silva et al. 2013).

$$\hat{\pi} = \frac{e^{\beta_0 + \beta_1 X_{1j} + \cdots + \beta_p X_{pj}}}{1 + e^{\beta_0 + \beta_1 X_{1j} + \cdots + \beta_p X_{pj}}} \tag{4.1}$$

This model can be adjusted using a non-linear regression. However, the traditional solution is to linearize the function presented in Eq. (4.1), using a logit transformation (Logit ($\hat{\pi}$))—Eq. (4.2).

$$\text{Logit}(\hat{\pi}) = \text{Ln}\left(\frac{\hat{\pi}}{1 - \hat{\pi}}\right) \tag{4.2}$$

Other important concept in logistic regression is the odds ratio ($\hat{\pi}/1 - \hat{\pi}$). This is the ratio between the probability of "success" ($\hat{\pi}$) and the probability of "not a success" (1-$\hat{\pi}$), as shown in Eq. (4.3).

$$\mathrm{Odds}(Y=1) = \frac{\hat{\pi}}{(1-\hat{\pi})} = \frac{[\#\{Y=1\}/n]}{[\#\{Y=0\}/n]} \tag{4.3}$$

In statistical models, the most common methods used in the parameters' estimation are: (i) the method of least squares (most common, usually used in regression analysis); (ii) and the method of maximum likelihood estimation (MLE). In logistic regression, the adjustment method used is the maximum likelihood method. MLE method allows estimating the regression coefficients (*B*) that maximize the probability of finding the realizations of the dependent variable sampled, i.e. which maximize the likelihood of these values. If all the *n* observations of *Y* are independent then the likelihood function (*L*) that gives the joint probability P ($Y = y_1$, $Y = y_2$, …, $Y = y_n$) of observing the values (y_1, y_2, …, y_n) is given by Eq. (4.4) (Maroco 2007).

$$L = P(Y=y_1) \times P(Y=y_2) \times \ldots \times P(Y=y_n) = \prod_{j=1}^{n} \pi_j^{y_j}\left(1-\pi_j\right)^{1-y_j} \tag{4.4}$$

When the MLE method is used for parameters' estimation, generally it is more convenient to use the negative logarithm of the likelihood function (Ljung 1987). The maximum likelihood function (*L*) is to $\partial L/\partial\beta = 0$ and $\partial 2L/\partial 2\beta < 0$. Since Ln (*L*) is a monotone transformation of *L*, the vector that maximizes Ln(*L*) is the same as that maximizing *L*. Thus, and to facilitate calculation, it is assumed that Ln(*L*) is the likelihood function—Eq. (4.5).

$$\begin{aligned}\mathrm{LL} = \mathrm{Ln}(L) &= \sum_{j=1}^{n} y_j\left(\frac{e^{x_j'\beta}}{1+e^{x_j'\beta}}\right) + \sum_{j=1}^{n}(1-y_j)\left(\frac{1}{1+e^{x_j'\beta}}\right)\\ &= \sum_{j=1}^{n}\left\{y_j\mathrm{Ln}(\hat{\pi}_j) - (1-y_j)\mathrm{Ln}(1-\hat{\pi}_j)\right\}\end{aligned} \tag{4.5}$$

However, it is not possible to obtain an analytical solution to this system, and it is necessary to use an algorithm that maximizes Ln(*L*). Therefore, the likelihood function for the model with qualitative variables, is now given by (4.6) (Hosmer and Lemeshow 2000; Gao and Shen 2007).

$$\mathrm{LL} = \sum_{j=1}^{J}\left\{y_j\mathrm{Ln}(\hat{\pi}_j) - (n_j - y_j)\mathrm{Ln}(1-\hat{\pi}_j)\right\} \tag{4.6}$$

where n_j represents the number of elements in each subpopulation; y_j the number of successes in each subpopulation; and $\hat{\pi}_j$ the probability of occurrence of these subpopulations.

4.1.1.2 Model's Goodness-of-Fit

The logistic regression parameters are obtained by an iterative process whose stopping criterion includes a maximum allowed number of iterations or other convergence criteria (Myung 2003). After parameter estimation, it is necessary to evaluate the significance and quality of the fitted model as well as the regression coefficients.

The likelihood ratio test is the most common test to evaluate the model's goodness-of-fit (Li 2007). This test is given by the ratio between the likelihood of the null model (model only with the constant) and the complete model (final model, with all the variables included in the analysis) (Suthar et al. 2010). In this test, two hypothesis are tested: (i) the null hypothesis (H_0): $\beta_1 = \beta_2 = \cdots = \beta_k = 0$, i.e. the final model is not statistically significant; (ii) the alternative hypothesis (H_1): $\exists\ i$: $\beta_i \neq 0$, i.e. there is at least one i where β_i is different from zero, which mean that the full model is statistically significant. If the null hypothesis is true, none of the independent variables included in the model is statistically significant in the explanation of the variability of the dependent variable. If the alternative hypothesis is true, then there is at least one independent variable that is appropriate to explain the behaviour of the dependent variable, and it is necessary to examine the statistical significance of each of the variables included in the model. To test these two hypotheses, the G-test of goodness-of-fit is generally used, which relates the likelihood ratio test between the null model (L_0) and the complete model (L_C). The smaller the test result ($\lambda = L_0/L_C$), the greater the results of G-test (Eq. (4.7)) (Agresti 2002).

$$G^2 = -2\mathrm{Ln}(\lambda) = -2\mathrm{Ln}\left(\frac{L_0}{L_C}\right) \tag{4.7}$$

This test is an "omnibus" test, since it is used to compare two nested models (a simpler model and a more complex model—including more variables), i.e. models that have a common structure but differ in terms of the causal relations represented (Hair et al. 2007).

To evaluate the fitting of the logistic regression model to the data it is used the Pearson chi-square test (χ^2) (Plackett 1983). The value of Pearson χ^2 is obtained as illustrated in Eq. (4.8) (Hosmer and Lemenshow 2000; Howell 2002).

$$\chi^2_{\mathrm{P}} = \sum_{j=1}^{J} \frac{\left(O_j - E_j\right)^2}{E_j} \tag{4.8}$$

where O_j represents the number of "successes" observed and E_j the number of "successes" predicted by the model.

The Pearson chi-square test (χ^2_{P}) has associated J-p-1 degrees of freedom and it is considered that, if the p-value is greater than the significance level considered, then the model presents a good fit to the data modelled. In this situation, it can be considered that the model fits the data, but the uncertainty regarding the existence of other models that could conveniently describes the sample's variability remains (Stelzl 1986; Lee and Hershberger 1990).

Besides the Pearson chi-square test, the Deviance test (also called log-likelihood ratio statistic) can also be applied in the analysis of the goodness-of-fit of the model to the data (Smyth 2003). This test is based on the logarithm of the ratio of the likelihood of the fitted or reduced model (L_C) and the saturated model (L_S)—model including all the independent variables and crossing their synergy effects. The deviance can therefore be estimated as indicated in Eq. (4.9).

$$D = -2\text{Ln}\left(\frac{L_C}{L_S}\right) \tag{4.9}$$

Newsom (2011) refers that, in logistic regression, the chi-square and deviance tests act like the determination coefficient (R^2) in regression analysis, measuring the goodness-of-fit of the model, analyzing the deviation between the values predicted by the model and the observed values. Thus, the deviation should be as small as possible. However, there is no threshold from which it can be said that the model fit is appropriate. For this reason, these tests should preferably be used to compare logistic regression models and are not suitable to be used as an indicator by itself.

To evaluate the model's goodness-of-fit, Akaike (1979, 1981, 1984) proposed the Akaike Information Criteria (AIC) and Schwarz (1978) proposed the Bayesian Information Criteria (BIC). These two criteria penalize the deviance test, the first depending on the number of model's parameters and the second depending on the sample size, ensuring the selection of the most parsimonious model. The lower the values of AIC and BIC, the better the model's fitting to the data is (Lewis et al. 2011).

It may not be suitable to use the chi-square test (χ^2) when the logistic regression model has more than one quantitative independent variable. To overcome this constraint, Hosmer and Lemeshow (1980) proposed other statistical test to evaluate the model's goodness-of-fit. To implement this test, predicted probabilities are grouped into L groups (usually ten). The test is calculated from the ratio between the frequency of observed values (O_g) and the average frequency of expected values (E_g) in each cell g (with $g = 1, 2, \ldots, 2G$), as indicated in Eq. (4.10).

$$\chi^2_{\text{HL}} = \sum_{g=1}^{2G} \frac{\left(O_g - E_g\right)^2}{E_g} \tag{4.10}$$

The chi-square test of Hosmer–Lemeshow follows an asymptotic chi-square distribution with G-2 degrees of freedom. However, some authors (Hosmer and Hjort 2002; Kuss 2002) refer the limitations of this test: it is a conservative test, enabling the identification of specific types of lack of fit of the data; it has a strong dependence on how data is clustered (when a small number of groups, less than five, is used, the test can lead to wrong conclusions about the goodness-of-fit of the data);

when the test indicates a lack of fit of the model, it becomes difficult to identify which cases are wrongly modelled. On the other hand, Xie et al. (2008) report that this test has been widely used due to its advantages, such as: it is intuitive and easy to compute and it is present in most available statistical software; finally, the popularity of this test largely results from the absence of better approaches.

4.1.1.3 Significance of Model Coefficients

The Wald test (1941, 1943) is usually used to evaluate the statistical significance of the model coefficients. If for a given explanatory variable the Wald test is significant, then it can be concluded that the parameters associated with these variables are not zero and can therefore be included in the model (Westergren et al. 2001). The Wald test statistic is calculated as indicated in Eq. (4.11).

$$W_j = \frac{\hat{\beta}_j}{\widehat{\mathrm{SE}}\left(\hat{\beta}_j\right)} \sim N(1,0) \tag{4.11}$$

where $\hat{\beta}_j$ represents the model coefficients and $\widehat{\mathrm{SE}}\left(\hat{\beta}_j\right) = \sqrt{\hat{\sigma}^2\left(\hat{\beta}_j\right)}$ the deviation or standard error of the coefficient $\hat{\beta}_j$.

To assess the significance of the model coefficients two hypotheses are tested: (i) the null hypothesis (H_0): $\beta_i = 0 | \beta_0, \beta_1, \beta_{i\text{-}1}, \beta_{i+1}, \beta_p$ ($i = 1, \ldots, p$); (ii) the alternative hypothesis (H_1): $\beta_i \neq 0 \mid \beta_0, \beta_1, \beta_{i\text{-}1}, \beta_{i+1}, \beta_p$ ($i = 1, \ldots, p$). When the null hypothesis is rejected (the *p*-value is less than the chosen significance level), it is possible to determine which coefficients should be included in the logistic regression model.

4.1.1.4 Pseudo R^2

The determination coefficient (R^2) is the most popular measure to evaluate a model's goodness-of-fit. The interpretation of this statistical measure is very simple, varying between zero and one, where zero represents a model without predictive power and one a model with a perfect fitting to the data (Hu et al. 2006). Hosmer and Lemeshow (2000) refer that, generally, in logistic regression, R^2 values are low, which can be a problem when reporting their values to an audience familiarized with linear regression values. In logistic and multinomial regression models, several studies have been made in order to define an adjusted determination coefficient for this type of models (McFadden 1973; Agresti 1986; Nagelkerke 1991; Cox and Wermuch 1992; Ash and Schwartz 1999). Various pseudo-R^2 are suggested, based on the comparison between the null model (model with no predictors) and the final model (including all the predictors). The most common pseudo-R^2 proposed in literature is: (i) Cox and Snell (1989), Eq. (4.12); (ii) Nagelkerke (1991); (iii) and McFadden (1973).

$$R^2_{\mathrm{CS}} = 1 - \mathrm{e}^{\frac{2(\mathrm{LL}_c - \mathrm{LL}_0)}{n}} \tag{4.12}$$

where LL_c represents the log-likelihood of the model with all predictors, LL_0 the log-likelihood of the null model and n the sample dimension.

The Cox and Snell pseudo-R^2 never reaches the maximum value of 1 (Wuensch 2011). Therefore, Nagelkerke (1991) proposed an improvement to the Cox and Snell pseudo-R^2, in order to obtain a statistical index varying within the range between 0 and 1, thus resulting in the Nagelkerke pseudo-R^2—Eq. (4.13).

$$R^2_{\mathrm{N}} = \frac{R^2_{\mathrm{CS}}}{1 - \mathrm{e}^{\frac{2\mathrm{LL}_0}{n}}} \tag{4.13}$$

Although varying between 0 and 1, the Nagelkerke pseudo-R^2 should also be carefully analysed; as mentioned by Mittlbock and Schemper (1996), the pseudo-R^2, unlike the determination coefficient in linear regression, does not represent the variability of the dependent variable that is explained by the model, since it only analyses the relationship between the null model and the complete model.

Shtatland et al. (2002) investigated the use of different pseudo-R^2 concluding that the one proposed by McFadden (1973) is the one whose interpretation is easier, since it measures the information gain obtained by adding the independent variables to the null model (Menard 2000). The pseudo-R^2 McFadden (1973) is determined as shown in Eq. (4.14).

$$R^2_{\mathrm{MF}} = 1 - \frac{\mathrm{LL}_C}{\mathrm{LL}_0} \tag{4.14}$$

4.1.1.5 Classification Table

In logistic regression, a classification table is usually used to analyse the accuracy of the model's classification. This table has four different scenarios: (i) true negatives; (ii) true positives; (iii) false positives; and (iv) false negatives. The cases correctly classified by the model (the predicted values are equal to the observed values) are designated true positives and true negatives. On the other hand, the case studies wrongly classified by the model are designated false positives and false negatives, meaning that the results predicted by the model do not correspond to the observed values. Table 4.1 illustrates a classification table, using as an example the service life prediction of façade claddings, answering the question "has the cladding reached the end of its service life?", with two possible outcomes: "yes" or "no".

In the definition of such classification tables, a cut value of 0.5 is normally used. With this threshold value, it is assumed that, if a case study has a probability equal

Table 4.1 Classification table

Observed values		Predicted values		Percentage correct
		"Has the cladding reached the end of its service life?"		
		No	Yes	
"Has the cladding reached the end of its service life?"	No	TN (true negatives)	FP (false positives)	Specificity (SPE)
	Yes	FN (salse negatives)	TP (true positives)	Sensitivity (SEN)
Overall percentage				Accuracy (ACC)

[a]The threshold value is 0.5

or greater than 50 % of having reached the end of its service life, then the answer is "yes" to the question "has the cladding reached the end of its service life?", otherwise the answer is "no". This threshold value can be changed in order to obtain a more or less restrictive classification criterion.

The classification table allows determining a set of statistical parameters that provide an indication of the model's classification accuracy. The accuracy of the classification is measured by its sensitivity and specificity. The sensitivity is the proportion of true positives that are correctly identified by the model (TP/(TP + FN)), which means that it represents the probability of the model to correctly identify the case studies that do in fact reach the end of their service life. On the other hand, the specificity is the proportion of true negatives correctly identified by the model (TN/(TN + FP)), i.e. this indicator provides the ability of the model to correctly predict the cases that have not reached the end of their service life. The accuracy is the proportion of true results (true positives and true negatives) correctly predicted by the model. A model with good classification capability shows sensitivity and specificity levels above 80 %; if between 50 and 80 %, it is considered that the model has a reasonable predictive ability; if less than 50 %, it is considered that the predictive ability of the model is poor (Maroco 2007).

To evaluate the classification ability of the model, it is usual to compare its accuracy with the percentage of case studies classified correctly by chance alone—Eq. (4.15).

$$\text{Case studies classified correctly by chance alone} = \sum_{i=1}^{k} \left(\frac{C_i^2}{N} \right) \times 100\ (\%) \tag{4.15}$$

where C_i represents the proportion of case studies for each category k of the dependent variable and N the sample dimension.

A model is considered to have a high classification capability when the cases correctly classified by the model are greater by at least 25 % than the proportion of case studies classified correctly by chance alone (White 2013).

4.1.2 *Probabilistic Analysis of the Degradation Condition in Relation to the Cladding's Age*

With logistic regression it is possible to evaluate the probability of each façade cladding to have a given degradation condition as a function of its age and its characteristics. The façade claddings analysed in this study are associated to a degradation scale (discussed in Chap. 2), which is defined based on the defects detected in the façades, their extension, severity and the costs associated with their repair. Defects are thus characterized and classified according to a scale of discreet variables that range from "A" (no visible degradation) to "E" (generalized degradation leading to immediate corrective action). Each degradation condition is associated with a qualitative scale (based on the evaluation of the physical and visual condition of the sample analysed) and a can be computed into a quantitative index that depicts the global performance of the façades. In this study a multinomial regression analysis is applied, in which the degradation condition of the façade's claddings is the dependent variable with five classes mutually exclusive: "Condition A"; "Condition B"; "Condition C"; "Condition D"; and "Condition E". In the next sections the probabilistic analysis of the degradation condition relative to the claddings' age is presented. The SPSS software was used to define the multinomial models.

4.1.2.1 Natural Stone Claddings

In natural stone claddings, none of the 203 case studies had the degradation condition "E" (generalized degradation). Therefore, in the multinomial regression performed, the degradation condition is the dependent variable corresponding to four mutually exclusive classes: "Condition A", "Condition B", "Condition C", and "Condition D". Multinomial regression thus calculates the probability that one cladding should belong to one the degradation conditions. Table 4.2 presents the percentage of case studies in each degradation condition. The majority of the case studies belong to conditions B and C.

Table 4.3 shows the likelihood ratio test between the null model (intercept only) and the final model. In this case it is found that the values of AIC, BIC and −2LL

Table 4.2 Statistical distribution of the sample according to their degradation condition

		Number of case studies	Percentage of case studies
Degradation condition	Condition A	9	4.4
	Condition B	114	56.2
	Condition C	57	28.1
	Condition D	23	11.3
	Condition E	0	0
Total		203	100

Table 4.3 Model fitting information

Model	Model fitting criteria			Likelihood ratio tests		
	AIC	BIC	−2 log likelihood	Chi-square[a]	Degrees of freedom (df)[b]	Significance[c]
Only with the constant	353.761	363.701	347.761			
Final	166.101	185.980	154.101	193.660	3	0.000

AIC—Akaike information criterion
BIC—Schwarz or Bayesian information criterion
−2 Log likelihood—minus twice the logarithm of the likelihood
[a]The chi-square (χ^2) statistics test is used to test the significance of the regression model; G^2 is a measure of the quality increment of the null model due to the addition of independent variables (Maroco 2007)
[b]The degrees of freedom are related to the available data for the statistical calculation
[c]The significance tests the probability of rejecting the null hypothesis when it is true (the null hypothesis is set aside whenever the value given in the table is higher than the significance level admitted). The null hypothesis in the case of the likelihood ratio test indicates that the model is not statistically significant; on the other hand, in the model adjustment test the null hypothesis indicates that the model adjusts itself to the data

are lower in the final model, which indicates a better fit to data than the model that only has the constant. The goodness-of-fit tests—$\chi^2_{\mathrm{P}}(165) = 133.603$ with $p = 0.965$ and *Deviance*: $\mathrm{D}(165) = 113.529$ with $p = 0.999$—reveal that the independent variable (age) is explanatory of the natural stone cladding's degradation condition. The proposed model presents values of the pseudo-R^2 of Cox and Snell ($R^2_{\mathrm{CS}} = 0.615$), Nagelkerke ($R^2_{\mathrm{N}} = 0.698$) and McFadden ($R^2_{\mathrm{MF}} = 0.448$), revealing a model with satisfactory quality.

The results of the likelihood ratio test for the independent variables included in the model are presented in Table 4.4. For the significance levels usually assumed (5 %), the likelihood ratio test reveals that the age is relevant to the degradation condition of natural stone claddings, with $G^2(4) = 193.66$ and $p < 0.000$.

Table 4.5 shows the models coefficients (*B*). As mentioned before, in this case, the dependent variable has four mutually exclusive classes and it was assumed that

Table 4.4 Likelihood ratio tests

Effect	Model fitting criteria			Likelihood ratio tests		
	AIC of reduced model	BIC of reduced model	−−2 log likelihood of reduced model	Chi-square	Degrees of freedom (df)	Significance
Constant	226.321	236.741	218.321	126.803	4	0.000
Age	268.638	279.059	260.638	169.121	4	0.000

The chi-square statistic is the difference in −2 Log-likelihoods between the final model and a reduced model. The reduced model is formed by omitting an effect from the final model. The null hypothesis is that all parameters of that effect are 0

Table 4.5 Estimates of the parameters of the model

Condition levels[a]		*B*	Standard error	Wald test[b]	Degrees of freedom (df)[c]	Sig.[d]	Exp (*B*)	95 % confidence range for Exp(*B*)	
								Lower bound	Upper bound
Condition B	Constant	0.472	0.629	0.563	1	0.453			
	Age	0.156	0.064	6.022	1	0.014	1.169	1.032	1.324
Condition C	Constant	−5.849	1.224	22.831	1	0.000			
	Age	0.288	0.067	18.272	1	0.000	1.334	1.169	1.523
Condition D	Constant	−16.956	2.903	34.112	1	0.000			
	Age	0.453	0.079	32.902	1	0.000	1.574	1.348	1.837

[a]The reference category is: condition A
[b]The Wald test is used to evaluate the significance of the model's coefficients. If, for a given explanatory variable, the Wald test is significant, then it can be concluded that the parameters associated with these variables are not nil and can be included in the model (Westergren et al. 2001)
[c]The degrees of freedom are related with the data available for the statistical procedure
[d]The significance tests the probability of rejecting the null hypothesis when it is true (the null hypothesis is set aside whenever the value given in the table is higher than the significance level admitted). The null hypothesis in the case of the likelihood ratio test indicates that the model is not statistically significant; on the other hand, in the model adjustment test the null hypothesis indicates that the model adjusts itself to the data

"Condition A" is the reference class. Once the model coefficients are statistically significant, it is possible to estimate the probability of each degradation condition according to cladding's age—Eqs. (4.16)–(4.19).

$$P(Y = \text{"Condition A"}) = \frac{1}{1 + e^{0.472 + 0.156\,\text{Age}} + e^{-5.849 + 0.288\,\text{Age}} + e^{-16.956 + 0.453\,\text{Age}}} \quad (4.16)$$

$$P(Y = \text{"Condition B"}) = \frac{e^{0.472 + 0.156\,\text{Age}}}{1 + e^{0.472 + 0.156\,\text{Age}} + e^{-5.849 + 0.288\,\text{Age}} + e^{-16.956 + 0.453\,\text{Age}}} \quad (4.17)$$

$$P(Y = \text{"Condition C"}) = \frac{e^{-5.849 + 0.288\,\text{Age}}}{1 + e^{0.472 + 0.156\,\text{Age}} + e^{-5.849 + 0.288\text{Age}} + e^{-16.956 + 0.453\,\text{Age}}} \quad (4.18)$$

$$P(Y = \text{"Condition D"}) = \frac{e^{-16.956 + 0.453\,\text{Age}}}{1 + e^{0.472 + 0.156\,\text{Age}} + e^{-5.849 + 0.288\,\text{Age}} + e^{-16.956 + 0.453\,\text{Age}}} \quad (4.19)$$

Figure 4.1 shows the probabilistic distribution of the degradation condition relative to cladding age. The model determines these probabilities based on the frequency of each degradation condition observed in each subpopulation. The probability of belonging to condition A decreases over time and is practically nil

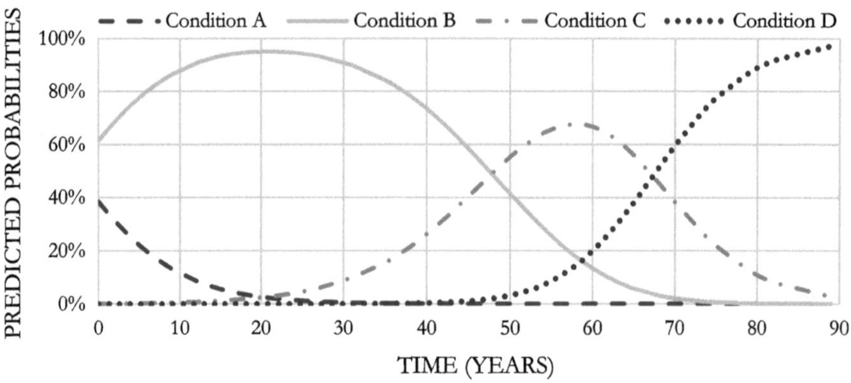

Fig. 4.1 Probabilistic distribution of the degradation condition according to the age of the case studies (stone claddings)

(under 3 %) at 20 years. The majority of the natural stone claddings analysed present the degradation condition B (56.2 %). The model reveals that there is a prevalence of the degradation condition B within the sample between age 0 and 48, reaching a maximum probability peak at 21 years ($P = 95$ %). Concerning degradation condition C, the maximum probability (67.8 %) is reached at 58 years. The probability of belonging to condition D increases with age and after 73 years it is higher than 70 % and after 86 years it is higher than 95 %. As expected, younger claddings are most likely to belong to lower degradation conditions.

Table 4.6 shows the probability of the claddings to belong to each degradation condition according to their age. The highest probability of transition between conditions B and C occurs after 48 years and the maximum probability of transition between conditions C and D occurs at age 67. It is important to note that condition A represents claddings with a degradation severity under 1 %; in fact, this range of degradation severity is much too small, allowing some misperception between conditions A and B (condition B is more comprehensive and includes all the case studies with a severity of degradation between 1 and 8 %). As a consequence, some case studies that belong to condition A are misclassified by the model (as shown in Table 4.7), which assumes that these cases belongs to condition B, which explains why the probability of belonging to condition B is higher than that of belonging to condition A in the range of 0–5 years.

Table 4.7 presents the classification table of the model, showing the percentage of cases correctly classified for each degradation condition, as well as the number of cases wrongly classified. The model obtained, where age is the only explanatory variable, classifies correctly 69.5 % of all cases analysed thus showing a reasonable predictive capacity. The percentage of case studies classified correctly by chance is given by the sum of the squares of the proportion of cases in each subgroup of the dependent variable (Bayaga 2010)—values shown in Table 4.2. In this case, the percentage of case studies classified correctly by chance it is given by $40.95\ \% = 100\ \% \times (0.044^2 + 0.562^2 + 0.281^2 + 0.113^2)$. The percentage of

Table 4.6 Probability of belonging to a condition level as a function of age

Range in years	Probability of belonging to a degradation condition			
	Condition A (%)	Condition B (%)	Condition C (%)	Condition D (%)
[0:5]	29.94	69.88	0.18	0.00
]5:10]	15.33	84.22	0.45	0.00
]10:15]	7.66	91.41	0.94	0.00
]15:20]	3.63	94.50	1.87	0.00
]20:25]	1.67	94.70	3.62	0.00
]25:30]	0.75	92.40	6.84	0.01
]30:35]	0.32	87.16	12.47	0.05
]35:40]	0.13	78.11	21.57	0.18
]40:45]	0.05	64.80	34.49	0.66
]45:50]	0.02	48.33	49.51	2.14
]50:55]	0.01	31.62	62.28	6.10
]55:60]	0.00	17.80	67.32	14.88
]60:65]	0.00	8.43	61.12	30.45
]65:70]	0.00	3.28	45.59	51.13
]70:75]	0.00	1.05	28.03	70.92
]75:80]	0.00	0.29	14.82	84.89
]80:85]	0.00	0.07	7.13	92.80
]85:90]	0.00	0.02	3.26	96.72

Table 4.7 Classification table

Observed	Predicted				
	Condition A	Condition B	Condition C	Condition D	Percentage correct
Condition A	0	9	0	0	0.0
Condition B	0	96	18	0	84.2
Condition C	0	15	33	9	57.9
Condition D	0	0	11	12	52.2
Overall percentage	0.0	59.1	30.5	10.3	69.5

cases correctly classified by the model is around 28.55 % higher than the percentage of random correct classifications, thus indicating that the model has good classifying properties.

4.1.2.2 Rendered Façades

Concerning the rendered façades, a sample comprising 100 case studies was used. In this case, the dependent variable (degradation condition) has five mutually exclusive classes: "Condition A" (14 % of the sample); "Condition B" (18 % of the sample); "Condition C" (31 % of the sample); "Condition D" (15 % of the sample); "Condition E" (22 % of the sample). The goodness-of-fit tests ($\chi^2_P(172) = 58.329$ with $p = 1.000$ and *Deviance*: $D(172) = 59.936$ with $p = 1.000$) indicate that the independent variable (age) is explanatory of the renders' degradation condition. The model applied to rendered façades has a pseudo-R^2 of Cox and Snell of 0.816, a pseudo-R^2 of Nagelkerke of 0.853 and a pseudo-R^2 of McFadden of 0.540, revealing a model with satisfactory quality. The likelihood ratio test, for a model only with age as explanatory variable, presents a $G^2(4)$ of 169.121 with $p = 0.000$, revealing that the age is statistically significant to the description of the façade's degradation condition.

The probability of each degradation condition according to renders' age is presented from Eqs. (4.20) to (4.24), assuming that "Condition A" is the reference class.

$$P(Y = \text{"Condition A"}) = \frac{1}{1 + e^{-1.788 + 0.597\text{Age}} + e^{-3.569 + 0.955\text{Age}} + e^{-10.032 + 1.427\text{Age}} + e^{-16.183 + 1.665\text{Age}}} \quad (4.20)$$

$$P(Y = \text{"Condition B"}) = \frac{e^{-1.788 + 0.597\text{Age}}}{1 + e^{-1.788 + 0.597\text{Age}} + e^{-3.569 + 0.955\text{Age}} + e^{-10.032 + 1.427\text{Age}} + e^{-16.183 + 1.665\text{Age}}} \quad (4.21)$$

$$P(Y = \text{"Condition C"}) = \frac{e^{-3.569 + 0.955\text{Age}}}{1 + e^{-1.788 + 0.597\text{Age}} + e^{-3.569 + 0.955\text{Age}} + e^{-10.032 + 1.427\text{Age}} + e^{-16.183 + 1.665\text{Age}}} \quad (4.22)$$

$$P(Y = \text{"Condition D"}) = \frac{e^{-10.032 + 1.427\text{Age}}}{1 + e^{-1.788 + 0.597\text{Age}} + e^{-3.569 + 0.955\text{Age}} + e^{-10.032 + 1.427\text{Age}} + e^{-16.183 + 1.665\text{Age}}} \quad (4.23)$$

$$P(Y = \text{"Condition E"}) = \frac{e^{-16.183 + 1.665\text{Age}}}{1 + e^{-1.788 + 0.597\text{Age}} + e^{-3.569 + 0.955\text{Age}} + e^{-10.032 + 1.427\text{Age}} + e^{-16.183 + 1.665\text{Age}}} \quad (4.24)$$

Figure 4.2 shows the probabilistic distribution of the degradation condition relative to rendered façades age. The probability of belonging to condition A decreases over time and is practically nil (less than 2 %) after 7 years. The probability of condition B increases initially reaching a peak at around 5 years

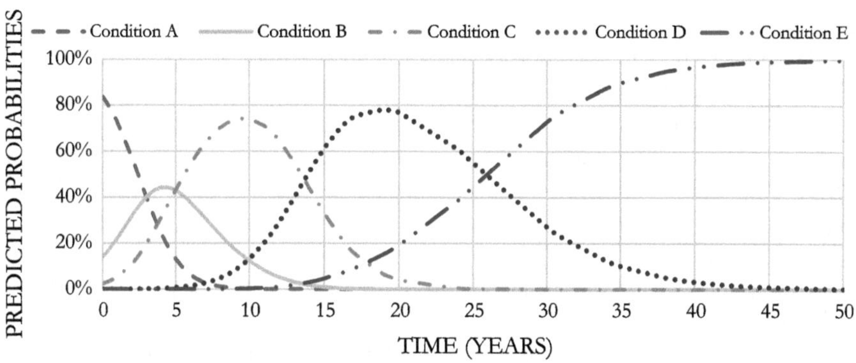

Fig. 4.2 Probabilistic distribution of the degradation condition according to the age of the case studies (rendered façades)

(P = 43 %) and then decreasing. Concerning condition C, the maximum probability (74.2 %) is reached at 10 years. For condition D, the peak is reached at around 19 years (probability of 78.3 %). Finally, the probability of condition E increases with age and after 30 years it is higher than 70 % and after 38 years it is higher than 90 %. The maximum probability of transition between conditions A and B occurs around 3 years. At year 5, the probability of any case study to be in condition B is similar to that of being in condition C (around 43 %). Between 13 and 15 years the probability of condition C is similar to that of condition D (between 40 % and 50 %). At 25 years the probability of a rendered façade to be in condition D of degradation is practically the same as being in condition E. These time intervals—when the probability of belonging to a given degradation condition is similar to that of the next condition—are also the ages when the probability of transition from one degradation condition to the next is greatest.

The multinomial model proposed for rendered façades classifies correctly 57.1 % of the cases belonging to degradation condition A, 22.2 % of those from condition B, 71.0 % of those from condition C, 60.0 % of those from condition D and finally 90.9 % of those from condition E. This model shows an overall percentage of cases correctly diagnosed (accuracy) of 63.0 %, thus showing a reasonable predictive capacity. The percentage of case studies classified correctly by chance it equal to 21.9 % [100 % × $(0.14^2 + 0.18^2 + 0.31^2 + 0.15^2 + 0.22^2)$]. The percentage of cases correctly classified by the model is thus 34.8 % higher than the percentage of case studies classified correctly by chance, indicating that the model has good classifying properties.

4.1.2.3 Painted Surfaces

Regarding painted surfaces, five mutually exclusive classes related to the degradation conditions, ranging from condition A to condition E, are also considered.

16.8 % of the sample belongs to condition A, 36.4 % to condition B, 22.7 % to condition C, 15.0 % to condition D and 9.1 % to condition E. The goodness-of-fit tests—$\chi^2_P(68) = 60.426$ with $p = 0.732$ and *Deviance*: $D(68) = 41.638$ with $p = 0.995$—reveal that the independent variable (age) is explanatory of the painted surfaces' degradation condition. The proposed model presents values of the pseudo-R^2 of Cox and Snell ($R^2_{CS} = 0.820$), Nagelkerke ($R^2_N = 0.862$) and McFadden ($R^2_{MF} = 0.568$), revealing a model with satisfactory quality. The model obtained conducts to a $G^2(4)$ of 376.832 with $p = 0.000$ (likelihood ratio test), revealing that the age is statistically significant to the description of the façade's degradation condition.

Equations (4.25)–(4.29) show the probability of each degradation condition according to paintings' age, assuming that "Condition E" is the reference class.

$$P(Y = \text{"Condition A"}) = \frac{e^{32.930-4.123\,\text{Age}}}{1 + e^{32.930-4.123\,\text{Age}} + e^{30.243-2.878\,\text{Age}} + e^{22.610-1.798\,\text{Age}} + e^{14.527-1.023\,\text{Age}}} \tag{4.25}$$

$$P(Y = \text{"Condition B"}) = \frac{e^{30.243-2.878\,\text{Age}}}{1 + e^{32.930-4.123\,\text{Age}} + e^{30.243-2.878\,\text{Age}} + e^{22.610-1.798\,\text{Age}} + e^{14.527-1.023\,\text{Age}}} \tag{4.26}$$

$$P(Y = \text{"Condition C"}) = \frac{e^{22.610-1.798\,\text{Age}}}{1 + e^{32.930-4.123\,\text{Age}} + e^{30.243-2.878\,\text{Age}} + e^{22.610-1.798\,\text{Age}} + e^{14.527-1.023\,\text{Age}}} \tag{4.27}$$

$$P(Y = \text{"Condition D"}) = \frac{e^{14.527-1.023\,\text{Age}}}{1 + e^{32.930-4.123\,\text{Age}} + e^{30.243-2.878\,\text{Age}} + e^{22.610-1.798\,\text{Age}} + e^{14.527-1.023\,\text{Age}}} \tag{4.28}$$

$$P(Y = \text{"Condition E"}) = \frac{1}{1 + e^{32.930-4.123\,\text{Age}} + e^{30.243-2.878\,\text{Age}} + e^{22.610-1.798\,\text{Age}} + e^{14.527-1.023\,\text{Age}}} \tag{4.29}$$

Figure 4.3 presents the probabilistic distribution of the degradation condition relative to painted surfaces age. The probability of a painted surface to belong to condition A is lower than 5 % after 4 years. The maximum probability to belong to condition B ($P = 87.9$ %) occurs at 4 years and the maximum probability to belong to condition C ($P = 68.7$ %) occurs at 9 years. Concerning condition D, the

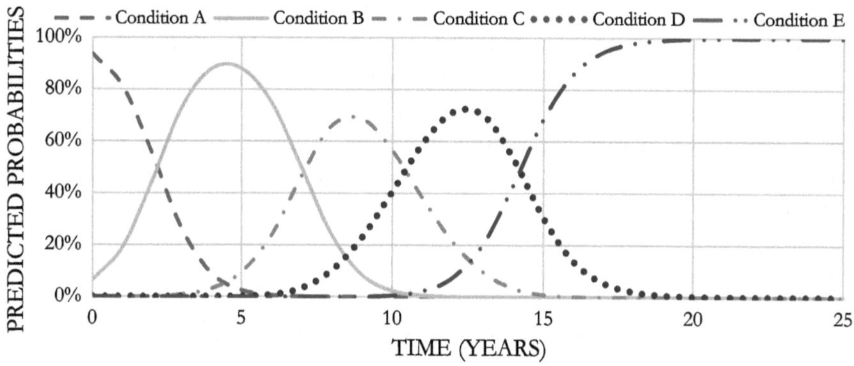

Fig. 4.3 Probabilistic distribution of the degradation condition according to the age of the case studies (painted surfaces)

maximum probability (71.3 %) is reached at 12 years. The probability of paintings to belong to condition E is less than 5 % at year 11 and is higher than 90 % after 16 years. The transition between the conditions A and B is more likely to occur between years 2 and 3. The maximum probability of transition between conditions B and C occurs between 7 and 8 years, the maximum probability of transition between conditions C and D occurs between 10 and 11 years and the maximum probability of transition between conditions D and E occurs between 14 and 15 years.

The model obtained for painted surfaces correctly classifies 86.5 % of the cases belonging to condition A of degradation, 77.5 % of those from condition B, 74.0 % of those from condition C, 48.5 % of those from condition D and finally 75.0 % of those from condition E. This model has an accuracy of 73.6 %. The percentage of case studies classified correctly by chance it equal to 24.3 % [100 % × ($0.168^2 + 0.364^2 + 0.227^2 + 0.15^2 + 0.091^2$)], revealing that the model has good classifying properties.

4.1.2.4 Ceramic Tiling Systems

For ceramic tiling systems, the dependent variable also has five mutually exclusive classes. 7.7 % of the sample belongs to condition A, 35.9 % to condition B, 32.8 % to condition C, 22.1 % to condition D and 1.5 % to condition E. Analyzing the likelihood ratio test between the null model (intercept only) and the final model, it is found that the values of AIC, BIC and -2LL are lower in the final model, which indicates a better fit to data than the model that only has the constant. The likelihood ratio test, for a final model with age as explanatory variable only, presents a $G^2(4)$ of 202.758 with $p = 0.000$, revealing that the age is statistically significant to the description of the cladding's degradation condition. The goodness-of-fit tests ($\chi^2_P(212) = 149.739$ with $p = 1.000$ and *Deviance*: $D(212) = 122.276$ with

$p = 1.000$) reveal that the independent variable (age) is explanatory of ceramic tiling systems' degradation condition. The model obtained leads to a pseudo-R^2 of Cox and Snell of 0.646, a pseudo-R^2 of Nagelkerke of 0.695 and a pseudo-R^2 of McFadden of 0.391, revealing a model with satisfactory quality.

The probability of each degradation condition according to ceramic tiling systems' age is presented in Eqs. (4.30)–(4.34), assuming that "Condition E" is the reference class.

$$P(Y = \text{"Condition A"}) = \frac{e^{28.241-0.659\,Age}}{1 + e^{28.241-0.659\text{ Age}} + e^{27.267-0.468\text{ Age}} + e^{24.681-0.385\text{ Age}} + e^{15.859-0.205\text{ Age}}} \tag{4.30}$$

$$P(Y = \text{"Condition B"}) = \frac{e^{27.267-0.468\text{ Age}}}{1 + e^{28.241-0.659\text{ Age}} + e^{27.267-0.468\text{ Age}} + e^{24.681-0.385\text{ Age}} + e^{15.859-0.205\text{ Age}}} \tag{4.31}$$

$$P(Y = \text{"Condition C"}) = \frac{e^{24.681-0.385\text{ Age}}}{1 + e^{28.241-0.659\text{ Age}} + e^{27.267-0.468\text{ Age}} + e^{24.681-0.385\text{ Age}} + e^{15.859-0.205\text{ Age}}} \tag{4.32}$$

$$P(Y = \text{"Condition D"}) = \frac{e^{15.859-0.205\,Age}}{1 + e^{28.241-0.659\text{ Age}} + e^{27.267-0.468\text{Age}} + e^{24.681-0.385\text{ Age}} + e^{15.859-0.205\text{ Age}}} \tag{4.33}$$

$$P(Y = \text{"Condition E"}) = \frac{1}{1 + e^{28.241-0.659\text{ Age}} + e^{27.267-0.468\text{ Age}} + e^{24.681-0.385\text{ Age}} + e^{15.859-0.205\text{ Age}}} \tag{4.34}$$

Figure 4.4 shows the probabilistic distribution of the degradation condition of ceramic tiling systems according to their age. The probability of belonging to condition A decreases over time, and is lower than 5 % after year 19. The probability of belonging to condition B reaches a maximum peak at year 16 ($P = 71.1$ %). In condition C, the maximum probability of belonging to this degradation condition is reached between 40 and 41 years ($P = 59.2$ %). Regarding condition D, the maximum probability (88 %) of belonging to this degradation condition occurs after 64 years. The probability of ceramic adherent claddings to belong to condition E is lower than 5 % at 64 years, having a probability higher than 95 % to belong to this condition after year 92. The highest probability of transition between conditions A and B occurs between 5 and 6 years, the transition

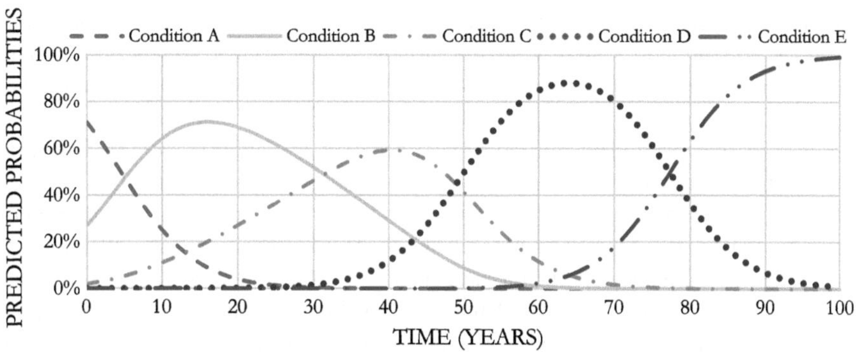

Fig. 4.4 Probabilistic distribution of the degradation condition according to the age of the case studies (ceramic tiling systems)

between condition B and C has a higher probability of occurrence between 31 and 32 years, while the transition between conditions C and D is more likely to occur between 48 and 49 years and the transition between conditions D and E are more likely to occur between 77 and 78 years.

The multinomial regression model applied to ceramic tiling systems classify correctly 60.0 % of the cases belonging to condition A of degradation, 58.6 % of those from condition B, 56.3 % of those from condition C, 76.7 % of those from condition D and finally 0.0 % of those from condition E. This model presents an accuracy of 61.0 %. The percentage of case studies classified correctly by chance is equal to 29.2 % [100 % × ($0.077^2 + 0.359^2 + 0.328^2 + 0.221^2 + 0.015^2$)], revealing that the model has good classifying properties.

4.1.3 Probabilistic Analysis of Façades Reaching the End of Their Service Life

Cladding systems present a large set of characteristics and are subject to a range of environmental agents that lead to their deterioration over time. As with other statistical methods, in logistic regression it is also possible to determine which variables are statistically significant in the explanation of the façade's degradation condition. Additionally, it is important to assess the probability of claddings to reach the end of their service life. To that purpose, a logistic regression analysis can be performed, which is used as the dependent variable the question "Has the cladding reached the end of its service life?" The dependent variable has thus two mutually exclusive classes: "yes" and "no". Using this analysis it also possible to evaluate which variables influence the moment in time at which the claddings reach the end of their service life. There are different methods for selecting the variables with more explanatory power. These selection methods are based on a stepwise technique, from two distinct options: first, the model starts with no explanatory variables and will add, step by step, the variables to be included (select Forward); in the second, the model starts with all variables included in the model and these are removed step by step, leaving only the variables that are statistically significant (Backward selection). The introduction or removal of variables is done based on well-defined selection criteria. In this study, the adopted selection method is based on the type of Forward ratio of likelihoods. In this type of selection, the addition of variables is performed based on the partial estimation of the maximum likelihood of the model. In the next sections, the influence of the claddings' characteristics on its probability of reaching the end of their service life is evaluated. For the façades claddings analysed in this study, it is considered that the end of the service life occurs when a façade meets or exceeds a degradation severity of 20 %.

4.1.3.1 Natural Stone Claddings

In natural stone claddings, only 11.3 % of the claddings have reached the end of their service life. Analyzing only the age of the cladding as explanatory variable, a model is obtained whose summary is presented in Table 4.8. The model summary and the Hosmer and Lemeshow test show there is an adequate fit of the model to the data ($-2LL = \chi^2$ (201) = 70.352, p = 1.00). The Hosmer and Lemeshow test also indicates that the estimated values are close to the observed values ($\chi^2_{HL}(8) = 5.019, p = 0.756$).

Using the Wald test, it is possible to analyse the validity of model coefficients. This test—$\chi^2_{Wald}(1) = 22.376$, with p = 0.000—confirms that age is statistically significant. Thus, it is possible to estimate the probability of a case study to reach the end of its service life, as described in Eq. (4.35).

$$P(Y = \text{"End of service life"}) = 1 - \frac{1}{1 + e^{-12.743 + 0.188\,\text{Age}}} \tag{4.35}$$

Figure 4.5 presents the probabilistic distribution of stone claddings to reach the end of their service life according to their age. Naturally, the probability of a stone cladding to reach the end of its service life increases over time; after 68 years, this probability is higher than 50 % and after 80 years, it is higher than 90 %. This model classifies correctly 90.1 % of the case studies.

Table 4.8 Model summary and Hosmer and Lemeshow test

Step	Model summary			Hosmer and Lemeshow test		
	−2LL (Log likelihood)	Cox and Snell R^2	Nagelkerke R^2	Chi-square	Degrees of freedom	Significance
1	70.352[a]	0.302	0.597	5.019	8	0.756

[a]Estimation stopped at iteration number 8 because parameter estimates changed by less than 0.001

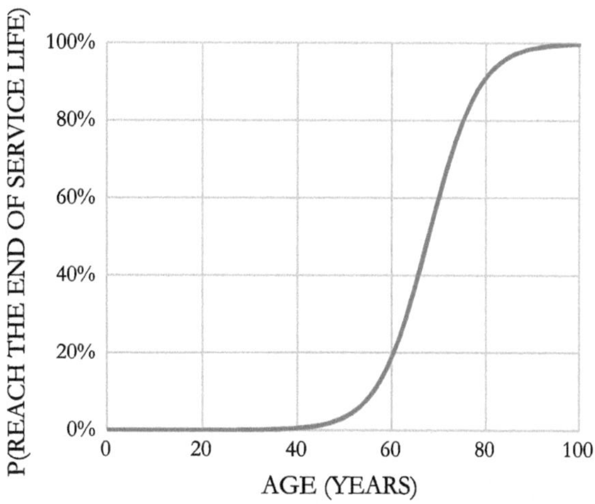

Fig. 4.5 Probabilistic distribution of the stone claddings to reach the end of their service life according to their age

Thus, a new logistic regression analysis is performed to identify the variables that significantly contribute to establish the probability of natural stone claddings to reach the end of their service life. For a 5 % significance level, that are three variables statistically significant: exposure to damp; distance from the sea; and claddings age. The pseudo-R^2 of Nagelkerke of the model is equal to 0.726, revealing a model with adequate quality (although this statistical indicator should be analysed with caution, as mentioned before). The value of −2 Log-likelihood indicates a proper fit of the model to the data ($-2LL = \chi^2(199) = 50.406$, $p = 1.00$). The Hosmer and Lemeshow test reveals that the estimated values are close to the observed values ($\chi^2_{\mathrm{HL}}(8) = 1.958, p = 0.982$). The model with three explanatory variables classifies correctly 95.1 % of the case studies (showing a higher classification power than the model only with age as explanatory variable). The probability of stone claddings to reach the end of their service life as a function of the environmental conditions considered is expressed by Eq. (4.36). The variable Age must be replaced in equation by claddings age, the variable Sea must be replaced by 1 if the cladding is located at less than 5 km from the sea and by 0 if the cladding more than 5 km from the sea; variable Damp should be replaced by 1 if the cladding has high exposure to damp and by 0 in claddings with low exposure to damp.

$$P(Y = \text{"End of service life"}) = 1 - \frac{1}{1 + \mathrm{e}^{-17.190 + 0.226\,\mathrm{Age} - 1.631\,Sea - 2{,}281\,\mathrm{Damp}}} \tag{4.36}$$

Figure 4.6 shows the correlation between the explanatory variables in the probability of natural stone claddings to reach the end of their service life.

Stone claddings in coastal areas are more likely to reach the end of their service life earlier than claddings located more than 5 km away from the sea. In fact, after

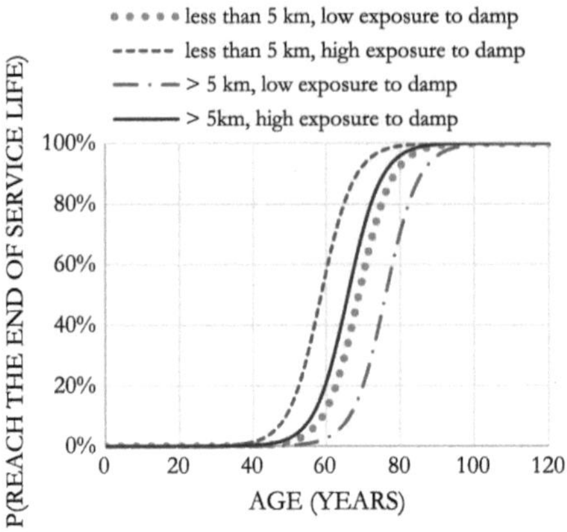

Environmental conditions	Age where a cladding achieves a probability of 90% of have reached the end of its service life
Less than5 km from the sea, low exposure to damp	79 years
Less than 5 km from the sea, high exposure do damp	69 years
> 5 km from the sea, low exposure to damp	86 years
> 5 km from the sea, high exposure to damp	76 years

Fig. 4.6 Probabilistic distribution of the stone claddings to reach the end of their service life according to their age, its distance from the sea and the exposure to damp

79 years, a stone cladding located at less than 5 km from the sea has a probability higher than 90 % of reaching the end of its service life, while a cladding located at more than 5 km only reaches this probability after 86 years. Concerning exposure to damp, claddings with low exposure to damp take longer to reach the end of their service life, with a probability of 90 % after 86 years; for claddings with high exposure to damp, this threshold is reached 10 years sooner. Figure 4.6 reveals that the most favourable situation corresponds to a natural stone cladding located more than 5 km away from the sea and with low exposure to damp; contrarily, a cladding located at less than 5 km and with high exposure to damp corresponds to the most unfavourable situation, with higher probability of reaching the end of its service life sooner.

4.1.3.2 Rendered Façades

Regarding rendered façades, in the sample analysed, 31 % of the cases had already reached the end of its service life. The first analysis concerns the probability of having reached the end of the service life as a function of age. The statistic relative to −2 Log of the likelihood indicates an adequate adjustment of the model to the data ($-2LL = \chi^2(98) = 26.434$, $p = 1.00$). The values of the pseudo-R^2 of Cox and Snell (R^2_{CS}) (0.622) and of the pseudo-R^2 of Nagelkerke (R^2_N) (0.876) indicate a model with adequate quality. The Hosmer and Lemeshow test reveals that the estimated values are close to the observed values ($\chi^2_{HL}(7) = 2.124, p = 0.953$). This model correctly classifies 96 % of the case studies. Wald test ($\chi^2_{Wald}(1) = 12.480$, $p = 0.000$) confirms that age is a statistically relevant variable. The probability of a rendering reaching the end of its service life as a function of age can be determined according to Eq. (4.37).

$$P(Y = \text{"End of service life"}) = 1 - \frac{1}{1 + e^{-7.767 + 0.467\,\text{Age}}} \tag{4.37}$$

Figure 4.7 presents the distribution of the probability of rendered façades reaching the end of their service life. After 17 years the probability is higher than 50 % and after 21 years it is higher than 90 %.

Based on the sample analysed, it is found that none of the variables analysed (render type, colour, height of the building, building shape, design level, eaves' protection, ground floor protection, platbands or balcony copings, façades orientation, distance from the sea, exposure to damp, distance from pollution sources and façade protection level) is considered statistically significant for usual significance levels (5 and 10 %). This means that the independent variables considered do not distinguish the probability of façades to belong to a given degradation condition. Independently of the statistical validity of this analysis, it is considered extremely relevant to evaluate the degradation condition of façades as a function of the type of mortar of which the rendering is made. Equation (4.38) expresses the probability of reaching the end of the service life as a function of the rendering's type of mortar. The variable type of mortar has four categories: B represents a lime–cement rendering; C a current cement mortar; Ma a mortar with crushed marble and Mo a single-layer mortar. To use the equation, one must replace the symbol of the category corresponding to the case study with 1 and the remaining categories' symbols with 0. To obtain the probability of reaching the end of the service life of a single-layer mortar Mo, which was used as the reference class, one must replace B, C and Ma with 0.

$$P(Y = \text{"End of service life"}) = 1 - \frac{1}{1 + e^{-7.794 + 0.628\,\text{Age} - 3.585\text{B} - 2{,}892\text{C} - 4.804\text{Ma}}} \tag{4.38}$$

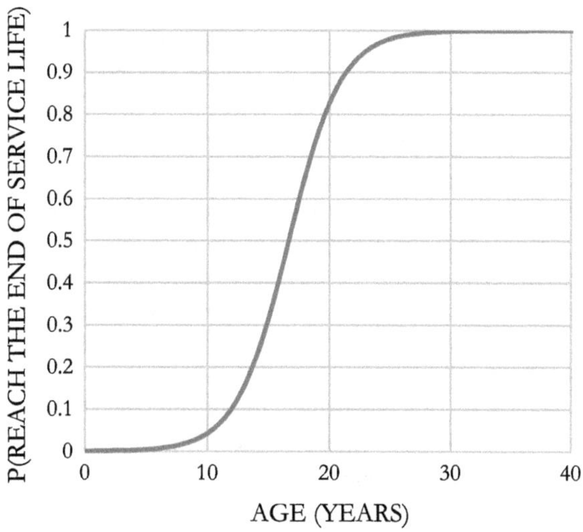

Fig. 4.7 Probabilistic distribution of rendered façades to reach the end of their service life according to their age (data sourced from Silva et al. 2013)

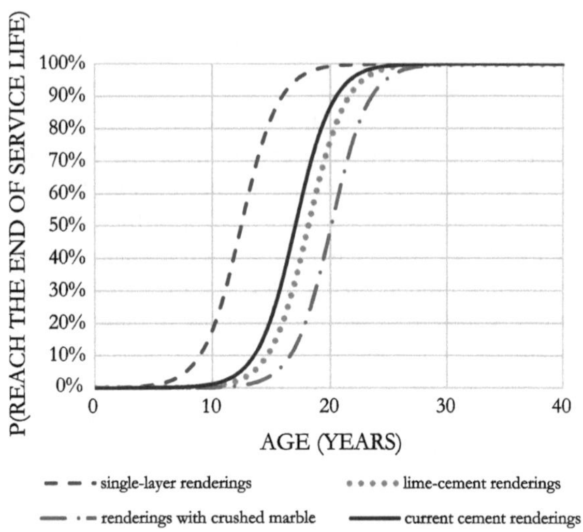

Fig. 4.8 Probabilistic distribution of rendered façades to reach the end of their service life according to their age and the type of mortar (data sourced from Silva et al. 2013)

Figure 4.8 presents the probability of rendered façades reach the end of their service life as a function of age and mortar type.

In the sample analysed the single-layer renderings have a higher probability of reaching sooner the end of their service life, with a probability higher than 90 % after 15 years. These conclusions slightly contradict expectations since this type of rendering is based on varying industrial premixed compositions with expectedly diversified performance levels (Veiga 2000). The lower durability of the single-layer renderings of the sample may be due to errors in tender specifications (i.e. mortars incorrectly prescribed for the in-service conditions). The sample analysed reveals that lime–cement renderings perform slightly better than the current cement renderings. In fact, current cement renderings have the lowest probability of remaining in the lowest degradation conditions. According to Kus and Nygren (2000), this type of rendering generally has a higher Young modulus (especially when compared to lime–cement renderings) and is therefore more prone to defects such as cracking or loss of adherence due to shear stresses generated between the mortar and the substrate. Lime–cement renderings have a probability higher than 90 % of reaching the end of their service life after 22 years, while cement based renderings only reach that value after 21 years. Kus et al. (2004) refer that lime–cement renderings display lower internal stresses as a result of slow changes between wetting and drying cycles, especially in dry and moderate climates. Thomasson (1982) also refers that cementitious renderings have drying shrinkage values about twice those of lime–cement renderings, which may justify the results obtained. In fact, renderings with lime in their composition have a higher flexibility than the ones made exclusively with cement and are capable of withstanding small movements without cracking (Mitchell 2007). However, due to their lower mechanical characteristics, lime–cement renderings are more friable than current cement renderings and may retain moisture for longer periods of time, especially if rendered with thick layers, which may somehow offset the previous effects (Silva et al. 2013). Renderings with crushed marble show the best performance, with a probability of 90 % of reaching the end of their service life after 24 years. Concerning this type of render, there are other factors that contribute to their durability: in the sample analysed, renderings with crushed marble are applied in buildings with a high design level, including joints in the rendered surfaces, drips, eaves and upper protections (Gaspar 2009).

4.1.3.3 Painted Surfaces

Regarding painted surfaces, 24.1 % of the case studies had already reached the end of their service life. The logistic model obtained reveals an satisfactory adjustment between the model and the sample ($-2LL = \chi^2(218) = 84.607$, $p = 1.00$). The Hosmer and Lemeshow test indicates that the values predicted by the model are similar to the observed values ($\chi^2_{HL}(7) = 2.921, p = 0.892$). The proposed model correctly classifies 91.4 % of the case studies. The age is an explanatory variable statistically significant in the probability of a painted surface reaching the end of its service life ($\chi^2_{Wald}(1) = 32.579$, $p = 0.000$)—Eq. (4.39).

$$P(Y = \text{"End of service life"}) = 1 - \frac{1}{1 + e^{-9.865 + 0.947\,\text{Age}}} \tag{4.39}$$

Figure 4.9 presents the probabilistic distribution of painted surfaces to reach the end of their service life according to their age. After 10 years, the probability of a painted surface to reach the end of its service life is higher than 50 % and, after 12 years, this probability is greater than 90 %.

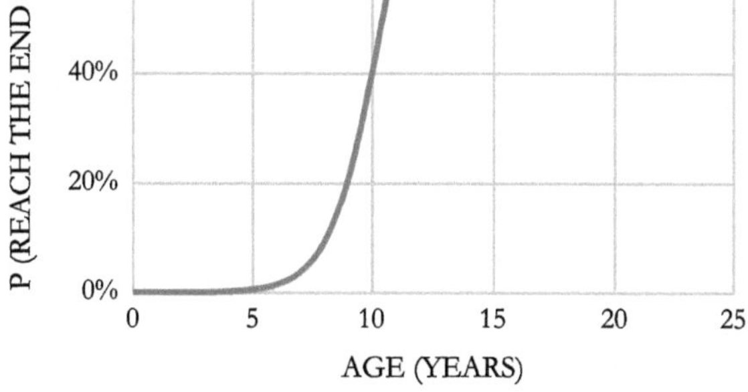

Fig. 4.9 Probabilistic distribution of painted surfaces to reach the end of their service life according to their age

It is then possible to evaluate the influence of other variables in the definition of the probability of painted surfaces to reach the end of their service life. Once more, for a 5 % significance level, none of the variables analysed is considered statistically significant. However, it seems relevant to analyse the influence of the distance from the sea and the exposure to damp on the probability of a painted surface reaching the end of its service life. As referred, the two new models have a low relevance from a statistical point of view; however, these models reveal an increased classification capability when compared with the model with only age as explanatory variable. Both models correctly classify 92.3 % of the case studies.

Equation (4.40) allows estimating the probability of a painted surface to reach the end of its service life according to its distance from the sea and Eq. (4.41) according to its exposure to damp.

$$P(Y = \text{"End of service life"}) = 1 - \frac{1}{1 + e^{-10.376 + 0.967\,\text{Age} + 0.781\,\text{Sea}}} \quad (4.40)$$

$$P(Y = \text{"End of service life"}) = 1 - \frac{1}{1 + e^{-9.638 + 0.966\,\text{Age} - 0.706\,\text{Damp}}} \quad (4.41)$$

The variable Age must be replaced in both equations by the claddings' age, the variable Sea must be replaced by 1 if the cladding is located at less than 5 km from the sea and by 0 if the cladding is located more than 5 km from the sea; variable Damp should be replaced by 1 if the cladding has low exposure to damp and by 0 in claddings with high exposure to damp. Figure 4.10 shows the probabilistic distribution of the painted surfaces to reach the end of their service life according to their age and their distance from the sea or their exposure to damp.

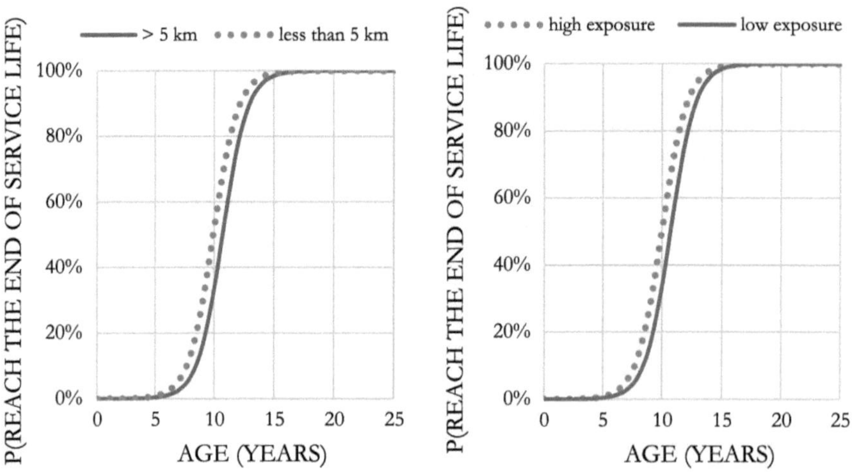

Fig. 4.10 Probabilistic distribution of the painted surfaces to reach the end of their service life according to their age and their distance from the sea (*left*) or their exposure to damp (*right*)

Concerning exposure to damp, after 10 years, a painted surface with high exposure to damp has a probability higher than 50 % of reaching the end of its service life and, after 12 years, this probability is higher than 90 %. On the other hand, for a painted surface with low exposure to damp, the probability of reaching the end of its service life is higher than 50 % only at 11 years and higher than 90 % at 13 years. Regarding distance from the sea, painted surfaces located at less than 5 km from the sea have a higher probability of reaching sooner the end of their service life than paintings located more than 5 km from the sea, with a probability higher than 50 % after 10 years and higher than 90 % after 13 years. Paintings located at more than 5 km have a probability higher than 50 % of reaching the end of their service life after 11 years and higher than 90 % after 14 years.

4.1.3.4 Ceramic Tiling Systems

Concerning ceramic tiling systems, 23.6 % of the cases had already reached the end of their service life, in the sample analysed. Using age as the only explanatory variable, a logistic regression model is obtained that shows an adequate adjustment to the data ($-2\text{LL} = \chi^2(193) = 90.951$, $p = 1.00$). The Hosmer and Lemeshow test reveals that the estimated values are close to the observed values ($\chi^2_{\text{HL}}(7) = 2.921, p = 0.892$). A pseudo-$R^2$ of Cox and Snell (R^2_{CS}) of 0.465 and a pseudo-R^2 of Nagelkerke (R^2_{N}) of 0.700 were obtained, revealing a model with satisfactory quality. The proposed model is able to correctly classify 89.2 % of the case studies. The Wald test (χ^2_{Wald} (1) = 35.712, $p = 0.000$) reveals that age is statistically relevant for the evaluation of the probability of a ceramic cladding to reach the end of its service life as a function of age, and the relationship between these two variables can be determined as described in Eq. (4.42).

$$P(Y = \text{"End of service life"}) = 1 - \frac{1}{1 + \mathrm{e}^{-10.059 + 0.201\,\text{Age}}} \tag{4.42}$$

Figure 4.11 shows the probabilistic distribution of a ceramic cladding to reach the end of its service life according to its age. After 50 years, ceramic has a probability higher than 50 % of reaching the end of its service life and after 61 years, this probability increases to more than 90 %.

Using the same methodology applied to the other claddings analysed, a multiple logistic regression is performed in order to identify the variables with an explanation power in the definition of the probability of ceramic claddings to reach the end of their service life. Again, using a 5 % significance level, none of the variables considered has a statistically significant relevance. However, a step by step analysis (without statistical relevance) is performed, identifying two variables that contribute to increase (even residually) the classification power of the logistic regression model. The selected variables are distance from the sea and exposure to damp. Two new models are defined, as described in Eqs. (4.43) and (4.44), using distance from

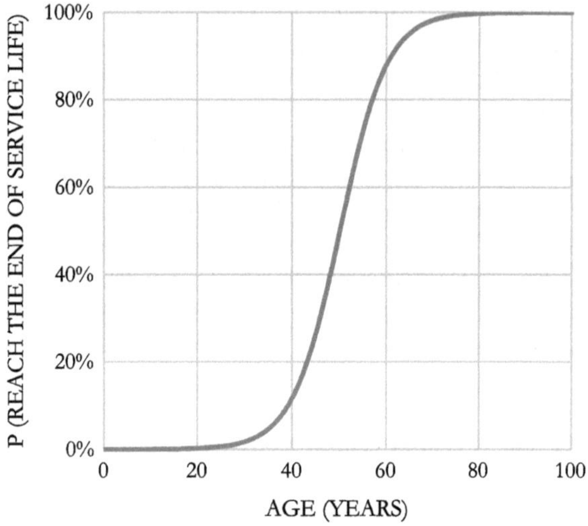

Fig. 4.11 Probabilistic distribution of ceramic claddings to reach the end of their service life according to their age

the sea and exposure to damp as explanatory variables, respectively. These models correctly classify 89.7 % of the case studies.

$$P(Y = \text{"End of service life"}) = 1 - \frac{1}{1 + e^{-9.549 + 0.192\text{Age} - 0.785\text{Sea}}} \quad (4.43)$$

$$P(Y = \text{"End of service life"}) = 1 - \frac{1}{1 + e^{-10.028 + 0.194\text{Age} - 0.561\text{Damp}}} \quad (4.44)$$

The variable Age must be replaced in both equations by the claddings' age, the variable Sea must be replaced by 1 if the cladding is located more than 5 km away from the sea and by 0 if the cladding is located less than 5 km from the sea; variable Damp should be replaced by 1 if the cladding has high exposure to damp and by 0 in claddings with low exposure to damp. Figure 4.12 shows the probabilistic distribution of ceramic claddings to reach the end of their service life according to their age and their distance from the sea or their exposure to damp.

Regarding exposure to damp, claddings with high exposure reach the end of their service life earlier than claddings with low exposure. The probability of a ceramic cladding with low exposure to damp of reaching the end of its service life is higher than 50 % after 51 years and higher than 90 % after 63 years. In the case of claddings with high exposure, the probability of reaching the end their service life is higher than 50 % after 48 years and higher than 90 % after 60 years. Concerning distance from the sea, claddings located at more than 5 km take longer to reach the end of their service life; after 53 years, this probability is higher than 50 % and after 65 years higher than 90 %. Ceramic claddings in coastal areas (less than 5 km from the sea) have a probability higher than 50 % of reaching the end of their service life after 49 years and higher than 90 % after 61 years.

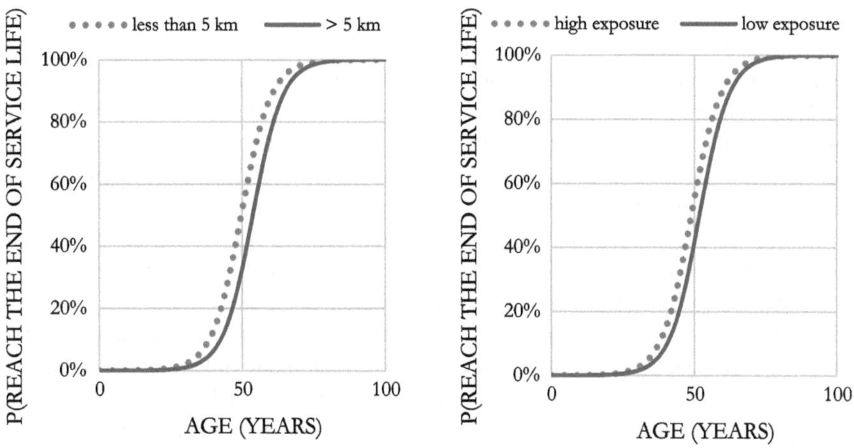

Fig. 4.12 Probabilistic distribution of ceramic claddings to reach the end of their service life according to their age and their distance from the sea (*left*) or their exposure to damp (*right*)

4.1.4 Conclusions

Stochastic models allow analyzing the service life of façade's claddings using probability distribution functions. Stochastic models usually are superior to purely deterministic models, since they are able to deal with the uncertainty associated with complex phenomena, such as buildings' degradation. In this chapter, a probabilistic analysis of the degradation condition of façade's claddings over time is performed. A logistic regression analysis is used to that purpose; this methodology has been successfully applied in several areas, including in the prediction of mortality in humans (and is to a certain degree related to the prediction of service life of building components).

This statistical tool allows predicting the probability of occurrence of each of various degradation conditions in each case study and defining an equivalent probabilistic model. Furthermore, using this model it is possible to describe the probability of transition from each degradation condition to the next one (more serious) and translate the data collected in the field work into statistical distributions functions allowing probabilistic analyses of the degradation phenomena.

The probability of each of the case studies of reaching the end of its service life can also be obtained through Logistic regression. In other words, it provides a probabilistic indication of the moment after which the façades must be intervened. Stone claddings have a probability higher than 50 % of reaching the end of their service life after 80 years. For ceramic tiling systems, this probability is reached at 50 years, for renderings at 17 years and for painted surfaces at 10 years.

For all claddings analysed, age is extremely relevant in the evolution of the degradation condition over time. For stone claddings, the degradation condition is also affected by distance from the sea and exposure to damp. For ceramic tiling systems and painted surfaces, the influence of distance from the sea and exposure to damp is also evaluated (although with little statistical relevance). As expected, claddings with unfavourable conditions (closer to the coast and with high exposure to adverse environmental agents) transit more rapidly to the most severe degradation conditions and are more prone to reaching sooner the end of their service life. For rendered façades, a probabilistic analysis of the degradation condition as a function of the mortar type shows that single-layer renderings have a higher probability of reaching sooner the end of their service life, followed by current cement renderings, lime–cement renderings and marble crushed renderings (the latter, applied in buildings that are generally better detailed).

The method proposed provides a probabilistic view of the evaluation of the service life of façades, with an indication of the probabilistic time range of transition between degradation conditions. This information is extremely useful and can be employed in various scopes of service life prediction and maintenance of constructions. For instance, it can be used as a technique to identify the parameters used in the factor method, currently used to estimate the service life of construction elements.

4.2 Markov Chains

4.2.1 Theoretical Foundations

Service life prediction models seek to extrapolate the future behaviour of buildings elements, giving an indication of the instant in time from which action is needed. However, degradation is a complex and dynamic process, wherein the agents and degradation mechanisms act synergistically and are in constant permutation. The current and future state of buildings involves a large number of uncertainties (Ellingwood 2005). These uncertainties are due to two main reasons: the limited ability to predict future events and the intrinsic randomness of natural phenomena (e.g. environmental agents) (Bocchini et al. 2013). To overcome this difficulty, it is sometimes necessary to use a stochastic approach in service life prediction (Frangopol et al. 2004). In general, this approach has a high complexity and is based on the assumption that the deterioration is a process governed by random variables (Moser 2003). Consequently, the degradation can be seen as a continuous process of loss of performance, which evolves probabilistically over time.

One method that evaluates, stochastically, the future condition of the façades claddings is Markov chains, initially proposed by Russian mathematician Andrei Andreyevich Markov (1856–1922). Markov chains can be used to emulate the evolution of the degradation condition of building elements, describing the probability of a future condition based only on the present condition, independently of previous deterioration history (Parzen 1962; Neves et al. 2006). This feature of the Markov chain is commonly referred as “loss of memory”. To better understand this concept, Uchwat and MacLeod (2012) make an analogy between the “loss of memory” of Markov chains and board games, where the next position of the pawn depends only on the current position and the rolling dice.

Markov chains have been successfully applied to model the deterioration state of bridges (Monika 2005; Morcous et al. 2003; Robelin and Madanat 2007), in the optimization of maintenance schedules (Lounis and Vanier 2000; Lacasse et al. 2008) and even in service life of building components (Mc Duling 2006). Augenbroe and Park (2002) refers that Markov chains are able to describe the randomness associated with buildings’ performance and can also be used in the decision-making process concerning the systematic replacement of building components.

4.2.1.1 Description

Currently, it is very common to find in the literature Markov models used to explain practical problems. Markov chains are practically ubiquitous in stochastic modelling, for two main reasons: (i) many models are naturally Markovian, as the fundamental laws of physics, in which the motion of a particle in a given time depends on his current position and speed, and it is irrelevant how it ended up in that position; (ii) the simple structure of Markov processes enables to develop powerful mathematical techniques and computer algorithms that would be untreatable by other methods (Handel 2008). Markov models are relatively simple to use and statistics on the degradation condition and service life of buildings, at any time of their life cycle, are easy to determine (Bogdanoff 1978; Lounis 2000).

In comparison with deterministic models, Markov chains have some advantages (Lounis et al. 1998; Robelin 2006; Uchwat and MacLeod 2012): (i) it is possible to develop a service life prediction model based exclusively on known degradation condition of the element in two periods of time, unlike what happens with regression models, where it is necessary to collect information on the degradation state of the elements over an extended period of time, thereby anticipating a performance trend; (ii) the ability to extrapolate the performance of building elements, even if information on the condition of these elements at a given time is omitted; (iii) they avoid regression techniques which, in some cases, include several factors that contribute to explain the degradation phenomena; (iv) models based on Markov chains can easily be adapted over time as further information relating to the degradation condition of building elements is being acquired; (v) they apply information regarding the degradation condition of building components in real in-use conditions, thereby offsetting some of the limitations of laboratory testing, and comprising, in some way, the interaction of different deterioration agents and the uncertainty and variability of degradation mechanisms in the buildings’ performance; (vi) they allow assessing the influence of the buildings characteristics on the transition rate between degradation conditions.

However, the use of Markov chains in the context of service life also have some limitations, especially the ones related with the information collection system, since the degradation of building elements are analysed based only on visual inspection, and therefore with some subjectivity associated (Das 1998; Frangopol and Neves 2004).

Markov chains are based on a set of discrete states which characterize the buildings performance. Garavaglia et al. (2004) refer that the degradation phenomena can be defined as a transition process through different condition states. There are various methods to evaluate the deterioration state of buildings and components that vary according to the hierarchical classification of building elements, hierarchical classification of defects and the extension of defects (Straub 2003). Various authors (Balaras et al. 2005; Shohet et al. 1999; Marteinsson and Jónsson 1999; Freitas et al. 1999; Brandt and Rasmussen 2002; Chew 2005) have established classification systems of defects or degradation scales in order to express the visual and physical degradation of the elements under analysis. These qualitative and quantitative deterioration scales ranges from the best possible condition (no visible degradation) to the highest degradation level (general degradation or loss of functionality).

In this study, in the definition of Markov processes there are two fundamental properties: (i) the "loss of memory" (mentioned before), where the future condition depends exclusively on the present condition; (ii) and their homogeneity, which ensures that the transition rates from one condition to another occurs in a constant way over time (Ng and Moses 1996).

To establish a Markov chain model it is necessary to define a matrix that describes the probability of transition between degradation conditions (Frangopol et al. 2004). Assuming that the various degradation states are given by $C = \{1,2, \ldots, n\}$, existing n discrete conditions, then the matrix P, of one step, will have a dimension $n \times n$, as presented in Eq. (4.45) (Li and Wang 2006).

$$P_{\Delta t} = \begin{bmatrix} P_{11} & P_{12} & \cdots & P_{1n} \\ P_{21} & P_{22} & \ddots & P_{2n} \\ \vdots & \vdots & \ddots & \vdots \\ P_{n1} & P_{n2} & \cdots & P_{nn} \end{bmatrix} \tag{4.45}$$

To apply the transition matrix P it is necessary to ensure that the following conditions are fulfilled (Baum and Petrie 1966; Horn and Johnson 1985; Madanat et al. 1995; Monika 2005; Baik et al. 2006; Liu 2010; Lethanh and Adey 2012):

(i) All the values within the transition matrix must be higher than zero (non-negativity rule);
(ii) Once it is a probability matrix, all the values of P_{ij} should range between 0 and 1 ($0 \leq P_{ij} \leq 1$, for i and $j \in C$);
(iii) Every row in the matrix P must represent a probability vector, which implies that $\sum_{j \in C} P_{ij} = 1$ for $i \in C$;
(iv) If in instant t the building component is in condition i, then in instant $t + 1$ this element can only remain in state i or change from state i to state $i + 1$;

this implies that over time the degradation condition of building components always tends to became more serious or severe (on the contrary, if the performance of components improves over time, it means that the element was subjected to some maintenance action, which cannot be modelled by Markov chain processes); condition $P_{ij} = 0$, for $i > j$ is then imposed;

(v) This matrix is an one-step transition matrix, which means that a building component can only transit from one degradation condition to the next one (Yokota et al. 2012); if in a period of time Δt the element moves from the most favourable condition to the most unfavourable condition, it is assumed that in the meantime it has passed through all the intermediate degradation states.

After a period of time Δt, the probability of a construction element remaining in state i is given by Eq. (4.46) (Morcous 2006). The elements of the main diagonal of the matrix P represents the probability of an element remaining in a given degradation condition (Roh and Xi 2000).

$$P_{ii}(\Delta t) = \Pr(X(\Delta t) = i|X(t) = i) \tag{4.46}$$

On the other hand, the conditional probability of a given element to transit from state i to the next state is given by Eq. (4.47) (Henry 1971; Dekker et al. 2008; Kobayashi et al. 2010).

$$P_{ij}(\Delta t) = \Pr(X(\Delta t) = j|X(t) = i) \tag{4.47}$$

Thus, it is possible to define the vector of probabilities $p(t)$, which describes for a given time t the probability of an element to belong to each of degradation states, as shown in expression (4.48) (Cargal 2013).

$$p(t) = [p_1 \quad p_2 \quad \ldots \quad p_n] \tag{4.48}$$

Concerning the time of transition between deterioration conditions, Markov processes can be divided in two common types: discrete and continuous. The discrete-time Markov chain is useful when transitions can only occur at specific instants (for example, when the building has a maintenance schedule). The continuous-time Markov chain is applied when transitions can occur at any time, as is in general the case of deteriorating performance. The uncertainty in the rate of transitions between the states is defined by a transition probability matrix (denoted P) for discrete-time processes and by an intensity matrix (denoted Q) for continuous-time processes (Kallen 2009). Various authors (Parnham 1997; Aikivuori 1999) refer that the decision to intervene is usually based on subjective and programmatic criteria. Besides, in some countries there is no well-defined policy regarding the periodicity of buildings inspections other than a maximum interval of time between repair works, in spite of the actual degradation level of buildings, and therefore it does not seem possible to model the performance of façade claddings through a discrete-time Markov process. Therefore, only continuous-time models are analysed in this study.

4.2.1.2 Continuous-Time Markov Processes

As mentioned before, in continuous-time Markov chain models, the transitions between degradation states occur at random time periods. In this case, the transition matrix is designated as infinitesimal generator matrix or intensity matrix (Q), and provides the transition rate between states (and not the transition probability, as for the discrete-time Markov chains). Matrix Q is composed by the elements q_{ii} and q_{ij} (where i $\neq$ j), which must comply with the following conditions (Johnson 1987):

(i) $\sum_{j \in C} q_{ij} = -q_{ii}$, i.e., the diagonal elements are defined to ensure that the sum of the elements in every row is zero;
(ii) $q_{ij} \geq 0$, for $j - i = 1$, which implies that all the diagonal elements are non-negative, except on the last row that represents the most unfavourable degradation condition;
(iii) $q_{ij} = 0$, for $i > j$; for $j - i > 1$; and for $i, j = n$.

A generic intensity matrix Q is given as shown in Eq. (4.49) (Singer 1981).

$$Q = \begin{bmatrix} -q_{0,1} & q_{0,1} & 0 & \cdots & \cdots & 0 \\ 0 & -q_{1,2} & q_{1,2} & \cdots & \ddots & 0 \\ 0 & 0 & -q_{2,3} & q_{2,3} & \ddots & 0 \\ 0 & 0 & \cdots & \cdots & \ddots & 0 \\ 0 & 0 & \cdots & \cdots & -q_{n-1,n} & q_{n-1,n} \\ 0 & 0 & \cdots & \cdots & 0 & 0 \end{bmatrix} \tag{4.49}$$

The transition between degradation states depends uniquely on the last recorded state and the transition rate between the current state and future state (given by matrix Q) (Silva et al. 2015). Using Chapman–Kolmogorov differential equation and based on the transition rates, it is possible to estimate the probability of transition between states of condition and the mean probability of permanence in these states (Norris 1997)—Eq. (4.50).

$$\frac{\mathrm{d}}{\mathrm{d}t} P(\Delta t) = Q \cdot P(\Delta t) \tag{4.50}$$

Equation (4.51) gives the solution of this system of differential equations (Carrington et al. 2005), where Δt represents the time interval considered. Therefore, it is possible to relate the infinitesimal generator matrix Q with the Markovian transition matrix P (Cox and Miller 1965; Bladty and Sorensen 2009).

$$P(\Delta t) = \exp(Q \cdot \Delta t) = \sum_{n=0}^{\infty} \frac{\Delta t^n \cdot Q^n}{n!} \tag{4.51}$$

In the literature there are different methods to calculate the intensity matrix Q. Optimization procedures are generally considered as a reliable and accurate methodology to estimate the intensity matrix Q, ensuring the efficiency of the degradation model (Silva et al. 2015). The optimization of matrix Q is based on the concept of maximum likelihood described by Kalbfleisch and Lawless (1985). The likelihood is defined as the predicted probability of occurrence of the observed transitions, as illustrated in Eq. (4.52), where i is the condition state in the initial instant, j is the condition state in the final instant, m is the number of elements, k is the number of intervals between inspections and P_{ij} is the probability of transition from the condition i to condition j.

$$L(Q) = \prod_{i=1}^{m} \prod_{i,j=1}^{k} P_{ij} \tag{4.52}$$

Gu and Kong (1998) suggest that matrix Q and the likelihood are computed simultaneously, and optimized in each iteration, until a given convergence criteria (previously established) is reached. Currently, a great variety of highly efficient optimization tools are available in commercial software. In this study, the optimization of the infinitesimal generator matrix Q is carried out using function *fmincon* (constrained non-linear optimization) available in MATLAB (Higham and Higham 2005).

The model's efficiency is evaluated through the comparison of the number of case studies belonging to a given degradation state according to the Markov chain model and the number of real values observed during fieldwork. Degradation models based on Markov chain models allow determining: (i) the mean time of permanence in each degradation condition, given by Eq. (4.53); (ii) the probability of belonging to each degradation condition over time and according to claddings characteristics (similar to the analysis performed by logistic regression).

$$T_i = \frac{1}{q_{ij}} \tag{4.53}$$

where T_i represents the mean time of permanence in the condition i (for $i = 1, 2, \ldots, n$) and q_{ij} the elements of intensity matrix (Q).

4.2.2 Application of Markov Chain Models to Façade Claddings

In this study, which intends to model the expected service life of façade coatings, through visual inspections of buildings in service conditions, all that is known is the degradation state in two instants: (i) initial condition, assuming that at time zero the degradation is equal to zero; (ii) and final condition, corresponding to the degradation state at inspection time. The degradation condition of the façade coatings is characterized and classified according to a range of discrete variables that vary from level "A" (no visible degradation) to level "E" (general deterioration, requiring immediate action). In the next sections, the probability of transition between degradation conditions over time is evaluated. The most relevant characteristics that explain the degradation of claddings are identified and Markov chain models are used to analyse the probability of belonging to each degradation condition over time according to the claddings' characteristics.

4.2.2.1 Natural Stone Claddings

The intensity matrix Q for natural stone claddings—Eq. (4.54)—is obtained through a Markov chain model. In the sample analysed, none of the case studies is in condition E and the transitions between degradation conditions occur faster for less deteriorated facades (changes from level A to level B occur in only 4.5 years). The transition between condition B and C occurs after 52.6 years and the transition between condition C and D after 87 years. Indeed, the transition between degradation conditions does not occurs homogeneously, since the condition states are determined through a numerical index (referred as severity degradation—in Chap. 2) that quantifies the overall degradation of façades and this index follows a scale which it is not continuous. Degradation condition A (most favourable) corresponds to a cladding with no visible deterioration (with a degradation severity value lower than 1 %); thus, for a stone cladding to move to the following condition it is only necessary to present changes in surface uniformity or stains. On the other hand, the transition between levels associated with higher degradation states implies the presence of a larger number of defects and/or more hazardous defects, including simultaneous occurrence of defects and synergies between them.

$$Q = \begin{bmatrix} q_{\mathrm{A,B}} \\ q_{\mathrm{B,C}} \\ q_{\mathrm{C,D}} \end{bmatrix} = \begin{bmatrix} 0.2210 \\ 0.0190 \\ 0.0115 \end{bmatrix} \tag{4.54}$$

Figure 4.13 shows the probabilistic distribution of the degradation condition of stone claddings over time, based on the Markov chain model. By default, transition times are exponentially distributed, for it is assumed that the probability of transition between states is constant in time; this is considered an acceptable simplification of a complex reality given the relatively small sample of available data (Silva et al. 2015). The probability of stone claddings to be in condition A is lower than 2 % after 8 years. The probability of stone claddings to be in condition B reaches a peak at around year 12 (probability of 79.2 %). Regarding condition C, stone claddings have a maximum probability (P = 46.1 %) of belonging to this condition between 71 and 73 years. The probability of stone claddings to belong to condition D increases over time, being lower than 10 % before year 41, and above 60 % after year 145.

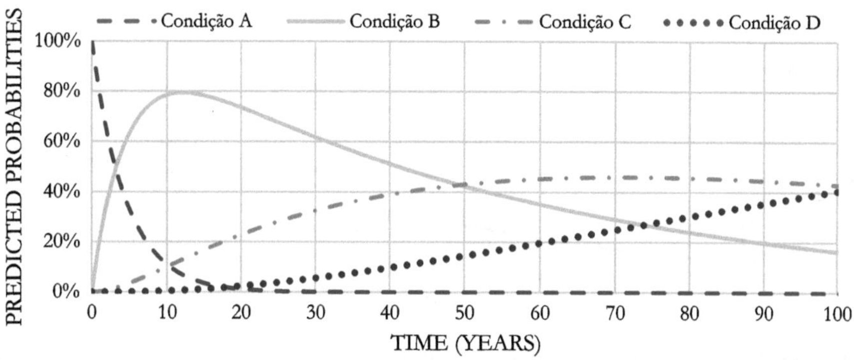

Fig. 4.13 Probabilistic distribution of the degradation condition for stone claddings according to the age of the case studies

Table 4.9 Classification accuracy of Markov chain model (stone claddings)

Degradation condition	Observed	Predicted	Mean percentage error (%)
Condition A	9	12.10	34.40
Condition B	114	105.42	7.53
Condition C	57	63.96	12.21
Condition D	23	21.53	6.41

Table 4.9 shows an analysis of the model's accuracy. In this analysis the number of cases belonging to each degradation state given by the model is compared with the number of cases within the equivalent degradation condition observed during fieldwork. The error obtained is relatively low, given the complexity of the degradation phenomena.

Regarding natural stone claddings, various analyses are performed to identify the characteristics of stone claddings that significantly affect their durability: (a) type of stone; (b) type of finishing; (c) size of stone plates; (d) exposure to damp; (e) and the distance from the sea. Therefore, five new Markov chain models were defined to analyse the probability of belonging to each degradation condition over time according to these characteristics.

Koch and Siegesmund (2004) refer that durability is an important issue when specifying stone panels as cladding materials for external façades, exposed to aggressive environmental agents. Therefore, it seems relevant to analyse the influence of type of stone in the stone claddings loss of performance process - Fig. 4.14. In the sample analysed, limestone claddings transit more rapidly between conditions A and B (after 2.2 years), followed by marble claddings (5.2 years) and granite claddings (6.0 years). Regarding condition B, in limestone claddings the maximum probability (P = 85.2 %) of belonging to this condition occurs at 7 years; for marble claddings, this peak (P = 73.7 %) is reached at 12 years and for granite coatings, this peak (P = 83.5 %) occurs after year 18. For condition C, marble coatings reach a maximum probability (P = 44.7 %) of belonging to this

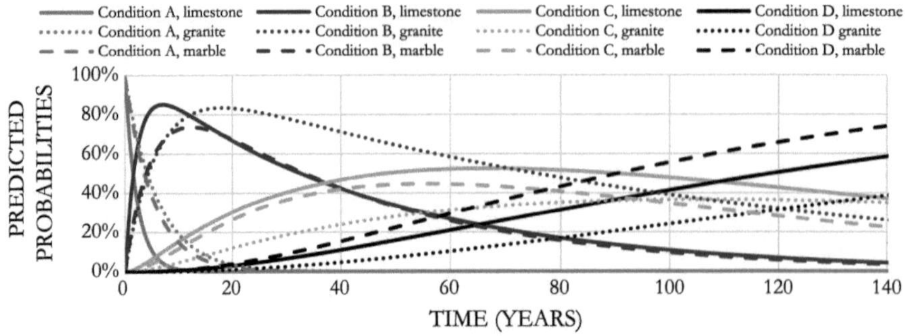

Fig. 4.14 Probabilistic distribution of the degradation condition for stone claddings according to their age and type of stone

condition between 55 and 56 years, limestone claddings reach this peak (P = 52.5 %) between 66 and 67 years and granite claddings (36.7 %) between 105 and 108 years (significantly later). The highest probability of transition between conditions B and C occurs: between 37 and 38 years for limestone claddings; between 40 and 41 years for marble claddings and around year 107 for granite claddings. Marble claddings are those with the lowest probability of belonging to the most favourable conditions (3.9 % of belonging to condition A and 42.9 % of belonging to condition B) and highest probability of being in the most unfavourable degradation condition (D) (19.5 %). Granite claddings have a very high probability (P = 94.4 %) of belonging to the more favourable conditions of degradation (A and B), and no probability of belonging to condition D. Based on the sample analysed, it is concluded that granite claddings are the most durable, followed by limestone claddings and marble claddings (the least durable) [these conclusions are similar to the results obtained by the study performed by Schouenborg et al. (2007)].

Figure 4.15 presents the probabilistic distribution of the degradation condition for stone claddings according to their age and to the size of the stone plates. Stone

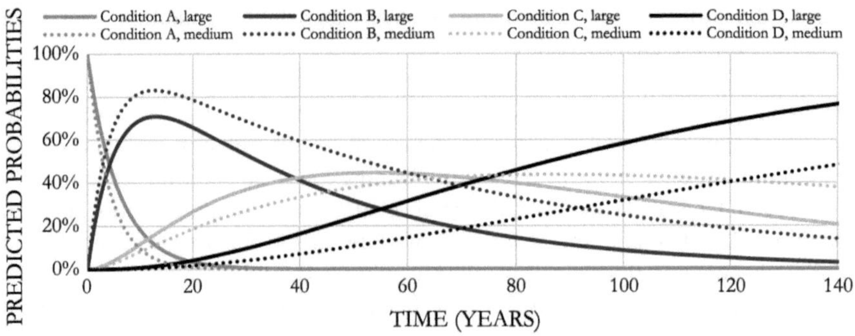

Fig. 4.15 Probabilistic distribution of the degradation condition for stone claddings according to their age and size of the stone plates

claddings with large plates reach a maximum probability of belonging to condition B (P = 71.1 %) at 13 years and reach a maximum probability of belonging to condition C (P = 44.6 %) between 53 and 54 years. These claddings have a maximum probability of transition between conditions B and C between 39 and 40 years; and between conditions C and D between 73 and 74 years. In turn, claddings with medium-sized plates (A < 0.4 m^2), reach a maximum probability of belonging to condition B (P = 83.2 %) at 13 years; and to condition C (P = 43.8 %) between 86 and 88 years. For these claddings, the transition between conditions B and C has a higher probability of occurrence between 64 and 65 years, while the transition between conditions C and D is more likely to occur between 121 and 122 years. Stone claddings with medium-sized plates have higher probability of belonging to more favourable conditions (71.3 % of probability of belonging to conditions A and B) and a low probability of belonging to condition D (3.9 %); claddings with larger plates have a higher probability of belonging to the most unfavourable conditions (24.3 % of probability of belonging to condition D). This suggests that larger plates reach the end of their service life sooner, especially when compared with medium-size plates; this can be explained by the lower relative area of the joints in the bigger plates and consequent higher concentration of stresses as well as the increased difficulty in applying uniformly to the back of the plate the adhesive mortar (Silva et al. 2012).

Concerning exposure to damp (Fig. 4.16), claddings with high exposure to damp transit more rapidly between the most favourable condition (A) and the following one. For this claddings, the probability of belonging to condition B reaches a peak (P = 83.3 %) at 8 years and the probability of belonging to condition C reaches a peak between 51 and 52 years (P = 38.7 %). On the other hand, a cladding with low exposure to damp has a maximum probability of belonging to condition B (P = 72.7 %) after 17 years and a maximum probability (P = 48.4 %) of belonging to the condition C at age 80. Claddings with low exposure to damp remain longer without degradation (condition A) and are more prone to belong to the lowest degradation levels (with a probability of belonging to conditions A and B of 3.4 %

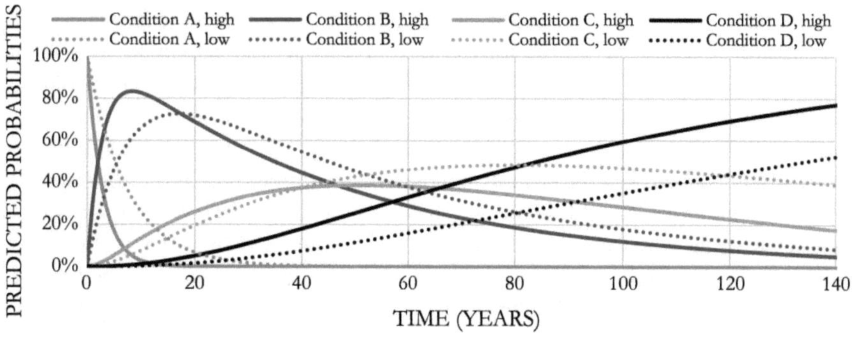

Fig. 4.16 Probabilistic distribution of the degradation condition for stone claddings according to their age and exposure to damp

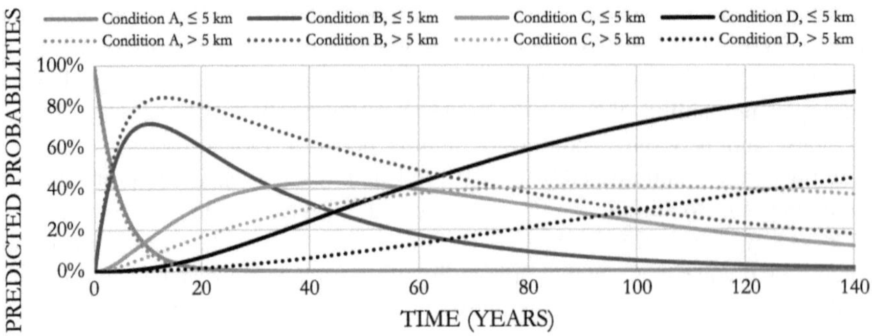

Fig. 4.17 Probabilistic distribution of the degradation condition for stone claddings according to their age and distance from the sea

and 52.9 %, respectively). Claddings highly exposed to damp are more likely to belong to the highest degradation condition (P = 14.3 %) when compared to claddings less exposed to damp (P = 9.2 %).

Figure 4.17 shows the probabilistic distribution of the degradation condition for stone claddings according to their age and to distance from the sea. The probability and mean time of permanence in the most favourable condition (A) is practically the same for façades located at more than 5 km and at less than 5 km from the sea. Façades in coastal areas present a maximum probability of belonging to condition B of 71.7 % at 10 years and a maximum probability of belonging to condition C of 43.1 % after 43 years. Claddings located at more than 5 km from the sea have a maximum probability of belonging to condition B of 84.6 % at 13 years and of belonging to condition C of 41.3 % between 92 and 94 years. Claddings in coastal areas have therefore a higher probability of belonging to the more unfavourable degradation condition (P = 26 %, substantially higher than those away from the coast, with 2.5 %). Claddings located at more than 5 km from the sea are more prone to belonging to degradation conditions A and B (P = 68.6 % as opposed to 51.9 % for coatings located less than 5 km away).

These results are consistent and credible from a physical point of view. The upper and lower limits of conditions C and D, respectively, correspond to a degradation severity of 20 %, thus corresponding to the maximum acceptable degradation level, assumed in this study as the threshold of the service life of façade claddings. Therefore, the maximum probability of transition between condition C and condition D marks the moment the end of stone claddings' service life is most likely to be reached. For the overall sample, the maximum probability of transition between conditions C and D occurs between 103 and 104 years (a relatively optimistic value when compared with the results obtained by other methods discussed in this book). Regarding the type of stone, this moment occurs:; between 77 and 78 years for marble claddings; at 109 years for limestone claddings; and between 131 and 132 years for granite claddings. The coatings with medium-sized plates reach condition D significantly later (between 121 and 122 years) than coatings with large plates (between 73 and 74 years), thus confirming the trend of claddings with large plates to deteriorate more quickly. Concerning exposure to damp, the transition between conditions C and D has a maximum probability of occurrence between 65 and 66 years for claddings highly exposed and between 118 and 119 years for claddings with low exposure, which suggest that claddings with high exposure are more prone to present higher degradation levels, reaching sooner the end of their service life. For distance from the sea, for claddings in coastal areas the transition between conditions C and D is more likely to occur between 57 and 58 years, while for claddings located at more than 5 km from the sea the transition between conditions C and D is more likely to occur between 124 and 125 years.

4.2.2.2 Rendered Façades

Regarding rendered façades, a Markov chain model is obtained, whose intensity matrix Q is given by Eq. (4.55). Figure 4.18 shows the probabilistic distribution of the degradation condition of rendered façades over time, based on the Markov chain model proposed. The probability of belonging to condition A decreases over time, with a probability lower than 2 % after 10 years. The model reveals that rendered façades go through the early stages of degradation in only 5 years (conditions A and B). The probability of belonging to condition B reaches a peak after 3 years ($P = 43.5$). The probability for levels C and D displays skewed distribution curves, with rapid increases and peaks at year 10 ($P = 51.4$ %) and 19 ($P = 38.5$ %) for levels C and D, respectively. The probability of belonging to condition E increases with age: it is practically nil (under 5 %) until 9 years, and above 75 % after 38 years.

$$Q = \begin{bmatrix} q_{A,B} \\ q_{B,C} \\ q_{C,D} \\ q_{D,E} \end{bmatrix} = \begin{bmatrix} 0.4016 \\ 0.2819 \\ 0.0994 \\ 0.0761 \end{bmatrix} \tag{4.55}$$

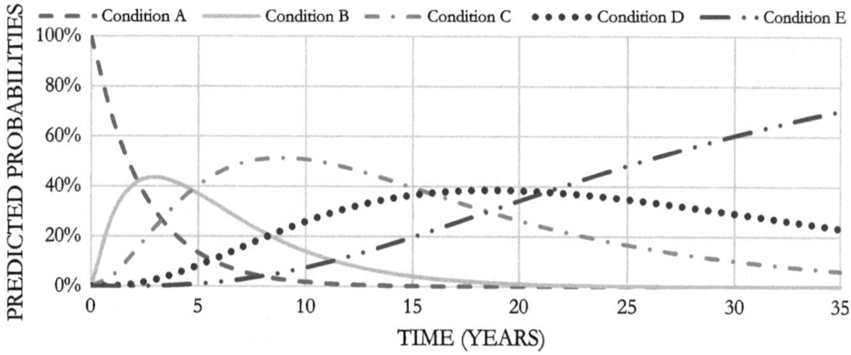

Fig. 4.18 Probabilistic distribution of the degradation condition for rendered façades according to the age of the case studies (data sourced from Silva et al. 2015)

Table 4.10 Classification accuracy of Markov chain model (rendered façades)

Degradation condition	Observed	Predicted	Mean percentage error (%)
Condition A	14	12.57	10.24
Condition B	18	17.77	1.27
Condition C	31	28.64	7.63
Condition D	15	17.29	15.27
Condition E	22	22.74	3.35

Table 4.10 presents an analysis of the model's accuracy, revealing an acceptable accuracy, with relatively low deviations (below 16 %) between the observed and the predicted values.

Rendered façades display significant differences in their deterioration patterns due to the variability of their exposure conditions. In this study, the degradation factors considered are: (a) distance from the sea; (b) exposure to damp; (c) exposure to the combined action to rain and wind; and (d) distance from pollutant sources. These analyses lead to the main conclusions:

- Maritime environments are particularly harmful to external renders, since wind carrying sea spray causes progressive deposition of salts on the external surfaces of the façades (Hossain et al. 2009), affecting their durability and reducing their service life (Soroka and Carmel 1987). Renderings in coastal areas are prone to belong to higher degradation conditions (probability of 26.1 and 30.4 % of belonging to conditions D and E, respectively). Façades located at more than 3.5 km from the sea remain longer in level A, with a probability of 50 % of being in level A at year 5. Renderings in coastal areas have a high probability of being in level B right from year 1 (P = 55.7 %); the probability of being in level C peaks at year 6 (P = 58 %) and the probability of being in level D reaches a peak at year 16 (P = 47 %). Renderings at more than 3.5 km from the sea reach: the maximum probability of belonging to condition B (P = 16.5 %) at year 4, the maximum probability of belonging to condition C (P = 52 %) at year 14, and of belonging to condition D (P = 38.5 %) at year 19;
- Exposure to damp is responsible for numerous defects in buildings such as: staining and soiling of the surface of the render, and biological growth (presence of algae and other microorganisms). Façades with low exposure to damp remain longer in the most favourable condition (A), with a maximum probability of belonging to condition B at year 4 (P = 42 %) while, under a severe exposure, this peak occurs at year 1 approximately (P = 35.6 %), and for moderate exposure it occurs at year 3 approximately, (P = 49.8 %). Renders with severe exposure to damp reach a maximum probability of belonging to condition C soon, at year 3 (P = 76.9 %); those with moderate exposure reach this peak at year 8 (P = 34.3 %), and for renders with low exposure, this peak is reached at year 10 (P = 57.2 %); regarding condition D, the results follow a similar trend;
- In the definition of the protection level of a façade the driven rain incidence and the existence of protections in surrounding area (for example, other buildings,

adjacent hills or vegetation) are considered. The combined action of wind and rain changes the trajectory of raindrops and consequently the pattern of runoff flow along the façade (Choi 1999). In the sample analysed, renderings with moderate exposure to wind-rain action are the ones that remain longer in degradation level A. Concerning condition B, renderings have a maximum probability of belonging to degradation level B at year 3, with a probability of P = 50.4, 40.3 and 39 % for unfavourable, standard and favourable conditions of exposure to driven-rain, respectively. Renders with poor and good levels of protection reach the maximum probability of belonging to condition C at 9 years (with a probability of 51.1 and 55.8 %, respectively), while renders with current level of protection reach this peak (44.1 %) at 8 years. Regarding condition D, renders with current level of protection reach this peak at year 18 (P = 44.7 %), renders with good levels of protection reach this peak at year 17 (P = 27.9 %). The corresponding value for renders with poor level of protection occurs at around 20 years (P = 41.6 %);

- The rain associated with atmospheric pollutants contributes to the physical and visual degradation of external renders (Kus and Nygren 2002). In the sample analysed, only six case studies have favourable conditions of exposure to pollutant sources and none of them presents degradation level E. These renders reach a maximum probability of belonging to condition B after year 2 (P = 69 %) while renders with unfavourable and normal exposure to pollutant sources have a maximum probability of belonging to this condition of 39.7 and 43 %, respectively, both at around year 3. Renders under unfavourable and favourable exposure conditions to pollutant sources have a maximum probability of belonging to condition C of P = 42.5 and 56 %, respectively, both at around year 8, while renders with normal exposure to pollutant sources reach maximum probability (P = 54.1 %) of belonging to this condition at year 9.

The results must be carefully analysed since in reality the environmental agents act simultaneously in rendered façades. Therefore, it seems relevant to analyse the synergy between the environmental factors considered. In this analysis, it was found that renders with severe exposure to damp, located in façades without protection are more prone to belong to the most severe degradation conditions. On the contrary, renderings with low exposure to damp are the ones that are more likely to belong to the less serious degradation of conditions (A and B). Façades located in coastal areas and unprotected are those with the highest probability (50 %) of belonging to the most unfavourable condition (E). In turn, renderings located in coastal areas are the ones that are less likely to belong to the less serious condition (condition A). Façades located at more than 3.5 km from the sea are more prone to belonging to conditions A and B, with an overall probability of 37 % for façades without protection, 50 % for façades in current situation and 30 % for façades with protection. Therefore, it is concluded that, although the façade protection level does not correctly model the effects of wind-rain action (since it models this effects indirectly), evidence shows that façades without protection are more exposed to the negative effects of the other environmental agents, while higher protection levels contribute to delay the effect of these agents in the performance of rendered façades.

Table 4.11 Analysis of the maximum probability of rendered façades belonging to degradation condition D (end of service life) (data sourced from Silva et al. 2015)

		Maximum probability of belonging to degradation level D (%)	Age at reaches the maximum probability of belonging to degradation level D (years)
Distance from the sea	Overall sample	38.47	19
	>3.5 km	5.38	16
	≤3.5 km	46.85	16
Exposure to damp	Low	27.34	19
	Moderate	47.12	17
	Severe	62.83	22
Protection level of the facades	Poor	41.64	20
	Current	44.72	18
	Good	27.90	17
Exposure to pollutants	Unfavourable	52.04	18
	Normal	32.36	18
	Favourable*	–	–

*In the sample analysed, there is no case studies belonging to degradation level E with favourable exposure to pollutants. For this reason, it is impossible to model the transition between levels D and E. The values are not shown because for favourable exposure to pollutants it is not possible to define a maximum probability of belonging to degradation level D, since the probability curve tends to 100% as age increases.0

Table 4.11 shows the maximum probability of belonging to degradation level D (which marks the end of service life). The overall sample reaches the maximum probability of belonging to level D at year 19 (38.5 %). Regarding the factor distance from the sea, renders located in coastal areas show a higher probability of being in the highest degradation levels (P = 56.5 % in either condition levels D or E), reaching the maximum probability of belonging to condition D after 10 year, while for renders at more than 3.5 km from the sea, this peak is reached only at 23 years.

Renderings with severe exposure to damp have a higher probability of being in degradation levels C and D, P = 91 %, and null probability of being in the lowest levels. Renderings with severe exposure reach a probability of 50 % of belonging to level D or E at year 10, sooner than facades with moderate and low exposure for which this probability is reached after 11 and 17 years, respectively.

In terms of the protection level, façades with good protection level reach the probability of 50 % of being in degradation level D or E after 15 years, while façades with current and poor exposure conditions reach this probability after 12 and 14 years, respectively. These results collide with the empirical perception that higher protection levels lead to lower degradation levels. In this study, the use of Markov chains do not allow correctly modelling the effects of wind driven rain (it was in fact the only environmental factor that was indirectly assessed), since this variable presents an indirect effect on the remaining environmental factors analysed. Silva et al. (2015) refer that lower levels of protection of the façade do enhance the hazardous effects of other environmental effects, while better protection levels delay the degradation effects of the latter. Concerning exposure to pollutants, service life variations are not as clear as the results for other environmental agents; nevertheless, it is possible to identify a pattern according to which buildings with low exposure to pollutants do not reach condition E. Renderings with unfavourable exposure reach a probability of 50 % of belonging to level D or E at 11 years, while façades with current exposure only reach this probability at 15 years and render subjected to favourable exposure reach this probability at 17 years.

4.2.2.3 Painted Surfaces

The intensity matrix Q obtained for the Markov chain model applied to painted surfaces is presented in Eq. (4.56).

$$Q = \begin{bmatrix} q_{A,B} \\ q_{B,C} \\ q_{C,D} \\ q_{D,E} \end{bmatrix} = \begin{bmatrix} 0.4868 \\ 0.1962 \\ 0.1524 \\ 0.1062 \end{bmatrix} \tag{4.56}$$

Figure 4.19 shows the probabilistic distribution of the degradation condition of painted surfaces over time, based on the Markov chain model proposed. The probability of belonging to condition A decreases from the outset and is lower than 5 % after 7 years. The maximum probability of belonging to condition B is reached at 3 years (P = 54.1 %). The probability of belonging to conditions C and D reaches a peak after 8 years (P = 38.5 %) and 16 years (P = 35.4 %), respectively. Finally, the probability of belonging to condition E is practically residual (less than 5 %) after 7 years, exceeding 90 % after year 41.

An analysis of the model's accuracy is presented in Table 4.12, revealing a model with an acceptable accuracy.

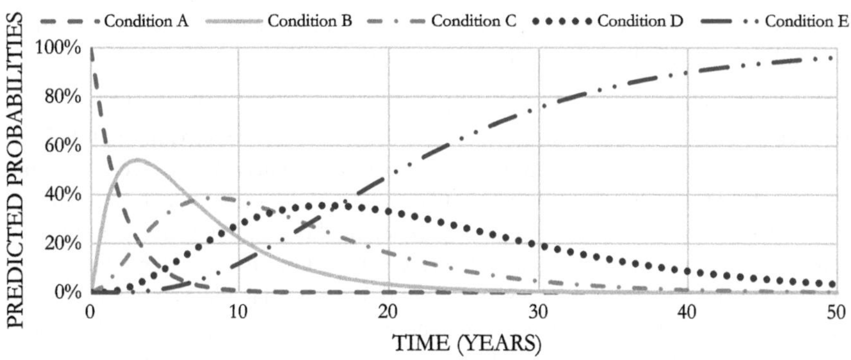

Fig. 4.19 Probabilistic distribution of the degradation condition for painted surfaces according to the age of the case studies

Table 4.12 Classification accuracy of Markov chain model (rendered façades)

Degradation condition	Observed	Predicted	Mean percentage error (%)
Condition A	37	32.54	12.05
Condition B	73	75.50	3.42
Condition C	48	58.86	22.63
Condition D	29	35.49	22.39
Condition E	20	17.61	11.97

For painted surfaces three environmental agents that affect façades durability are analysed: (a) exposure to damp; (b) distance from the sea; (c) and proximity to pollution sources. These models lead to the following results:

- Façades with low exposure to damp reach a maximum probability of belonging to condition B after 3 years (P = 54.4 %); to condition C after 8 years (P = 42.7 %) and to condition D at year 14 (P = 32.1 %). For facades with high exposure to damp, the maximum probability is reached after 4 years for condition B (P = 55.9 %), after 9 years for condition C (P = 34 %) and after 19 years for condition D (P = 45.4 %);
- Painted surfaces located more than 5 km from the sea reach the maximum probability of belonging to condition B after 3 years (P = 53 %); of belonging to condition C after 8 years (P = 43.2 %) and of belonging to condition D after 15 years. (P = 31.8 %) For painted surfaces in coastal areas, the maximum probability of belonging to condition B is reached after 4 years (P = 56.7 %), after 9 years for condition C (P = 32.9 %) and after 19 years for condition D (P = 45.8 %);
- Painted surfaces with current exposure to pollutants, reach a maximum probability of belonging to condition B after 3 years (P = 56.3 %), a maximum probability of belonging to condition C after 8 years (P = 36.9 %) and of belonging to condition D after 17 years (P = 38.5 %). For façades with favourable exposure to pollutants, the maximum probability of belonging to conditions B is reached after 3 years, with P = 44.9 %, the maximum probability of belonging to conditions C is reached after 8 years (P = 43.8 %) and for condition D, this peak is reached after 14 years (P = 29.7 %).

As in the previous claddings, in the case of painted surfaces the transition between conditions C and D also establishes the instant of the maximum probability of the claddings to reach the end of their service life. Analyzing the overall sample, it is found that this threshold occurs between 12 and 13 years, which is within the expected service life values obtained by other methods applied in this book. The analysis of the probability of painted surfaces to belong to each degradation condition according to the variables considered reveals that the sample is slightly biased. This particularity of the sample is not relevant to the application of the other models (namely the deterministic or computational ones). Contrary to what would be expected, paintings subjected to unfavourable exposure conditions are more prone to belong to the less severe degradation conditions (conditions A and B) and paintings subjected to favourable environmental conditions are more likely to belong to the most severe degradation conditions (D and E). This is justified by the specific nature of the sample, since the most recent façades are the ones exposed to the most unfavourable environmental conditions, while the oldest façades (with higher degradation rates) are subjected to favourable exposure conditions. In the sample analysed, and considering exposure to damp, it was found that painted surfaces with high exposure have an average age of 5.4 years (about half of the estimated service life this type of coatings) while those with low exposure have an average age of 8.7 years (near the end of the service life of painted surfaces). This skew of the data results in a Markov chain model that associate the unfavourable environmental conditions to the lower degradations rates (based on the sample supplied to the model). Despite the sample features, considering distance from the sea, it is found that painted surfaces in coastal areas reach their service life limit between years 10 and 11, substantially earlier than coatings more than 5 km away from the sea, whose transition occurs between years 14 and 15. Regarding exposure to damp, coatings with low exposure transit between conditions C and D between 13 and 14 years, while coatings highly exposed transit between these conditions earlier, between 11 and 12 years. This result reveals that coatings subject to adverse environmental conditions tend to have shorter service lives.

4.2.2.4 Ceramic Tiling Systems

In ceramic tiling systems the same methodology was adopted. Equation (4.57) presents the intensity matrix Q obtained for the Markov chain model applied to ceramic claddings.

$$Q = \begin{bmatrix} q_{\mathrm{A,B}} \\ q_{\mathrm{B,C}} \\ q_{\mathrm{C,D}} \\ q_{\mathrm{D,E}} \end{bmatrix} = \begin{bmatrix} 0.1519 \\ 0.0403 \\ 0.0252 \\ 0.0100 \end{bmatrix} \tag{4.57}$$

Figure 4.20 presents the probabilistic distribution of the degradation condition of ceramic claddings over time, based on the Markov chain model proposed.

The probability of a ceramic cladding to belong to condition A decreases over time, with a probability lower than 2 % after 26 years. The probability of belonging to condition B reaches a maximum peak (P = 61.9 %) at 12 years. The maximum probability of belonging to condition C is reached after 39 years (P = 44.4 %). The probability of belonging to condition D is almost residual before year 16 (less than 5 %), reaching a peak between 98 and 100 years (P = 49.5 %). Finally, is very low until year 44 (less than 5 %) and slowly increases over time, to around 30 % at year 100.

Table 4.13 shows an analysis of the model's accuracy. The model is less accurate than the previous ones, and is barely able to describe the cases belonging

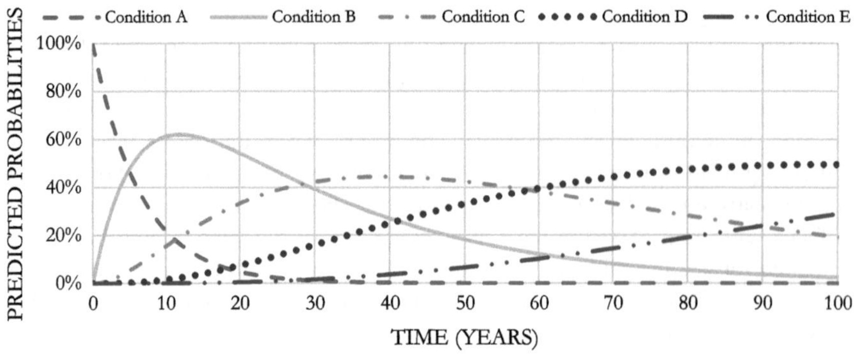

Fig. 4.20 Probabilistic distribution of the degradation condition for ceramic claddings according to the age of the case studies

Table 4.13 Classification accuracy of Markov chain model (ceramic claddings)

Degradation condition	Observed	Predicted	Mean percentage error (%)
Condition A	15	16.11	7.41
Condition B	70	64.78	7.46
Condition C	64	67.27	5.11
Condition D	43	39.29	8.62
Condition E	3	7.54	151.47

to condition E; however those cases comprise 1.5 % of the sample only (3 case studies) and as such do not allow the model to draw firm conclusions about the probability distribution of this condition. Even so, the deviation obtained seems reasonable since the model intends to explain a complex phenomenon, such as is the degradation condition of buildings' elements.

In the case of adherent ceramic tiles, some characteristics of environmental exposure, relevant to the degradation of ceramic tiles, were analysed: (a) exposure to damp; (b) distance from the sea; (c) exposure to wind-rain action. The following results can be drawn:

- Claddings less exposed to damp are more prone to remain undamaged (A and B), with $P = 54.9$ %; none of the case studies analysed less exposed to damp belong to the most unfavourable condition (condition E). The probability of belonging to condition B reaches a maximum probability at 14 years ($P = 65.6$ %) for façades with low exposure to damp, while in façades with high exposure, this peak occurs earlier, at 9 years, with $P = 59.3$ %. Concerning condition C, façades with low exposure have a maximum probability of belonging to this condition after 48 years ($P = 41.6$ %), while once again for a high exposure this peak occurs earlier, at 30 years ($P = 51.3$ %). Regarding condition D, the maximum probability of belonging to this condition occurs between 114 and 115 years ($P = 45.4$ %) for coatings with low exposure and between 86 and 87 years (P = 52.4 %) for coatings with high exposure. Finally, claddings highly exposed to damp are more prone to belong to the most unfavourable conditions, and have a relatively small probability of belonging to conditions A and B (P = 28.5 %);

- Claddings located in coastal areas have higher probability (P = 67.7 %) of reaching higher degradation levels (conditions C, D and E), transiting more rapidly from the most favourable conditions to the most serious conditions. For coatings in the coastal area, the maximum probability of belonging to condition B occurs at 9 years (P = 66.5 %), for condition C at 33 years (P = 47.2 %) and for condition D (P = 51.1 %) between 91 and 92 years. Regarding coatings located more than 5 km from the sea, the maximum probability of belonging to condition B occurs at 18 years (P = 68.4 %), for condition C between 76 and 77 years (P = 50.7 %) and for condition D (P = 26.1 %) after 150 years. Claddings located more than 5 km from the sea have a probability of 71.4 % of belonging to the most favourable conditions (A and B);
- Claddings with severe exposure to the combined action of wind and rain are the most prone to undergo higher degradation (P = 37.8 % of belonging to conditions D and E) and have the lowest probability of belonging to the most favourable conditions. Claddings with low exposure to wind-rain action reach the maximum probability of belonging to degradation condition B at 10 years (P = 73.9 %); this peak also occurs at 10 years for claddings with severe exposure (P = 56.8 %) and at 13 years for claddings with moderate exposure (P = 61.7 %). The maximum probability of belonging to condition C occurs after 45 years for claddings with low exposure conditions (P = 43.9 %), after 44 years (P = 44 %) for moderate exposure and after 32 years (P = 48.3 %) for severe exposure. Regarding condition D, the maximum probability of occurrence occurs between 113 and 115 years (P = 44.5 %) for low exposure, after 113 years (P = 48 %) for moderate exposure and between 86 and 87 years (P = 53.2 %) for severe exposure.

As in previous claddings, the transition between conditions C and D marks the instant of the end of the service life of ceramic claddings (according to the assumptions made here). From the analysis of the overall sample, it is found that the maximum probability of transition between conditions C and D occurs between 58 and 59 years. This result is similar to those obtained by other methods used in this book and to the values contained in the literature. Concerning exposure to damp, ceramic claddings with low exposure transit between conditions C and D between 70 and 71 years, while claddings with high exposure transit between these conditions earlier, between 51 and 52 years, confirming a lower durability of claddings subjected to unfavourable exposure conditions. For distance from the sea, the results shows that coatings located in coastal areas (less than 5 km from the sea) transit between conditions C and D between 53 and 54 years, substantially earlier than the ceramic claddings more distant from the shore. Concerning exposure to wind-rain action, claddings with low exposure transit between conditions C and D between 73 and 74 years, while claddings with moderate exposure transit between these conditions between 62 and 63 years, and claddings with severe exposure make this transition between 49 and 50 years. These results confirm the empirical perception that claddings with adverse exposure to wind-rain action are more prone to present higher degradation indexes, reaching the end of their service life earlier.

4.2.3 Conclusions

This chapter intended to model the probability distribution of the degradation condition of façade claddings over time. In service life prediction models, it is relevant to know the rate of degradation transition between degradation condition, anticipating the element's behaviour under analysis and predicting the point at which it will be unable to meet the requirements for which it was designed. To accomplish this, a Markov chain based model was used, which allows predicting the probability of each degradation condition for the samples analysed.

In this chapter, it was performed a relatively thorough study regarding the influence of the various claddings' characteristics and degradation agents in the description of the loss of performance of façade coatings over time. The proposed methodology provides information on the synergy between degradation agents and allows understanding the probabilistic distribution of façade's degradation condition over time according to the surrounding environmental features. This methodology also allows knowing the mean time of permanence in each degradation condition and the most probable instant for the transition to the next degradation condition. However, Markov chains model can lead to biased results, especially in samples in which the most recent façades are the ones exposed to the most unfavourable environmental conditions, while the oldest façades (with higher degradation rates) are subjected to favourable exposure conditions. Samples with these characteristics, as occurs in this study, can skew the results achieved by Markov chain model, leading to the incorrect conclusion that the unfavourable environmental conditions correspond to lower degradations rates. This is the main disadvantage of this model, and the remaining proposed models, there is all the models react to the data.

The results reveal that environmental agents contribute to establish the end of service life of façades claddings. The information concerning the degradation effect on the durability of claddings can be applied to the implementation of maintenance schedules, rationalizing these interventions. The results thus obtained can be used to assess the risk of claddings not performing as expected due to degradation. Therefore, a definition of a threshold of acceptable risk can be proposed, so that a stakeholder may visualize the need for repair based on the probabilistic analysis of a set of data. To illustrate this concept, it can be considered that "high", "average" and "low" probability of a given condition correspond to "high", "average" and "low" risk associated to the consequences of defects and the cost of repair and thus produce an indication of the urgency to maintenance and repair actions. It is considered that claddings with a low risk (no action required apart from monitoring) are those that present a $P > 75\ \%$ of belonging to either condition levels "A" or "B". Claddings with high risk (a need to undertake extensive repair) are considered to correspond to $P > 25\ \%$ of belonging to condition levels D (for stone and ceramic claddings) and "E" (for renderings and paintings). An average risk corresponds to the intermediate states. Based on the results obtained in this chapter and based on these assumptions, some recommendations can be made:

- For stone claddings: (i) monitor them until year 20; (ii) perform maintenance actions until year 70; (iii) consider their substitution after year 70, subject to on-site confirmation of their condition state;
- For renderings: (i) monitor the render facade until year 5; (ii) perform light maintenance actions until year 10; (iii) perform repair works until year 15; and (iv) consider replacement of the render from year 15 onwards, subject to on-site confirmation of its condition;
- For painted surfaces: (i) they should be monitored every 3 years; (ii) maintenance actions should be performed around year 13; (iii) repainting should be considered after year 13 subject to on-site confirmation of their condition state;
- For ceramic claddings: (i) monitor the façade until year 13; (ii) maintenance actions should be performed before year 40; (iii) full replacement should be considered after 40 years subject to on-site confirmation of their condition state.

The application of stochastic models in service life prediction is extremely useful, with numerous practical applications. For example, they can be applied in the context of insurance policies, since they model uncertainty and provide information about the risk of failure and the most probable failure time of building elements based on their characteristics. Additionally, this information can be used in the definition of maintenance strategies throughout the building life cycle and in the establishment of joint maintenance strategies for different parts and elements of the building.

References

Agresti A (1986) Applying R^2 type measures to ordered categorical data. Technometrics 28 (2):133–138

Agresti A (2002) Categorical data analysis, 2nd edn. Wiley, New York

Aikivuori AM (1999) Critical loss of performance—what fails before durability. In: 8th international conference on durability of buildings materials and components, DBMC, Vancouver, Canada, pp 1369–1376

Akaike HA (1979) Bayesian extension of the minimum AIC procedure of autoregressive model fitting. Biometrika 66(2):237–242

Akaike HA (1981) Likelihood of a model and information criteria. J Econometrics 16(1):3–14

Akaike HA (1984) New look at the statistical model identification. IEEE Trans Autom Control 19 (6):716–723

Ash A, Shwartz M (1999) R2: a useful measure of model performance when predicting a dichotomous outcome. Stat Med 18(4):375–384

Augenbroe GLM, Park C-S (2002) Towards a maintenance performance toolkit for GSA. Interim Report submitted to GSA, Georgia Institute of Technology, Atlanta, USA
Baik HS, Jeong HS, Abraham DM (2006) Estimating transition probabilities in Markov chain-based deterioration models for management of wastewater systems. J Water Resour Plan Manage 132(1):15–24
Balaras A, Droutsa K, Dascalaki E, Kontoyiannidis S (2005) Deterioration of European apartment buildings. Energy Build 37(5):515–527
Baum LE, Petrie T (1966) Statistical inference for probabilistic functions of finite state Markov chains. Ann Math Stat 37(6):1554–1563
Bayaga A (2010) Multinomial logistic regression: usage and application in risk analysis. J Appl Quant Methods 5(2):288–297
Bladty M, Sorensen M (2009) Efficient estimation of transition rates between credit ratings from observations at discrete time points. Quant Finance 9(2):147–160
Bocchini P, Saydam D, Frangopol DM (2013) Efficient, accurate, and simple Markov chain model for the life-cycle analysis of bridge groups. Struct Saf 40:51–64
Bogdanoff JL (1978) A new cumulative damage model- Part 1. J Appl Mech 45(2):246–250
Brandt E, Rasmussen M (2002) Assessment of building conditions. Energy Build 34(2):121–125
Cargal JM (2013) Discrete mathematics for neophytes: number theory, probability, algorithms, and other stuff. Accessed in 31 Jan 2013. Available online in http://www.cargalmathbooks.com/lectures.htm
Carrington PJ, Scott J, Wasserman S (eds) (2005) Models and methods in social network analysis, 1st edn. Cambridge University Press, New York
Chew M (2005) Defect analysis in wet areas of buildings. Constr Build Mater 19(3):165–173
Choi EC (1999) Wind driven rain on building facades and the driving rain index. Wind Eng Ind Aeronaut 79(1–2):105–122
Choi M, Lee G (2010) Decision tree for selecting retaining wall systems based on logistic regression analysis. Autom Construct 19(7):917–928
Cox DR, Miller HD (1965) The theory of stochastic processes, 1st edn. Chapman and Hall, London
Cox DR, Snell EJ (1989) The analysis of binary data, 2nd edn. Chapman and Hall, London
Cox DR, Wermuch N (1992) A comment on the coefficient of determination for binary responses. Am Stat 46(1):1–4
Das PC (1998) New developments in bridge management methodology. Struct Eng Int 8(4):299–302
Dekker R, Nicolai RP, Kallenberg LCM (2008) Maintenance and Markov decision models. Encyclopedia of statistics in quality and reliability. Wiley, London
Ellingwood BR (2005) Risk-informed condition assessment of civil infrastructure: state of practice and research issues. Struct Infrastruct Eng Maintenance Manage Life-Cycle Des Perform 1 (1):7–18
Frangopol DM, Neves LC (2004) Probabilistic maintenance and optimization strategies for deteriorating civil infrastructures. In: Topping BHV, Mota Soares CA (eds) Progress in computational structures technology. Saxe-Coburg Publishers, Stirling, pp 353–377
Frangopol DM, Kallen M-J, Noortwijk JMV (2004) Probabilistic models for life-cycle performance of deteriorating structures: review and future directions. Prog Struct Mat Eng 6 (4):197–212
Freitas VP, Sousa M, Abrantes V (1999) Survey of the durability of facades of 4000 dwellings in northern Portugal. In: 8th DBMC, international conference on the durability of building materials and components; Ottawa, Canada, pp 1040–1050
Gao S, Shen J (2007) Asymptotic properties of a double penalized maximum likelihood estimator in logistic regression. Stat Probab Lett 77(9):925–930
Garavaglia E, Gianni A, Molina C (2004) Reliability of porous materials: two stochastic approaches. J Mater Civ Eng 16(5):419–426

Gaspar P (2009) Service life of constructions: development of a method to estimate the durability of construction elements. Application to renderings of current buildings (in Portuguese). Doctor thesis in sciences of engineering, Instituto Superior Técnico, University of Lisbon, Portugal

Gu MG, Kong FH (1998) A stochastic approximation algorithm with Markov chain Monte Carlo method for incomplete data estimation problems. Nat Acad Sci USA 95(13):7270–7274

Hair JF, Black WC, Babin B, Anderson RE, Tatham RL (2007) Multivariate data analysis, 6th edn. Prentice-Hall Publishers, Englewood Cliffs

Handel RV (2008) Hidden Markov models. Lecture Notes, Princeton University, New Jersey, USA

Henry NW (1971) The retention model: a Markov chain with variable transition probabilities. J Am Stat Assoc 66(334):264–267

Higham D, Higham N (2005) MATLAB guide, 2nd edn. Society for Industrial Mathematics, Philadelphia

Horn R, Johnson C (1985) Matrix analysis. Cambridge University Press, Cambridge

Hosmer DW, Lemeshow S (1980) Goodness of fit tests for the multiple logistic regression model. Commun Stat Part A Theory Methods 9(10):1043–1069

Hosmer DW, Lemeshow S (2000) Applied logistic regression, 2nd edn. Wiley, New York

Hosmer DW, Hjort NL (2002) Goodness-of-fit processes for logistic regression: simulation results. Stat Med 21(18):2723–2738

Hossain KMA, Lachemi M, Şahmaran M (2009) Performance of cementitious building renders incorporating natural and industrial pozzolans under aggressive airborne marine salts. Cement Concr Compos 31(6):358–368

Howell DC (2002) Statistical methods for psychology, 5th edn. Wadsworth Publishing, Pacific Grove

Hu B, Shao J, Palta M (2006) Pseudo-R^2 in logistic regression model. Statistica Sinica 16(3):847–860

Johnson JT (1987) Continuous-time, constant causative Markov chains. Stoch Process Appl 26:161–171

Kalbfleisch JD, Lawless JF (1985) The analysis of panel data under a Markov assumption. J Am Stat Assoc 80(392):863–871

Kallen MJ (2009) A comparison of statistical models for visual inspection data. Safety, reliability and risk structures, infrastructures and engineering systems. In: Tenth international conference on structural safety and reliability (ICOSSAR), Osaka, Japan, pp 3235–3242

Kobayashi K, Do M, Han D (2010) Estimation of Markovian transition probabilities for pavement deterioration forecasting. KSCE J Civil Eng 14(3):343–351

Koch A, Siegesmund S (2004) The combined effect of moisture and temperature on the anomalous expansion behaviour of marble. Environ Geol 46(3–4):350–363

Kus H, Nygren K (2000) Long-term exposure of rendered autoclaved aerated concrete: measuring and testing programme. In: RILEM/CIB/ISO international symposium—integrated life-cycle design of materials and structures, Helsinki, Finland, pp 415–420

Kus H, Nygren K (2002) Microenvironmental characterization of rendered autoclaved aerated concrete. Build Res Inf 30(1):25–34

Kus H, Nygren K, Norberg P (2004) In-use performance assessment of rendered autoclaved concrete walls by long-term moisture monitoring. Build Environ 39(6):677–687

Kuss O (2002) Global goodness-of-fit tests in logistic regression with sparse data. Stat Med 21 (24):3789–3801

Lacasse MA, Kyle BR, Talon A, Boissier D, Hilly T, Abdulghani K (2008) Optimization of the building maintenance management process using a markovian model. In: 11th international conference on the durability of building materials and components, Istanbul, Turkey, pp 1–9 (T72)

Lee S, Hershberger S (1990) A simple rule for generating equivalent models in covariance structure modeling. Multivar Behav Res 25(3):313–334

Lethanh N, Adey B (2012) A hidden Markov model for modeling pavement deterioration under incomplete monitoring data. In: International conference on mathematical, computational and statistical sciences, and engineering 2012, Stockholm, Sweden, pp 722–729
Lewis KN, Heckman BD, Himawan L (2011) Multinomial logistic regression analysis for differentiating 3 treatment outcome trajectory groups for headache-associated disability. Pain 152(8):1718–1726
Li P (2007) Hypothesis testing in finite mixture models. Doctoral thesis in Statistics, University of Waterloo, Ontario, Canada
Li Z, Wang W (2006) Computer aided solving the high-order transition probability matrix of the finite Markov chain. Appl Math Comput 172(1):267–285
Liu T (2010) Application of Markov chains to analyze and predict the time series. Mod Appl Sci 4 (5):162–166
Ljung L (1987) System identification—theory for the user. Prentice-Hall, Englewood Cliffs
Lounis Z, Vanier DJ (2000) A multi-objective and stochastic system for building maintenance management. Comput Aided Civil Infrastruct Eng 15(5):320–329
Lounis Z, Lacasse MA, Siemes AJM, Moser K (1998) Further steps towards a quantitative approach to durability design. Materials and technologies for sustainable construction, construction & environment. CIB World Building Congress, Gävle, Sweden, pp 315–324
Madanat S, Mishalani R, Ibrahim WHW (1995) Estimation of infrastructure transition probabilities from condition rating data. J Infrastruct Syst 1(2):120–125
Maroco J (2007) Statistical analysis using SPSS. (in Portuguese), 3rd edn. Sílabo Editions, Lisbon
Marteinsson B, Jónsson B (1999) Overall survey of buildings—performance and maintenance. In: 8th DBMC, international conference on the durability of building materials and components, Ottawa, Canada, pp 1634–1654
Mc Duling JJ (2006) Towards the development of transition probability matrices in the Markovian model for the predicted service life of buildings. PhD thesis in civil engineering, Faculty of Engineering, Built Environment and Information Technology, University of Pretoria, Pretoria
McFadden D (1973) Conditional logit analysis of qualitative choice behavior. In: Zarembka P (ed) Frontiers of econometrics. Academic Press, New York, pp 105–142
Menard S (2000) Coefficients of determination for multiple logistic regression analysis. Am Stat 54(1):17–24
Mitchell DS (2007) The use of lime & cement in traditional buildings. Technical Conservation, Research and Education Group, Historic Scotland, Longmore House, Salisbury Place, Edinburgh, Scotland, 8p
Mittlbock M, Schemper M (1996) Explained variation for logistic regression. Stat Med 15 (19):1987–1997
Monika SO (2005) Statistical inference and hypothesis testing for Markov chains with interval censoring—application to bridge condition data in the Netherlands. Master thesis in civil engineering, Delft University of Technology, Delft
Morcous G (2006) Performance prediction of bridge deck systems using Markov chains. J Perform Construct Facil 20(2):146–155
Morcous G, Lounis Z, Mirza MS (2003) Identification of environmental categories for Markovian deterioration models of bridge decks. J Bridge Eng 8(6):353–361
Moser K (2003) Engineering design methods for service life planning—state of the art. In: WMDBP 2003, international workshop on management of durability in the building process, Politecnico di Milano, Milan, Italy, paper 40
Myung IJ (2003) Tutorial on maximum likelihood estimation. J Math Psychol 47(1):90–100
Nagelkerke NJD (1991) A note on a general definition of the coefficient of determination. Biometrika 78(3):691–693
Neves LC, Frangopol DM, Cruz PJS (2006) Lifetime multi-objective optimization of maintenance of existing steel structures. In: 6th international symposium steel bridges, European Convention for Construction Steelwork, Prague, Czech Republic, pp 206–215
Newsom JT (2011) Logistic regression. Course of Data analysis II, Institute on Aging, School of Community Health, Portland State University

Ng S, Moses F (1996) Prediction of bridge service life using time-dependent reliability analysis. In: 3rd International conference on bridge management, University of Surrey, Guildford, UK, pp 26–33

Norris JR (1997) Markov chains, 1st edn. Cambridge University Press, Cambridge

Parnham P (1997) Prevention of premature staining of new buildings. E & FN Spon, London

Parzen E (1962) Stochastic processes. Holden Day, San Francisco

Plackett RL (1983) Karl Pearson and the Chi-squared test. Int Stat Rev 51(1):59–72

Robelin C-A (2006) Facility-level and system-level stochastic optimization of bridge maintenance policies for Markovian management systems. PhD thesis in civil and environmental engineering, University of California, California, USA

Robelin CA, Madanat S (2007) History-dependent bridge deck maintenance and replacement optimization with Markov decision process. J Infrastruct Syst 13(3):195–201

Roh Y-S, Xi Y (2000) A general A general formulation for transition probabilities of Markov model and the application to fracture of composite materials. Probab Eng Mech 15(3):241–250

Schouenborg B, Grelk B, Malaga K (2007) Testing and assessment of marble and limestone (TEAM) - Important results from a large European research project on cladding panels. ASTM Int 4(5):10–22

Schwarz GE (1978) Estimating the dimension of a model. Ann Stat 6(2):461–464

Shohet I, Rosenfeld Y, Puterman M, Gilboa E (1999) Deterioration patterns for maintenance management—a methodological approach. In: 8th international conference on durability of building materials and components, Vancouver, Canada, pp 1666–1678

Shtatland ES, Kleinman K, Cain EM (2002) One more time about R^2 measures of fit in logistic regression. In: 15th annual Nesug conference; statistics, data analysis and econometrics, Issue M, pp 1–6

Silva A, de Brito J, Gaspar PL (2012) Application of the factor method to maintenance decision support for stone cladding. Autom Constr 22(3):165–174

Silva A, de Brito J, Gaspar P (2013) Probabilistic analysis of the degradation evolution of stone wall cladding (directly adhered to the substrate). J Mater Civil Eng 25(2): 227–235

Silva A, Neves LC, Gaspar PL, de Brito J (2015) Probabilistic transition of condition: render facades. Build Res Inf doi:10.1080/09613218.2015.1023645

Singer B (1981) Estimation of nonstationary Markov chains from panel data. Sociol Methodol 12:319–337

Smyth GK (2003) Pearson's goodness of fit statistic as a score test statistic. In: Goldstein DR (ed) Science and atatistics: a festschrift for terry speed. IMS lecture notes—monograph series, vol. 40. Institute of Mathematical Statistics, Hayward, California

Soroka I, Carmel D (1987) Durability of external renderings in a marine environment—durability of building materials. Elsevier Science Publishers, Amsterdam, pp 61–72

Stelzl I (1986) Changing causal relationships without changing the fit: some rules for generating equivalent LISREL models. Multivar Behav Res 21(3):309–331

Straub A (2003) Using a condition-dependent approach to maintenance to control costs and performances. Facil Manage 1(4):380–395

Suthar V, Tarmizi RA, Midi H, Adam MB (2010) Students' beliefs on mathematics and achievement of university students: logistics regression analysis. In: International conference on mathematics education research (ICMER 2010). Procedia—Social and Behavioral Sciences, vol 8, pp 525–531

Thomasson F (1982) Les enduits monocouches à base de liants hydrauliques. UNITECNA, France, Paris, 129p

Trexler JC, Travis J (1993) Nontraditional regression analysis. Ecology 74(6):1629–1637

Uchwat C, MacLeod D (2012) Case studies of regression and Markov chain models. In: Conference of the transportation Association of Canada, New Brunswick, Canada, Session: pavement performance case studies (SES), 19p

Veiga MR (2000) Methodology to evaluate the cracking susceptibility of mortars. Selection criteria of rendering and repointing mortars for ancient buildings. Seminar "Malte a vista com sabie locali nella conservazione degli edifici storici", Torino, Italy, Politecnico di Torino, July, pp 1–16

Wald A (1941) Asymptotically most powerful tests of statistical hypotheses. Ann Math Stat 12 (1):1–19

Wald A (1943) Tests of statistical hypotheses concerning several parameters when the number of observations is large. Trans Am Math Soc 54(3):426–482

Wang Y (2005) A multinomial logistic regression modeling approach for anomaly intrusion detection. Comput Secur 24(8):662–674

Westergren A, Karlsson S, Andersson P, Ohlsson O, Hallberg IR (2001) Eating difficulties, need for assisted eating, nutritional status and pressure ulcers in patients admitted for stroke rehabilitation. J Clin Nurs 10(2):257–269

White JL (2013) Logistic regression model effectiveness: proportional chance criteria and proportional reduction in error. J Contemp Res Educ 2(1):4–10

Witten IH, Frank E (2005) Data mining: Practical machine learning tools and techniques, 2nd edn. Morgan Kaufman, Boston

Wuensch KL (2011) Binary logistic regression. East Carolina University, Department of Psychology

Xie XJ, Pendergast J, Clarke W (2008) Increasing the power: a practical approach to goodness-of-fit test for logistic regression models with continuous predictors. Comput Stat Data Anal 52(5):2703–2713

Yokota H, Furuya K, Hashimoto K, Hanada S (2012) Reliability of deterioration prediction with Markov model for mooring facilities. In: IALCCE—International Association for Life-Cycle Civil Engineering, 3rd international symposium on life-cycle civil engineering, Vienna, Austria, pp 523–530, 3–6 Oct 2012

Chapter 5
Computational Models

5.1 Artificial Neural Networks

5.1.1 Theoretical Foundations

The concept of intelligent machines that emulate human behaviour is not new. Artificial neural networks (ANNs) resulted from a combination between mathematical models and the study of neurological phenomena (Abraham 2002). In 1943, McCulloh and Pitts addressed the mathematical modelling of a neuronal network with a binary nature, which assumed that the formal model of a neuron followed a law of "all or nothing". The authors demonstrated that a network containing an appropriate number of these simple units with well adjusted synaptic connections would, in principle, be able to calculate any computable function (Haykin 1999). The model of neural network proposed by McCulloh–Pitts was able to process data but unable to learn from them. Thereafter, several studies have emerged in the field of ANNs, among which the studies of Hebb (1949), Minsky (1954), Von Neumann (1958) and McClelland et al. (1986) stand out. However, it was the study of Rosenblatt (1958) that gave an important impulse to this area, creating the "perceptron": a network completely connected, powered forward, able to learn and generalize from given patterns by changing the weights of the connections. The network proposed by Rosenblatt had two significant limitations though: the first one related with the assumptions made and the second one with the resources available at that time (McClelland 2009). Only 25 years later Rumelhart et al. (1986) were able to overcome these two limitations.

In the past decades, ANNs had been successfully used to solve complex problems in various application fields, such as: classification (e.g. pattern recognition), function approximation (e.g. time series analysis), optimisation (e.g. constraint

A. Silva et al., *Methodologies for Service Life Prediction of Buildings*,
Green Energy and Technology, DOI 10.1007/978-3-319-33290-1_5

satisfaction), data processing (e.g. data mining) and control systems (e.g. robotics) (Haykin 1999). The adaptive capacity of learning with ANNs (they learn the behaviour of a given phenomenon directly from data) enables complex problem-solving that would be hard to deal with analytically (Rafiq et al. 2001). ANNs benefit from special characteristics such as low error sensitivity, massive parallel processing, distributed information storage, generalization capability and adaptability to new data (Schalkoff 1997; Silva et al. 2011).

ANNs have been widely used in different areas of science and engineering (Suhartono et al. 2006; Wang and Xu 2010) and they have proven to be a powerful and efficient tool that can be used in particular to solve problems related with civil engineering and construction (Moselhi et al. 1991; Garret 1994; Chao and Skibniewski 1998; Leung et al. 2001). In recent years, neural networks have been applied in service life prediction of building components (Usman and Resdiansyah 2008) and new materials used in Civil Engineering (Freitag et al. 2009), and in decision support of maintenance operations (Moselhi et al. 1993).

5.1.1.1 Description

Generally, ANNs are emulations of biological neural systems (Rezeki et al. 2006) and are inspired in the basic mechanisms of the functioning of the human brain, gathering information through a learning process (Kazanasmaz et al. 2009) and, for this reason, its fundamental processing element is usually called neuron. Haykin (1999) argues that a neural network is a massive parallel distributed processor consisting of simple processing units which have a natural tendency to store experimental knowledge and make it available to be used. ANNs are thus typically composed by hundreds of simple processing units—neurons—linked together by a complex network of communication (Lippmann 1987; Awodele and Jegede 2009). Essentially a neuron receives inputs, (through its synapses), combines them in some way (involving the synaptic weights), executes a generally non-linear operation on the result and provides this result as its output (Fig. 5.1). Usually the output of a neuron feeds several synapses, each of which establishes a connection with another neuron within the network.

The multilayer perceptron (MLP) is the most common type of ANNs (Fig. 5.2) and consists in a network with three or more layers of neurons. Typically, a MLP comprises an input layer, one or more intermediate or hidden layers and an output layer. Nowadays, it is accepted that ANNs of the MLP type can approximate any measurable function with a given arbitrary accuracy (Hornik et al. 1989; Johannet et al. 2007).

MLPs are fully connected feedforward networks: each neuron in each layer is linked to all neurons in the next layer and to no others within the network (Amirikian 1999). It is usual for the neurons to have an extra input (called bias)

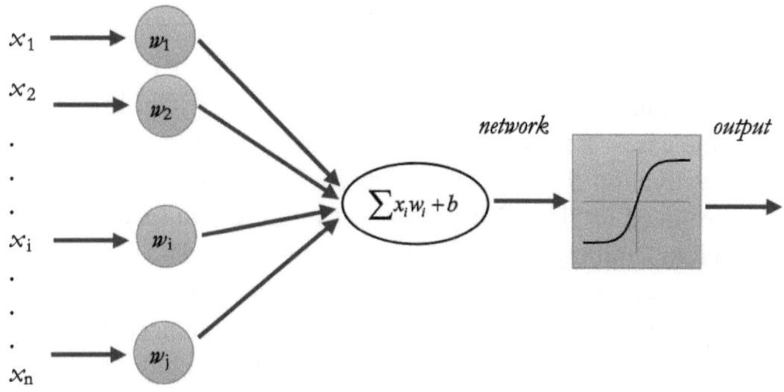

Fig. 5.1 Typical artificial neuron

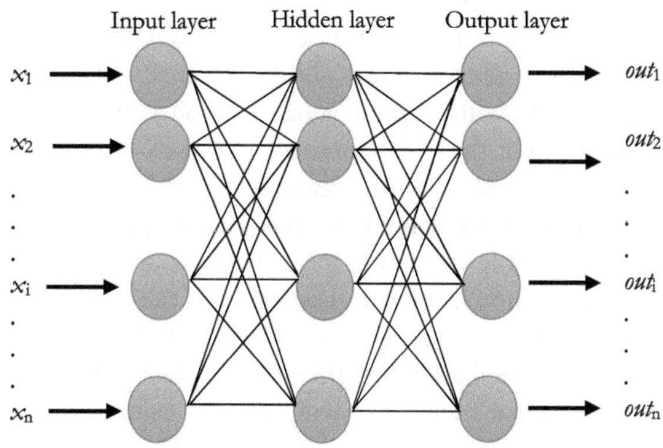

Fig. 5.2 Typical multilayer perceptron

unrelated to the other neurons, which normally improves the network's performance (Dias et al. 2014). Haykin (1999) describes the operation of a typical MLP, referring that the input layer acquires the network inputs (a single input per neuron) and distributes them to the hidden layer; in the other layers neurons get input data from their synapses and combine them through Eq. (5.1).

$$\text{net}_j = \sum_{i=1}^{n} (w_{ij} x_i) + b_j \tag{5.1}$$

where net_j is the global input for the jth neuron of the layer, n is the number of neurons in the preceding layer, w_{ij} is the weight of the connection (synapse) to the ith neuron of the preceding layer, x_i is the output of the ith neuron of the preceding layer and b_j is the bias of the neuron (Topçu and Sarıdemir 2008). The output of the neuron, out_j, is then determined through an activation function, generally a sigmoid function (Liu 2002), such as a hyperbolic tangent, as provided, for example, in Eq. (5.2).

$$\text{out}_j = \tanh(\beta\,\text{net}_j) \tag{5.2}$$

where β is a constant used to control the gradient of the semi-linear region of the function. In some cases, it may be convenient for the neurons of the output layer to use a different activation function (sometimes even linear) (Silva et al. 2011).

MLPs are designed to "learn" from a set of input and output patterns belonging to a given problem (training sample). The training of MLPs is an iterative process that consists of presenting a representative set of patterns to a version of the network and determining the errors associated with the differences between the actual outputs and the target values (Silva et al. 2011). The weights in the network are computationally adjusted until the desired input–output mapping occurs (Gardner and Dorling 1998). However, excessively complex networks, where the learning process is based only on the minimization of the errors of the training set, may suffer from a problem of overfitting of synaptic weights (Costa et al. 2003), as happens when the neural network memorizes the patterns of the training set, thus losing their generalization ability. To ensure that the training of the network ceases in a model with a good generalization capability, a cross-validation (CV) set is usually used; the CV set can also be used to monitor the evolution of the errors in patterns not used during training (Stone 1974). Once training is concluded, the network is tested using new data (test sample) from the same problem. The network should be able to predict the behaviour of these new data (seen for the first time) based on knowledge acquired through the learning process (Boussabaine 1996).

The typical learning algorithm for MLPs is a backpropagation algorithm (BPA) (Rumelhart et al. 1986; Cigizoglu and Kişi 2005). The BPA intends to minimize the error function in weight space using the method of gradient descent (Rojas 1996). The BPA established the use of MLPs to solve problems with practical interest (Kamarthi and Pittner 1999). Additionally, the MLPs with BPA become a powerful alternative to model arbitrary data, such as the data collected in the field work, used in this book to model the degradation of claddings' over time. The output of any neuron is determined by their weights (bias included). Many alternatives to BPA have been developed over the years, but its advantages usually lie more in the training swiftness than the generalization capability (Dias and Silvestre 2011).

The choice of an adequate MLP to represent a given mapping $R^n \rightarrow R^m$ generally implies a trial and error selection process involving various network architectures (Silva et al. 2011). The difference in architectures normally concerns the number of neurons in each layer. Generally, the number of neurons in input and output layers is established by the problem: the input layer of ANNs has a number of neurons equal to the number of explanatory or independent variable and the output layer has one neuron (dependent or desired variable). Regarding the number of neurons in hidden layer(s), there are no established rules to define these numbers (Ayat et al. 2013). Usually an hidden layer has the same number of neurons as the input layer (this value may vary between half and twice the neurons of the input layer). In fact, Lawrence (1994) refers that normally a network containing few hidden neurons generalizes better than other with more neurons. Although a MLP may have more than one hidden layer, theoretical studies showed that more than one hidden layer is rarely necessary (Cybenco 1989; Hornik et al. 1989).

Regardless of the architecture used, there are two key parameters in the use of neural network models: (i) the selection of the appropriate variables; (ii) the quality of the sample—it is important to ensure the acquisition of a sufficient amount of training examples to allow the neural network to correctly model the problem under analysis (Zahedi 1996; Walczak 2001). To create a ANN model, the first step is to collect a representative data set in the shape of patterns, i.e. pairs $\{x^p, t^p\}$, where x^p and t^p are vectors with, respectively, n input values and m target values for pattern p. The minimum amount of data to be collected is determined by empirical rules. There are no studies that indicate the size of the optimal sample. Atkinson and Tatnall (1997) refer that the training of neural networks requires a large amount of data and that researchers are often concerned with finding the minimum size of data set necessary (Hepner et al. 1990). In fact, the size of training sets varies considerably according to the researchers (Pathak and Dikshit 2005). ANNs react to data modification, so that their ability to generalize increases, in principle, with the size of available data. The sample size and its relevance and representativeness are extremely important in ANNs-based models. It is important to provide the network with a representative data set, ensuring that the network can interpolate for other cases within the population. If the data are inadequate, the network may fail in the prediction of the behaviour of new data. Besides that, it is important to be aware that the data only represent a sample from the entire population (some of the characteristics of the entire population remain unknown) (Olden et al. 2004). ANNs can be put to use even with limited amount of initial data and later be adapted (through learning processes) to incorporate more information.

The selection of input variables remains a key issue in the definition of ANNs' models (Hill et al. 1996; Zhang et al. 1998; Kourentzes and Crane 2008). ANNs have been labelled as "black box" models since they do not provide information on the contributions of the independent variables in the prediction process (the true importance of the variables is unknown) (Olden et al. 2004). Tetko et al. (1996) demonstrated that the selection of the model variables can be ascertained directly through ANNs, without a previous knowledge of the model (for example, through a sensitivity analysis). However, it was found that this procedure would consume too much time. For this reason, some authors (Basak et al. 2000; Beck et al. 2000) have stated that the most efficient way to get the set of variables to be included in the ANNs' model is through the stepwise method (multiple linear regression). The stepwise method is an effective method that allows obtaining the set of significant variables to explain the desired model; however, this tool does not take into account non-linear relationships between variables (González-Arjona et al. 2002). The identification of significant inputs (or variables) allows obtaining more accurate results of the network, reducing the procession time; furthermore, less data are required to obtain an adequate performance of the network (Sung 1998).

Typically, the process of conceiving and evaluating a specific MPL architecture can be divided into three steps: (i) divide the data (according to empirical rules) into three parts: training, cross-validation and testing; (ii) select the MLP architecture (number of hidden layers and of neurons in each layer)—architectures are usually selected by trial and error; (iii) initialization of the network's parameters (weights and biases). The learning process is composed of two steps (Silva et al. 2011): (i) in the first one, a learning algorithm (e.g. BPA) is used to change the network parameters based on the errors with the training set, thus reducing the output errors; this stage evolves through several epochs, i.e. through several presentations of all patterns of the training set to the learning algorithm; (ii) in the second step, the performance of the network version at the end of stage one is measured based on its errors with the CV set. This stage serves only to evaluate the network generalization capability. The learning iterative process goes on until there is a reduction of the network performance with the CV set; at this point learning ceases and the network is tested (with the test set) to evaluate its performance. The whole process is repeated until the best architecture is achieved.

5.1.2 Application to Façade Claddings

In the last decades, ANNs have become an extremely useful and versatile computational tool, being commonly accepted to model complex real-world problems (Basheer and Hajmeer 2000). There are various studies where structural degradation of buildings and their elements is modelled using ANNs' models (Szewczyk and Hajela 1994; Worden 1997; Chang et al. 2000). Friswell et al. (1998) reported that ANNs can easily model the degradation phenomena, since they exhibit non-linear behaviour; however, the author states that, for ANNs to be applied successfully, it is necessary to provide information on essential features of the component under analysis, e.g. in stone wall claddings it is important to analyze the influence of the type of stone in the façade's degradation condition. Therefore, it seems viable to use ANNs' models in service life prediction of the façades coatings.

In this chapter, ANNs are used to model the coatings' degradation, yielding the equivalent mathematical formula that also allows the estimation of the expected service life of each case study within the sample.

5.1.2.1 Variables Selection

The selection of variables may be performed in two ways. The first one is based on a multiple linear regression (MLR) (discussed in Chap. 4), which allows identifying the explanatory variables to include in the model for each type of exterior cladding, giving a priori knowledge of which variables should be included in the model produced by ANNs. The second way is based on the direct use of the ANNs in the identification of the significant variables; a sensitivity analysis is thus performed to determine which variables best describe the degradation severity of the façades analyzed. Thus, it is possible to analyze whether both methods select the same variables or instead propose different variables. In some occasions, for small samples, the sensitivity analysis is unable to clearly identify the significant variables. In fact, while some variables are unquestionably relevant, others may be so or not, leading to the suspicion that the introduction of new cases could alter the relevance of introducing these variables in the model. However, in this study, and concerning the size of the samples used, it is chosen not to introduce in the proposed models too many variables, which would make the model too complex and lead to a decrease of its statistical significance.

5.1.2.2 ANNs Architectures Selection

In this study, a MLP is used to develop a formula to approximately estimate the degradation severity of the façades analyzed. For the definition of MLP a NeuroSolutions software was used (Lefebvre et al. 2003). In all MLPs created, the activation functions were the hyperbolic tangent for the intermediate layer neurons, and the linear function for the output layer neurons. All channels (the inputs and the output) were normalized so that all corresponding values would be within the range [−0.9, 0.9]. Various architectures were tested for the networks and 10 runs were performed per architecture. In this book only the architecture with the best overall result is shown. The MLPs created include only one hidden layer, which means an architecture *N*-*H*-1, where *N* is the number of cells in the input layer and *H* the number of cells in the hidden layer. The BPA was used for training. The mean square error (MSE) (for both training and cross-validation) was used as stoppage criterion. In each run the cross-validation error was evaluated after each 50 epochs and training was concluded as soon as this error ceased to decrease or when a maximum of 65,000 epochs was reached (typically many epochs are required to train the neural network). The best run was selected for each architecture (according to the cross-validation criterion) as being representative, after which the architecture was tested for the following statistical parameters: mean percentage error normalized with respect to the maximum ($\bar{\varepsilon}$), normalized mean square error (NMSE), and Pearson correlation coefficient (r). Considering a test set with P patterns, where for each pattern p, t_p is its target value, and z_p is the actual network output (corrected to consider the negative values as zero), and where t_{max} is the greatest target value within the set, the statistical parameters referred to above are defined by Eqs. (5.3) to (5.7) (Silva et al. 2011).

$$\varepsilon_p = \left| \frac{z_p - t_p}{t_{max}} \right| \cdot 100\,\% \tag{5.3}$$

$$\bar{\varepsilon} = \frac{1}{P} \sum_{p=1}^{P} \varepsilon_p \tag{5.4}$$

$$\text{MSE} = \sqrt{\frac{1}{P} \sum_{p=1}^{P} (z_p - t_p)^2} \tag{5.5}$$

$$\text{NMSE} = \frac{\text{MSE}}{\bar{t}}, \quad \bar{t} = \frac{1}{P}\sum_{p=1}^{P} t_\text{p} \tag{5.6}$$

$$r = \frac{\sum_p (z_\text{p} - \bar{z})(t_\text{p} - \bar{t})}{\sqrt{\sum_p (z_\text{p} - \bar{z})^2 \sum_p (t_\text{p} - \bar{t})^2}}, \quad \bar{z} = \frac{1}{P}\sum_{p=1}^{P} z_\text{p} \tag{5.7}$$

Furthermore, for clarity's sake (Dias and Silvestre 2011; Silva et al. 2011), the best runs were also tested for the maximum normalized percentage error ($\varepsilon_{\text{max.}}$) and for the percentage of patterns where ε is greater than x = 5 %, x = 10 %, x = 20 % and x = 30 % ($PP_{\varepsilon>x}$).

Several authors (Bebis and Georgiopoulos 1994; Bebis et al. 1997; Rezeki et al. 2006) stated that one of the most critical problems in the definition of an ANNs' model is the selection of the size of the network so that it is suitable for the intended application. In this study, once the different architectures are tested, the selection of the best one is made by comparing the statistical parameters mentioned, also taking into consideration the parsimony principle, i.e. in the presence of two networks with similar performance, the network with fewest elements is chosen.

Once these procedures are completed, the explicit formulation of the degradation severity is obtained directly from the selected MLP using their activation functions, their parameters (weights and biases) and their normalization factors. The formulation is simply a closed form expression for MLP (i.e. that can be analytically expressed through a certain finite number of well known functions).

Once the assumptions of the models are defined, it is possible to apply ANNs in the service life prediction of external claddings.

5.1.2.3 Natural Stone Claddings

In natural stone claddings, in all runs, the global set of patterns was divided into three groups: training (60 % of the sample—84 case studies), cross-validation (15 %—21 case studies) and test (25 %—35 case studies). According to the MLR performed in Chap. 2 of this book, the variables with more statistical relevance in the explanation of the degradation severity of stone claddings are age, distance from the sea, type of finish and size of the stone plates. A sensitivity analysis was performed to determine the variables that best describe the degradation severity; this analysis shows that the four variables listed before are the most relevant and no others show statistical significance that justifies their addition to the model.

To use ANNs' models it is necessary to codify the variables: age and size of the stone plates are numerical and their real values were used; the other two variables (type of stone finish and distance from the sea) have only two possible arbitrary "values" according to the data collected during the field work that are: (i) type of stone finish—smooth stone finish, value 1; rough stone finish, value −1; (ii) distance of the building from the sea—less than 5 km, value 1; otherwise, value −1.

Since four variables in the input layer were used, the MLP produced has a 4-4-1 architecture.

The degradation severity (S_w) is a function of the type of finish, distance from the sea, size of the stone plates and age of the stone cladding, as seen in Eqs. (5.8) and (5.9).

$$S_w = h_0 + \sum_{i=1}^{4} h_i H_i \tag{5.8}$$

$$H_i = \tanh(c_{i0} + c_{i1} f + c_{i2} d + c_{i3} s + c_{i4} a) \tag{5.9}$$

where variables f, d, s and a are the type of finish, distance from the sea, size of the stone plates and age of the stone cladding, respectively; coefficients h_0 to h_4 and c_{i0} to c_{i4} are listed in Table 5.1.

For clarity's sake, an example of how to calculate the degradation severity of a case study is given in Table 5.2. This example shows a case study with 5 years, with a size of the stone plates equal to 0.4 m^2, located less than 5 km from the sea and with rough finish. The degradation severity observed in field work is equal to 0.015 (1.5 %), similar to the value estimated by ANNs' model (0.0168). It is possible to use the same procedure for the remaining cases within the test sample.

To analyze the applicability of the model, the statistical parameters presented before were used. The mean absolute percentage error ($\overline{\varepsilon}$) and the maximum absolute percentage error ($\varepsilon_{max.}$) are relatively small: 11.04 and 37.55 %, respectively. Analyzing the percentage of patterns with ε greater than $x = 5$ %, $x = 10$ %, $x = 20$ % and $x = 30$ % ($PP_{\varepsilon > x}$), it is found that the model proposed presents error percentages relatively low (57.1, 48.6, 17.1 and 8.6 %, respectively). In fact, the model obtained leads to relatively small errors and shows a relatively high Pearson correlation coefficient ($r = 0.783$). Pearson correlation coefficient (r) evaluates the linear dependency between the degradation severity values from the sample and the corresponding values predicted by each model, showing a strong correlation between the values observed and predicted by the ANNs' model (Piaw 2006). These results indicate that the model is statistically valid and can describe the reality observed in field work.

After proving the statistical validity of the model, the values of the degradation severity can be used to determine the estimated service life of each case study within the test sample (corresponding to 35 case studies). For this, it was stipulated

Table 5.1 Coefficients of the proposed formula (data sourced from Silva et al. 2011)

i	h_i (−)	c_{i0} (−)	c_{i1} (−)	c_{i2} (−)	c_{i3} (m^{-2})	c_{i4} ($year^{-1}$)
0	1.368E−01					
1	−7.461E−02	1.040E+00	2.079E−01	−2.551E−01	−1.675E−01	−1.294E−02
2	−6.571E−02	5.543E−02	−1.161E−01	−1.838E−01	−3.709E−01	−6.621E−03
3	8.618E−02	7.720E−01	5.671E−02	−2.963E−01	−1.034E+00	4.278E−03
4	−1.190E−01	2.495E+00	1.620E−01	−2.387E−01	−1.596E+00	−2.333E−02

Table 5.2 Example of the calculation of the degradation severity using the ANNs model for a case study

Case study	Calculus
	$H_1 = \tanh(1.04 + 0.208 \cdot -1 - 0.255 \cdot 1 - 0.168 \cdot 0.4 - 0.013 \cdot 5) \approx 0.418$ $H_2 = \tanh(0.055 + 0.116 \cdot -1 - 0.184 \cdot 1 - 0.371 \cdot 0.4 - 0.007 \cdot 5) \approx -0.191$ $H_3 = \tanh(0.772 + 0.057 \cdot -1 - 0.296 \cdot 1 - 1.034 \cdot 0.4 - 0.004 \cdot 5) \approx 0.027$ $H_4 = \tanh(2.495 + 0.162 \cdot -1 - 0.239 \cdot 1 - 1.596 \cdot 0.4 - 0.023 \cdot 5) \approx 0.871$ $S_w = 0.1368 - 0.075 \cdot 0.418 - 0.066 \cdot -0.191 + 0.086 \cdot 0.027 - 0.119 \cdot 0.871 \approx 0.0168$

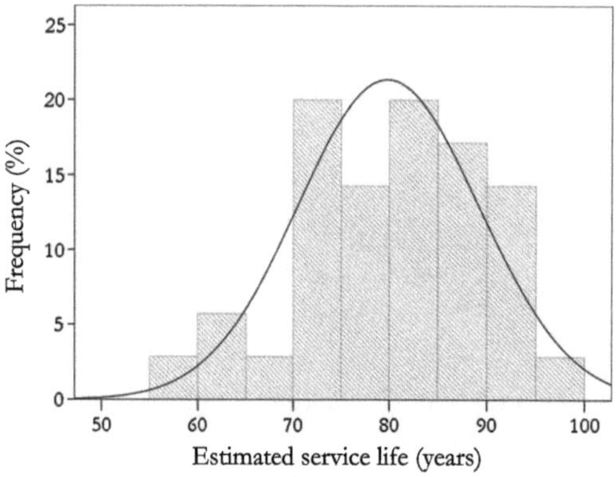

Fig. 5.3 Histogram of the estimated service life for ANN-based model

that a coating with a degradation severity level of 20 % has reached the end of its service life. Therefore, and using the proposed mathematical model, the point in time (age of the coating since its last major repair) when the degradation severity reaches 20 % was determined for each case study. In Fig. 5.3 the histogram of the estimated service life for ANN-based model is presented. A normal distribution was adapted to this histogram. The average estimated service life obtained by ANNs' model is equal to 79.81 years, with a standard deviation of 9.34 years. A study performed by Shohet and Paciuk (2004) about service life prediction of claddings subjected to normal conditions leads to an estimated service life for stone claddings (directly adhered to the substrate) which equals to 64 years (with a possible range of 59–70 years). These values apply only to normal conditions and it is expected that the scope of results may broaden if substandard and exceptional conditions are to be considered. The value obtained in the present study is close to the values estimated by other authors and it seems realistic taking into account the claddings' characteristics. It is found that the service life estimated by the model lies within a range between 58.81 and 95.77 years. The minimum estimated service life corresponds to the worst case scenario, i.e. a cladding located less than 5 km from the sea, with rough finish and a large size of stone plates (about 0.4 m^2). On the other hand, the maximum estimated service life corresponds to a case study with favourable conditions, i.e. located more than 5 km from the sea, with smooth finish and a small size of the stone plates (approximately 0.25 m^2). These results are consistent with the empirical feedback from real life case studies. A study performed by Silva et al. (2012) suggest that larger plates ($A \geq 0.4$ m^2) reach the end of their service life sooner, especially when compared with medium sized plates ($A < 0.4$ m^2), for three main reasons: (i) higher concentration of stresses in bigger plates; (ii) larger plates are thicker/heavier per unit area of façade; (iii) the larger the contact area of each plate the more difficult it is to guarantee that the adhesive material is evenly spread on its back.

Figure 5.4 shows an analysis of the average estimated service life according to the variables included in the ANNs-based model. Once again, the consistency of the

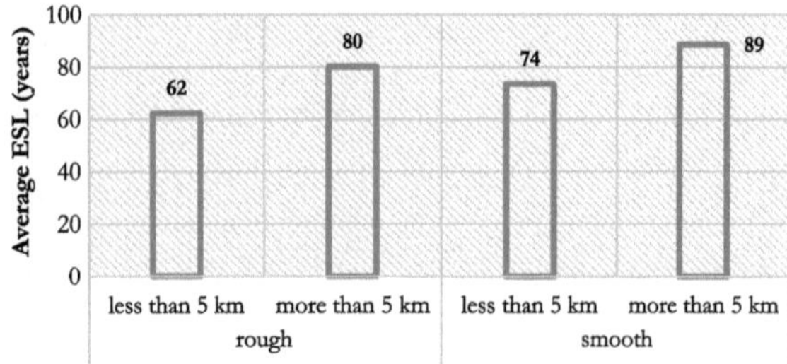

Fig. 5.4 Analysis of the mean estimated service life according to the variables included in the ANNs-based model

model is verified and it is possible to conclude that: (i) claddings with smooth finish are more durable than claddings with rough finish, at the same distance from the sea; (ii) claddings located closer to the coast are more prone to degradation, with shorter service lives, for the same type of finish. Analyzing the results from a physical point of view, the model seems to be able to portray reality; it is easily understandable that stone claddings located near the coast are subjected to strong winds that carry moisture and soluble salts (Rivas et al. 2010), which drastically accelerate their deterioration. Concerning the type of finish, rough finish is more prone to degradation, which may be due to the fact that this type of finish has a greater area of stone exposed to adverse atmospheric conditions.

5.1.2.4 Rendered Façades

In rendered façades, it was the same procedure as in stone claddings. The first step is to perform a sensitivity analysis in order to define the variables that should be included in the model. In the sensitivity analysis, the age of the render and its type are unquestionably relevant to explain the degradation of rendered façades. Furthermore, exposure to damp and the façade's protection level were found to be relevant to explain the degradation severity. These variables were also chosen in the MLR analysis performed in Chap. 2 of this book. An 11-4-1 architecture for the model was adopted. The 11 entries are: age of the render, type of render (with four entries: marble agglomerate, monomass, cement and lime mortar and cementitious mortar), exposure to damp (with three entries: favourable, normal and unfavourable exposure), and façade protection level (with three entries: poor, average and good protection). In all runs the global set of patterns was divided into three groups: training (60 % of the sample—60 case studies), cross-validation (15 %—15 case studies) and test (25 %—25 case studies).

In the ANNs-based model proposed the degradation severity was a function of the render type, exposure to damp, the façade's protection level and render age, as seen in Eqs. (5.10) and (5.11).

$$S_w = h_0 + \sum_{i=1}^{4} h_i H_i \tag{5.10}$$

$$H_i = \tanh\left(c_{i0} + \sum_{n=1}^{11} c_{in} V_n \right) \quad (5.11)$$

where V_1 represents age of the render, V_2 marble agglomerate, V_3 monomass, V_4 cement and lime mortar, V_5 cementitious mortar, V_6 a favourable exposure, V_7 an unfavourable exposure, V_8 a normal exposure, V_9 a poor façade's protection, V_{10} a good façade's protection and V_{11} an average façade's protection. The coefficients h_0 to h_4 and c_{i0} a c_{i11} are presented in Table 5.3.

The model proposed was applied to the test sample (25 case studies) to determine the predicted degradation severity. To evaluate the efficiency of the model, the predicted values were compared with the values obtained through the field survey and the errors between predicted and observed values were estimated. The errors obtained were relatively low; the maximum absolute percentage error ($\varepsilon_{max.}$) were about 28 % and the mean absolute percentage error ($\bar{\varepsilon}$) were equal to 5.3 %. The percentage of patterns with ε greater than $x = 5$ %, $x = 10$ %, $x = 20$ % and $x = 30$ % ($PP_{\varepsilon > x}$) were, respectively, 32, 16, 4 and 0 %. These results demonstrate that the model leads to small errors. Furthermore, the Pearson correlation coefficient obtained is certainly high ($r = 0.948$), showing a very strong correlation between the values observed and predicted by the ANNs' model.

Using the ANNs-based model it is possible to predict the estimated service life for rendered façades. In Fig. 5.5 the histogram of the estimated service life for the rendered façades and the normal distribution curve associated are presented. The average estimated service life is 17.50 years, with a standard deviation of 2.74 years. The estimated service life of rendered façades ranges between 12 and 23 years and with 16 years occurring most often. These results are coherent with the values obtained previously by other authors: Gaspar (2009) predicted a service life of 15–20 years for rendered façades for a maximum acceptable degradation level between 20 and 30 % (using a graphical method); Shohet and Paciuk (2004) obtained an estimated service life of 23 years (with a range of 19–27 years) for a lower level of requirement. A service life of 12 years applies to situations of cement and lime mortar with severe exposure to damp and medium façade's protection (strong exposure to the environmental agents). On the other hand, a service life of 22 years concerns a marble agglomerate mortar with favourable exposure to damp and good façade's protection.

Figure 5.6 shows an analysis of the average estimated service life according to the variables included in the ANNs-based model. It seems that the results are logical and physically adjusted to the reality observed in the field work. When subjected to the same conditions, the façades with unfavourable exposure to damp have a lower estimated service life than the ones with favourable exposure, as expected. Regarding the façade's protection level, the trend is also as expected, i.e. for the same render's characteristics a façade with good protection has higher values than the other situations. As for the render type, it is found that under the same exposure conditions, the marble agglomerate mortars have the highest service lives, followed by single-layer and cementitious mortars, while cement and lime mortars have the lowest service lives.

Table 5.3 Coefficients of the proposed formula (data sourced from Silva et al. 2013)

i	h_i (−)	c_{i0} (−)	c_{i1} ($year^{-1}$)	c_{i2} (−)	c_{i3} (−)	c_{i4} (−)	c_{i5} (−)	c_{i6} (−)	c_{i7} (−)	c_{i8} (−)	c_{i9} (−)	c_{i10} (−)	c_{i11} (−)
0	5.522E−1												
1	−1.633E−3	6.595E−1	6.829E−3	−1.451E−2	−3.665E−1	5.274E−1	−4.389E−1	−4.684E−1	7.275E−1	−5.817E−1	−2.956E−1	−7.405E−1	−2.010E−1
2	2.456E−1	−6.857	6.033E−2	2.119	1.695	2.129	−1.065	−1.183	1.730	1.219	6.100E−1	−4.122E−2	1.188
3	1.131E−1	−2.096	2.590E−2	1.167E−1	−3.695E−3	5.269E−1	−8.988E−2	6.885E−1	1.058E−2	3.334E−1	4.102E−2	1.138	5.284E−1
4	−2.683E−1	1.012	−5.563E−2	4.303E−1	2.145E−1	2.586E−1	1.545E−1	−6.401E−2	−3.298E−1	−1.451E−1	−1.417E−1	4.584E−1	3.639E−2

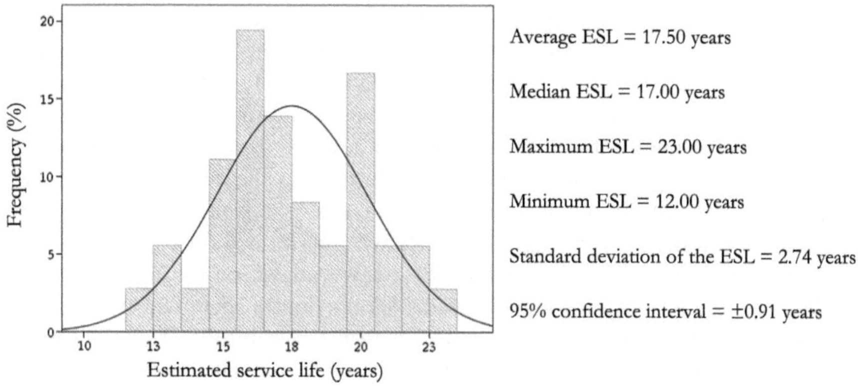

Fig. 5.5 Histogram of the estimated service life for ANNs-based model (data sourced from Silva et al. 2013)

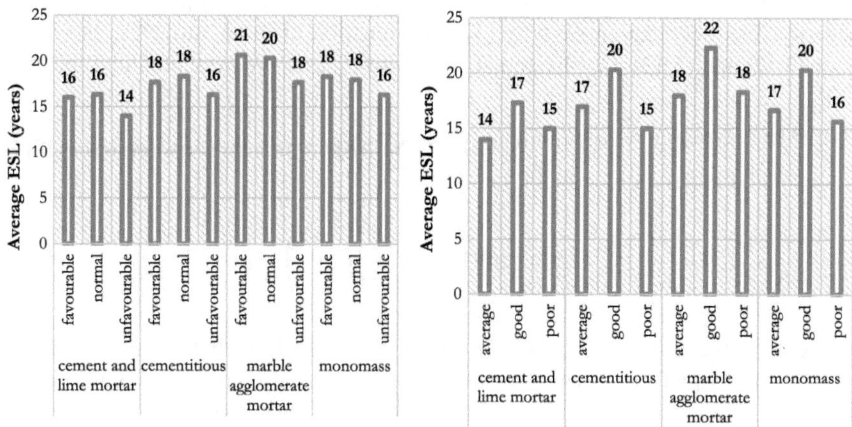

Fig. 5.6 Analysis of the mean estimated service life according to the variables included in the ANNs-based model

5.1.2.5 Painted Surfaces

Concerning painted surfaces, various architectures were tested for the networks and three runs were performed per architecture. In all runs the global set of patterns was divided into two groups: training (85 % of the sample—187 case studies) and cross-validation (15 %—33 case studies). The samples used for cross-validation were also used to test the networks. Like in the other claddings analyzed, in this case a sensitivity analysis was also performed to determine the variables (degradation factors) that best describe the degradation severity of painted façades. ANNs cannot be executed with symbolic (i.e. qualitative) values in the input layer. Therefore, to perform this analysis it was necessary to quantify the qualitative variables. Since the variable age is numerical their real values were used. The other variables were split into binary variables. For example, the variable orientation has four possible outcomes: North, East, South and West. Therefore, four new variables were created, each of which has two possible values: 1, if the characteristic is present, or 0, in the opposite case. The MLR analysis performed in Chap. 2 of this book shows that the most influential factors to describe the degradation of painted surfaces are the age of the façade, its distance from the sea and the façade's orientation. In the sensitivity analysis performed, a new variable (added one by one) is introduced in the network, and three runs were performed for each model, thus analyzing the influence of the addition of a new variable on the accuracy of the model. This analysis shows that the variables previously identified by MLR are definitely the ones that lead to the model with the best results. Thus, the model proposed in this study includes only the variables age of painted surfaces, façade's orientation and distance from the sea. In this model, an 8-4-1 architecture was adopted. The eight entries are: age of painted surfaces, distance from the sea (three entries: more than 5 km, less than 5 km, less than 1 km) and façades orientation (with four entries: North, South, East and West), and the output is degradation severity. The degradation severity (S_w) is a function of these variables, as seen in Eqs. (5.12) and (5.13). Coefficients h_0 to h_4 and c_{0i} a c_{8i} are presented in Table 5.4.

$$S_w = h_0 + \sum_{i=1}^{4} h_i H_i \tag{5.12}$$

$$H_i = \tanh\left(c_{0i} + \sum_{n=1}^{8} c_{ni} V_n \right) \tag{5.13}$$

where V_1 represents a façade facing north, V_2 a façade facing east, V_3 a façade facing south, V_4 a façade facing west, V_5 a building located less than 1 km from the sea, V_6 a building located more than 1 km and less than 5 km from the sea, V_7 a building located more than 5 km from the sea and V_8 the age of the painted surfaces.

The mean absolute percentage error ($\overline{\varepsilon}$) and the maximum absolute percentage error ($\varepsilon_{max.}$) are very small: 5.3 and 15.79 %, respectively. Analyzing the percentage of patterns with ε greater than $x = 5$ %, $x = 10$ %, $x = 20$ % and $x = 30$ % ($PP_{\varepsilon > x}$), it is found that the model proposed presents error percentages relatively low (42.2, 15.2, 0 and 0 %, respectively). Based on the errors obtained, the model seems to be accurate and useful to be used in service life prediction of painted surfaces. The Pearson correlation coefficient ($r = 0.979$) reveals a very strong correlation between the values observed in field work and predicted by the ANNs' model.

Figure 5.7 presents the histogram of the estimated service life for the painted surfaces obtained by the ANNs' model. The average estimated service life obtained is equal to 9.49 years, with a standard deviation of 0.63 years. The results obtained agree with previous studies: based on the same sample and using a graphical method, Chai et al. (2014) predicted a service life of 10 years for painted façades

Table 5.4 Coefficients of the proposed formula (data sourced from Dias et al. 2014)

i	h_i (−)	c_{0i} (−)	c_{1i} (−)	c_{2i} (−)	c_{3i} (−)	c_{4i} (−)	c_{5i} (−)	c_{6i} (−)	c_{7i} (−)	c_{8i} (year^{-1})
0	3.87E−01									
1	−3.84E−02	4.29E−01	7.64E−02	7.48E−01	−6.74E−01	5.90E−01	−1.01E−01	−1.33E−01	−4.63E−01	−5.09E−02
2	1.58E−01	−1.59E+00	−5.08E−01	4.30E−01	−1.72E−01	9.88E−02	−1.07E−01	2.21E−01	−4.60E−01	1.47E−01
3	9.31E−02	−2.39E+00	1.06E+00	−4.67E−01	3.69E−01	8.44E−01	5.63E−01	3.73E−01	4.37E−01	9.34E−02
4	1.67E−01	−3.08E+00	6.35E−01	7.34E−01	1.12E+00	7.14E−01	2.19E−01	−6.87E−01	8.87E−02	1.73E−01

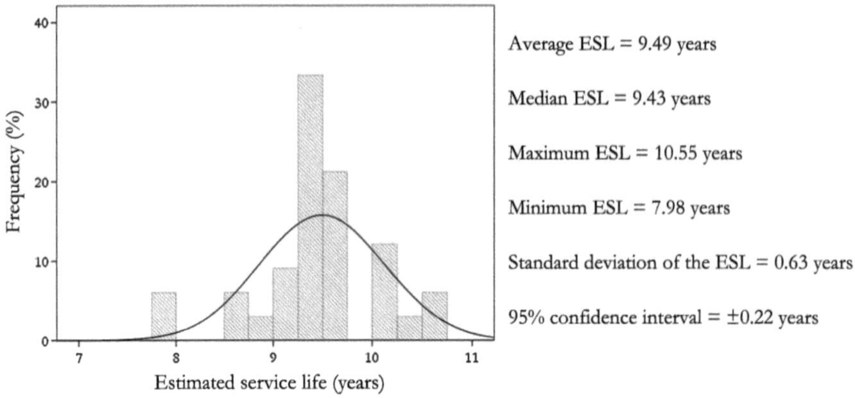

Fig. 5.7 Histogram of the estimated service life for ANNs-based model (data sourced from Dias et al. 2014)

for a maximum degradation level of 20 %. Using a MLR and for the same maximum degradation level, Chai et al. (2015) obtained an estimated service life of 8.5 years with a standard deviation of 0.54 years. The values obtained are consistent with the existing perception relative to the durability of painted coatings.

Figure 5.8 shows an analysis of the average estimated service life according to the variables included in the ANNs-based model. From the analysis of Fig. 5.8 it is possible to notice that, when subjected to the same conditions, façades located less than 1 km from the sea have a lower estimated service life than those located at less than 5 km. The façades located more than 5 km from the sea are generally the most durable ones. These results are coherent from a physical point of view; since it is known that façades near to coastal areas are more prone to degradation, with lower service lives due to the effects of the sea spray in the coatings.

Concerning the orientation of the façades, Fig. 5.8 does not reveal a clear pattern about the relationship between façades' degradation and their orientation.

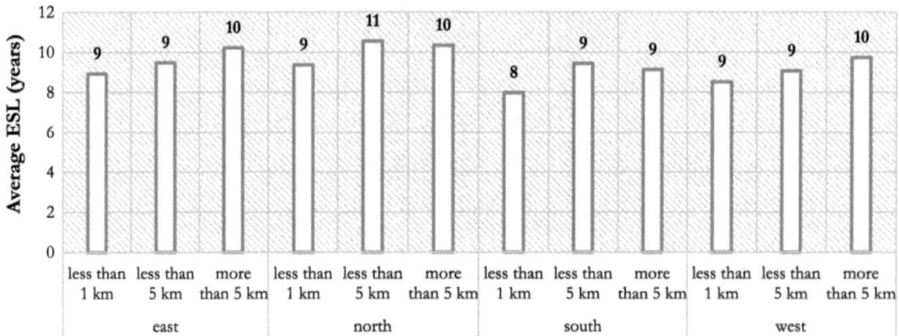

Fig. 5.8 Analysis of the mean estimated service life according to the variables included in the ANNs-based model (data sourced from Dias et al. 2014)

Gaspar (2009) reports that the most aggressive directions in Portugal are usually North, because greater humidity is combined with fewer hours of sunshine, and West, because of strong solar exposure leading to temperatures that may affect the walls. The data obtained partially agree with this pattern. In fact the sample shows that there is generally a higher incidence of cracking, detachment, colour/brightness changes and chalking in façades facing South and West and of biological stains in façades facing North (Chai et al. 2014). In terms of estimated service life as a function of solar orientation, differences less than 1 year are found between South- and North-facing façades. For that reasons, it is not possible to present clear and objective findings relative to solar orientation.

5.1.2.6 Ceramic Tiling Systems

The procedure adopted to model the service life of adhesive ceramic claddings was the same as for painted façades. In all runs the global set of patterns was divided into two groups: training (85 % of the sample—166 case studies) and cross-validation (15 %—29 case studies). The sample used for cross-validation was also used to test the networks. The first step is to determine the significant variables, to be included in the model. The MLR analysis performed in Chap. 2 of this book revealed that the most significant variables to describe the degradation of ceramic tiling systems are the claddings' age, distance from the sea and tiles' size. A sensitivity analysis was also performed but indicates that no other variable should be included in the model. To use ANNs' models it is necessary to codify the categorical variables; the variables distance from the sea and the size of tiles have only two possible arbitrary "values" according to the data collected during the field work, which are: (i) distance of the building from the sea—less than 5 km, value 1; otherwise, value −1; (ii) tiles' size—less than 20 cm in at least one of the sides of the tile, value −1; otherwise, value 1. In this model, a 3-4-1 architecture was adopted. The three entries are: age of ceramic claddings, distance from the sea and tiles' size and the output is degradation severity. The degradation severity (S_w) is a function of these variables, as seen in Eqs. (5.14) and (5.15). Coefficients h_0 to h_4 and c_{0i} a c_{8i} are presented in Table 5.5.

$$S_w = h_0 + \sum_{i=1}^{4} h_i H_i \tag{5.14}$$

$$H_i = \tanh\left(c_{0i} + \sum_{n=1}^{3} c_{ni} V_n\right) \tag{5.15}$$

where V_1 represents the age of the ceramic cladding, V_2 distance from the sea and V_3 tiles' size.

The model proposed was applied to the test sample (29 case studies) to determine the predicted degradation severity. In the evaluation of the accuracy of the

Table 5.5 Coefficients of the proposed formula

i	h_i (−)	c_{0i} (−)	c_{1i} (−)	c_{2i} (−)	c_{3i} (−)
0	2.39E−01				
1	2.58E−02	3.02E−01	3.06E−03	−4.74E−01	3.74E−01
2	9.01E−02	−9.66E−01	2.86E−02	4.72E−01	−3.41E−01
3	1.61E−01	−5.20E+00	1.01E−01	1.62E−01	5.83E−01
4	−7.39E−02	4.15E−01	−7.53E−03	1.69E−01	−2.45E−01

model, the errors between the predicted and the values obtained through the field survey are estimated. The errors obtained are small, with a mean absolute percentage error ($\bar{\varepsilon}$) equal to 5.2 % and a maximum absolute percentage error ($\varepsilon_{max.}$) equal to 17.3 %. The percentage of patterns with ε greater than x = 5 %, x = 10 %, x = 20 %, and x = 30 % ($PP_{\varepsilon > x}$) was respectively 37.9, 17.2, 0 and 0 %. There is a very strong correlation between values observed during the field work and the values predicted by the ANNs' model, with a Pearson correlation coefficient equal to 0.930.

Figure 5.9 presents the histogram of the estimated service life for the rendered façades and the normal distribution curve associated. The average estimated service life is 50.2 years, with a standard deviation of 3.52 years. The estimated service life of rendered façades ranges between 39.8 and 53.5 years, with high prevalence of values between 50 and 55 years (86 % of the sample). These results are coherent with previous studies: a study performed by Galbusera et al. (2014) quantifies the expected service life of external ceramic claddings as 50 years.

Figure 5.10 presents an analysis of the average estimated service life according to the variables included in the ANNs-based model. Once again, the results confirm that claddings located nearer the coast are more prone to degradation, with lower service lives. Regarding the tiles' size, a study performed by Medeiros (2000) shows that one of the most unfavourable conditions for detachment of tiles stems

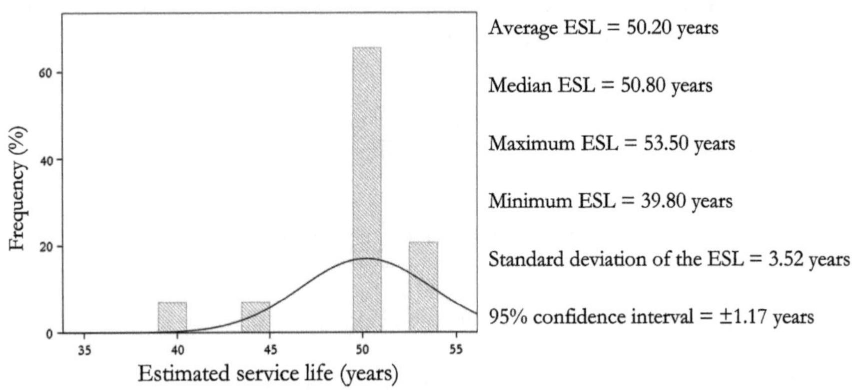

Fig. 5.9 Histogram of the estimated service life for ANN-based model

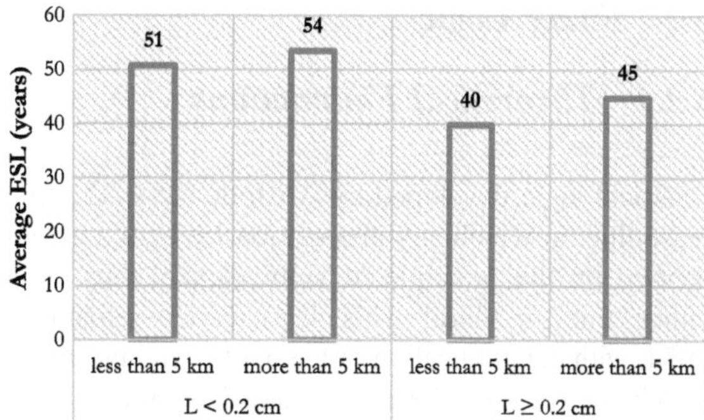

Fig. 5.10 Analysis of the mean estimated service life according to the variables included in the ANNs-based model

from the use of ceramic tiles greater than or equal to 20 cm in any direction. Thus, in this book, the tiles' size was categorized according to the lateral sizes of the tile; if at least one size is greater than 20 cm, then the cladding belongs to category "$L \geq 20$ cm". This study confirms that façades with tiles larger than 20 cm seem to deteriorate faster than the others.

5.1.3 Conclusions

In the last decades, ANNs have been successfully applied in different areas of knowledge. In fact, ANNs are found to have many advantages. They are able to learn from a data set related to a given problem. They use part of the available sample to define the model, another part to validate and adjust it and a third one to test it with case studies that were unknown until then. These methods are thus capable of learning and generalizing based on experience and examples, adapting to new situations, an ability that is extremely important since it means that complex problems can be solved, which are difficult to solve either analytically or numerically.

However, like in any other statistical method, ANNs' models react to data changes and are directly related to the size and quality of data available. Furthermore, the main disadvantages of neural networks are related to the perception that they function as a "black box" and do not provide clear explanations on how to solve those problems to which they are applied. In fact one does not intuitively grasp how a network functions and it is impossible to find out how it reaches its conclusions (the decision criteria of the network are unknown); additionally, it uses thousands of synaptic weights that are not subject to a logical interpretation.

Independently of their advantages or disadvantages, it is shown in this chapter that ANNs-based models are able to describe conveniently the degradation of external wall claddings. In fact, the models proposed in this chapter lead to realistic values, with physical sense, low errors and a high correspondence between the values obtained during the field work and the values estimated by the models.

5.2 Fuzzy Logic

5.2.1 Theoretical Foundations

Classical logic developed by Aristotle (384–322 BC) is based on a set of rules of reasoning, in which any proposition leads to a given conclusion. Classical logic is centred on the principle of ambivalence, thus assuming that any proposition is either "true" or "false". This classical view remained unchanged for many years. In 1954, with the publication of the book "An investigation of the laws of thought—on which are founded the mathematical theories of logic and probabilities", George Boole gives a significant enhancement to the evolution the concept of logic. This book led to the founding of the designated Boolean algebra, which is essentially a formalism used in the systematic treatment of logic (i.e. it is a mathematization of logic providing the basis to set theory). There are other relevant studies for the development of mathematical logic, such as: (i) the study by Frege (1879, 1882) that in a different view of Boole, sought to use the formal language and logic in arithmetic's foundation; (ii) the study by Whitehead and Russell (1910, 1912, 1913) who, in the book Principia Mathematica, encode the boolean logic, turning it into a symbolic logic; the study of Tarski (1956) that sought to implement the consistent use of the notion of truth, thus avoiding paradoxical issues; and the study of Gödel (1992), which showed that all logical system is, in some way, incomplete, showing that it is not possible to demonstrate the consistency of arithmetic.

Often, classical logic is unable to model reality, since it confines the truth in two possible values, e.g. "0/1", "yes/no", "truth/lie". Also the conventional computer models (based on boolean logic) work this way and are unable to deal with ambiguous variables (Mukaidono 2001). In this approach, computers manipulate facts, turning them into zeros and ones. However, the world is not binary and the human brain rationalizes reality through vague expressions, involving value judgments (Kosko and Isaka 1993). In addition, the available knowledge about a given reality can be neither absolutely true nor absolutely false and can be sometimes inaccurate, incomplete, fragmented, unrealistic, contradictory, deficient, vague, indefinite, among many other hypotheses (Klir and Yuan 1995).

In 1920, Jan Lukasiewicz, proposes an improvement to binary logic, by adding the uncertainty principle, assuming that there could be three degrees of truth (0, ½, 1), thus adding an intermediate term which defines the situations that through unknowingly, can not be classified in neither false nor true. Later, this concept was developed by Zadeh (1965) formalizing the mathematical principles of fuzzy logic, which allow transforming information with vague nature in numerical data. The study Zadeh (1965) made the basis of mathematical modelling of the approximate reasoning, thus encompassing the notions of chance and uncertainty (notions that are very relevant in intuitive or decision processes). Thus was created the concept of fuzzy logic, which states that a given proposition can be partially true and partially

Fig. 5.11 Illustrative example of the classical logical view (on the *left*) and fuzzy logic view (on the *right*)

false, with a given degree of belonging (or membership) to each of the conditions. Klir and Yuan (1995) refers that in fuzzy logic, the truth is a matter of degree.

Fuzzy logic is distinguished from boolean logic since it is able to get closer to reality where there are not only extreme answers (Cox 1994). For example, it can be said that the Boolean logic sees reality in "black" and "white" not contemplating any other colour of the spectrum, while the fuzzy logic is able to view the nuances between the extremes of colour scale (Fig. 5.11). Kosko and Isaka (1993) distinguish boolean logic and fuzzy logic using the example of the glass half empty or half full; in Boolean logic, the glass is either "full" or is "empty", while in fuzzy logic the amount of water in the glass can be measured by an infinite set of possibilities, from "half full", "half empty", "medium full", "overflowing", among others. In fact, human language is full of imprecise terms. Zadeh (1973) reports that, as the complexity of a system increases, the ability to make accurate and meaningful statements about system behaviour decreases until it reached a limit from which the accuracy and relevance become features mutually exclusive (principle of incompatibility). In conclusion, one of the advantages associated with models based on fuzzy logic is related to their ability to deal with uncertainty, being able to model subjective concepts.

Fuzzy logic models, and neural networks, are known to be able to function as universal approximators (Wang 1992; Kosko 1994; Castro 1995). Fuzzy logic has been successfully applied in the construction area, including: the risk analysis of construction projects (Tah and Carr 2000); in multi-criteria analysis for supporting selection of coating materials (Athanasopoulos et al. 2009); in life cycle analysis and ranking of the environmental performance of various types of floors (Kucukvar et al. 2014); among other applications. This type of models is especially interesting when the problem modelled is subject to uncertainties. The extensive application of fuzzy logic models is mainly due to (McNeill and Thro 1994; Jang and Gulley 1997): (i) its flexibility; (ii) its tolerance to inaccurate data; (iii) its ability to model systems naturally vague and difficult to describe; (iv) models can be developed based on the experience of experts; (v) accept as input data information based on human observation; (vi) be able to model non-linear functions extremely complex; (vii) generally simplify and expand the scope and capabilities of conventional monitoring methods; (viii) are based on linguistic (human communication). Usually, fuzzy logic models have better performance and accuracy than classical linear models (Vieira et al. 2015).

5.2.1.1 Description

One of the key ideas in fuzzy logic relates to the notion of fuzzy sets. These are a generalization of the crisp sets, in which, given a set A and an element x, there are only two well defined options: $x \in A$ or $x \notin A$. Fuzzy sets propose a comprehensive characterization; in fuzzy logic, a given element can belong to more than one set. The partial assignment to a particular set can be described numerically by a membership function which can take an infinite number of values in the range [0, 1] (Şen 1998). The membership function $\mu A(x)$ shows the degree of compatibility between the component x and the set A, where high values show a high degree of membership to the set. Therefore, fuzzy sets introduce an additional hypothesis to traditional sets: $x \in A$; x partially belongs to A ($0 < \mu A(x) < 1$) or $x \notin A$.

Vieira et al. (2005) refers that fuzzy modelling allows approximating non-linear systems when there is no prior knowledge about the phenomenon modelled. Generally, fuzzy modelling follows three stages: (i) model structure identification; (ii) the estimation of model parameters; (iii) and validation of the model. The fuzzy logic models are defined based on a set of independent systems of multiple inputs and a single output (Mendel 1995). The output can be represented as a non-linear model for multiple entries given by $y = f(x)$, where $\mathbf{x}$ is the vector of input data. In this case, the vector $\mathbf{x}$, for each sample k can be obtained from the inputs of the process (Eq. 5.16).

$$\mathbf{x}(k) = [\mathbf{x}_1(k), \mathbf{x}_2(k), \ldots, \mathbf{x}_n(k)] \tag{5.16}$$

where the parameter n is the number of available variables, and $k = 1, \ldots, N$, where N is the number of samples.

To estimate the degradation of the façades coatings, the model can be represented by Eq. (5.17).

$$\hat{y}(k) = f(\mathbf{x}(k)) \tag{5.17}$$

The vector $\mathbf{x}$ is called the regressor (or explanatory variable) and the predicted outputs $\hat{y}$ the regressand (depend variable).

In the present study, a first order Takagi–Sugeno (TS) fuzzy model (Takagi and Sugeno 1985) was used. This type of model uses precise functions as consequent variables, i.e. variables that result in the application of logic rules of the model. Therefore, this type of models can be seen as a combination of a linguistic modelling and mathematical regression analysis. This model is a set of fuzzy rules that describe local relationships of input–output. When the TS fuzzy models are used, each discriminant function is given by a set of rules of the type (Vieira et al. 2015):

$$\begin{aligned} \text{Rule } R_i : \ & \text{If } x_1 \text{ is } A_{i1} \text{ and} \ldots \text{and } x_n \text{ is } A_{in} \\ & \text{then } y_i(x) = f_i(x), \quad i = 1, 2, \ldots, C \end{aligned}$$

where f_i is the consequent function for rule R_i and C is the number of rules. The degree of activation of the ith rule is given by Eq. (5.18). It can be said that the degree of activation of each rule is obtained as a function of the degree of membership of each fuzzy set associated with each of the variables included in the model.

$$\beta_i = \prod_{j=1}^{n} \mu_{A_{ij}}(x) \tag{5.18}$$

where $\mu_{A_{ij}}(x)$: $\mathbb{R} \rightarrow [0, 1]$ is the membership function of the fuzzy set A_{ij}.

The output of the model (S_w) (in the present study, the severity of degradation of façades coatings) is obtained by adding the individual contribution of rules, represented by the Eq. (5.19):

$$S_w = \frac{\sum_{i=1}^{C} \beta_i f_i(x)}{\sum_{i=1}^{C} \beta_i} \tag{5.19}$$

In Takagi–Sugeno fuzzy models, the discriminant function is defined as shown in Eq. (5.20) (Silva et al. 2016).

$$f_i(x) = a_i x + b_i, \quad i = 1, 2, \ldots, C \tag{5.20}$$

Given N available input–output data pairs (x_k, y_k), the n-dimensional pattern matrix $X = [x_1, \ldots, x_n]^T$, and the corresponding output vector $y = [y_1, \ldots, y_N]^T$ are constructed.

The number of rules C, the antecedent fuzzy sets A_{ij}, and the consequent parameters a_i and b_i are determined in this step, by means of fuzzy clustering in the product space of the input and output variables (Sousa and Kaymak 2002). This technique of unsupervised learning is extremely useful in data classification and pattern recognition (Backer 1995), with the aim of allocating a data set into various clusters based on similarity measures. The objective of this analysis is to ensure that the data belonging to a given cluster are as similar as possible to each other (high degree of similarity), and as diverse as possible of data in the remaining clusters (Salgado and Igrejas 2007). Thus, the data set Z to be clustered is composed by X and y (Eq. 5.21).

$$Z = [X, y]^T \tag{5.21}$$

Given the training set Z and the number of clusters C, various data clustering algorithms can be used (Xie and Beni 1991; Pal and Bezdek 1995; Krishnapuram et al. 1995). Clustering algorithms are generally classified according to how they define the different classes, thereby distinguishing themselves in hierarchical and partitional algorithms (Jain et al. 1999; Halkidi et al. 2001). The hierarchical algorithms are based on a repetitive process, where the data set is organized in the form of a dendrogram (or tree diagram). On the other hand, the partitional algorithms seek to group the input data into a number of preset function sets by optimizing an objective function (Carvalho 2007). In traditional techniques (crisp clustering), the data are grouped into disjoint sets, where each data is associated with one, and only one cluster. In opposite, in fuzzy clustering algorithms, the data are grouped into clusters overlap, that is, each data belong to multiple clusters with a given degree of membership (Hartigan and Wong 1979).

In the present study, the fuzzy algorithm C-means (FCM) was used (Ruspini 1969; Dunn 1973; Bezdek 1981, 1992) to compute the fuzzy partition matrix U. The matrix Z provides a description of the system in terms of its local characteristic behaviour in regions of the data identified by the clustering algorithm, and each cluster defines a rule (Vieira et al. 2015). The FCM algorithm can group data in multiple clusters, making use of a distance measurement. In this case, Euclidean distance to measure the distance between data point x_j and cluster centre v_i and is denoted by d_{ij}, which is a squared inner-product distance given by Eq. (5.22).

$$d_{ij}^2(x_j, v_i) = (x_j - v_i)^{\mathrm{T}} A (x_j - v_i) \tag{5.22}$$

where A is a norm-inducing matrix. The clustering criterion is expressed via a cost function called C-means functional, which has to be minimized in order to obtain optimal cluster solutions (Eq. 5.23) (Silva et al. 2016).

$$J_f(X, U_f, V) = \sum_{i=1}^{C} \sum_{j=1}^{N} \mu_{ij}^m d_{ij}^2(x_j, v_i) \tag{5.23}$$

The $C \times N$ matrix $U_f = (\mu_{ij})$ is called a fuzzy partition matrix. Parameter $m \in [1, \infty[$ is the weighting exponent that determines the degree of "fuzziness" of the clustering. In order to use a FCM clustering it is necessary to define two parameters: (i) the number of clusters, which are reflected in the number of fuzzy rules; (ii) the degree of fuzziness of the clustering m, which is the weighting exponent of the clustering algorithm (Bezdek 1981). According Bezdek et al. (1984), there are not computational or theoretical evidence to define optimal value of m, referring that the best way to discover the value m is through experimentation. When m is close to 1, the matrix U tends to be rigid (all data belonging to one single set); in turn, when the value of m tends to infinity, this means that all the sample points have the same degree of membership of all sets. Pal and Bezdek (1995) suggest adopts a value of m equal to 2. The fuzzy sets in the antecedent of the rules are obtained from the partition matrix U, whose ikth element $\mu_{ik} \in [0, 1]$ is the membership degree of the data object z_k in cluster i. One-dimensional fuzzy sets A_{ij} are obtained from the multidimensional fuzzy sets defined point-wise in the ith row of the partition matrix by projections onto the space of the input variables x_j (Eq. 5.24) (Silva et al. 2016).

$$\mu_{A_{ij}}(x_{jk}) = \mathrm{proj}_j^{\mathbb{N}_{n+1}}(\mu_{ik}) \tag{5.24}$$

where proj is the point-wise projection operator (Klir and Yuan 1995). The point-wise defined fuzzy sets A_{ij} are approximated by suitable parametric functions in order to compute $\mu_{Aij}(x_j)$ for any value of x_j.

The consequent parameters for each rule are obtained as a weighted ordinary least square estimate. Let $\theta_i^T = f_i$, X_e denote the matrix [X, 1] and W_i denote a diagonal matrix in $\mathbb{R}^{N \times N}$ with the degree of activation, $\beta_i(x_k)$, as its kth diagonal element. Assuming that the columns of X_e are linearly independent and $\beta_i(x_k) > 0$ for $1 \leq k \leq N$, the weighted least squares solution of $y = X_e\, \theta + \varepsilon$ becomes in Eq. (5.25) (Vieira et al. 2015).

$$\theta_i = \left[X_e^T W_i X_e\right]^{-1} X_e^T W_i y \tag{5.25}$$

Rule bases constructed from clusters can be redundant due to the fact that the rules defined using the multidimensional antecedents are overlapping in one or more dimensions. A possible approach to solve this problem is to reduce the number of features of the model (Vieira et al. 2015).

5.2.2 Application to Façade Claddings

Using the methodology described in the previous sections, in this study, fuzzy models were used to define a model to estimate the degradation severity of façade claddings. In the present study, the termination tolerance of the clustering algorithm was 0.01, i.e. it stops when the partition matrix U has a variation of less than 0.01 between two iterations (clustering stability). In this analysis it is applied a multi-dimensional clustering, which include all the explanatory variables in the data clustering. In all the models produced, façades' degradation are predominantly influenced by age, and for that reason the façades' age are the most relevant parameter in data clustering. For the coatings analyzed, it was tested various models with different clustering sets (ranging from two to ten clusters). It was found that, concerning the data used in the present study, two clusters are sufficient to conveniently group the data. The degree of activation of each rule, i.e. the result of the degree of activation of the combination of membership values for all variables included in the model are obtained through a toolbox for Matlab (Fuzzy Identification Toolbox for Matlab).

5.2.2.1 Natural Stone Claddings

In natural stone claddings, the global set of patterns was divided into a training set (75 % of the sample—254 case studies) and a test set (25 % of the sample—51 case studies). The variables included in the model are age, distance from the sea, type of finish and size of the stone plates; the same variables used in neural networks model (introduced in Sect. 5.1 of this chapter). In fuzzy logic models, it is necessary to codify the variables; since age and size of the stone plates are numerical, their real values were used; variables type of stone finish and distance from the sea are codified in the following way: (i) type of stone finish—smooth stone finish, value −1; rough stone finish, value 1; (ii) distance of the building from the sea—less than 5 km, value 1; otherwise, value −1.

The fuzzy rules describing the local input–output relation are presented in Eqs. (5.26) and (5.27).

Rule 1: If u_1 is A_{11} and u_2 is A_{12} and u_3 is A_{13} and u_4 is A_{14} then

$$y_1(k) = 4.25 \cdot 10^{-4} u_1 + 6.00 \cdot 10^{-4} u_2 - 2.78 \cdot 10^{-2} u_3 + 1.38 \cdot 10^{-3} u_4 + 1.56 \cdot 10^{-2} \quad (5.26)$$

Rule 2: If u_1 is A_{21} and u_2 is A_{22} and u_3 is A_{23} and u_4 is A_{24} then

$$y_2(k) = 1.04 \cdot 10^{-2} u_1 + 1.29 \cdot 10^{-2} u_2 + 1.09 \cdot 10^{-1} u_3 + 5.24 \cdot 10^{-3} u_4 - 1.96 \cdot 10^{-1} \quad (5.27)$$

where u_1 represent the type of finishing, u_2 the distance from the sea, u_3 the size of stone plates and u_4 the age of stone cladding.

In Table 5.6 is presented the clusters centres of the model. The membership functions for each of the four input variables are shown in Fig. 5.12.

The continuous lines (blue) represent the membership functions obtained from cluster 1, and the dashed lines (red) represent the membership functions obtained from cluster 2. The analysis of the variables included in the model reveals that age is the most important variable in defining the severity of degradation of stone claddings. Figure 5.12 shows that the clustering clearly divides the data between newer and older buildings. In the variable "type of finishing", only cluster number two can discriminate the different types of finishing, thus revealing that the type of finishing has little significance to the degradation of the newer buildings. Concerning the "distance from the sea", only cluster number one can discriminate the different locations of the façades, i.e. from a certain age, the distance from the sea ceases to influence the degradation of stone claddings. With regard to the size of stone plates, it is also possible to verify that there is a distinction between the two clusters for this variable.

For clarity's sake, an example of how to calculate the degradation severity of a case study is given in Table 5.7. This example shows a case study with 42 years, with a size of the stone plates equal to 0.45 m^2, located at more than 5 km from the

Table 5.6 Cluster centres of the model proposed (natural stone claddings)

Cluster	u_1	u_2	u_3	u_4
1	−2.44E−01	3.56E−02	3.31E−01	1.28E01
2	7.66E−02	−2.10E−01	4.41E−01	5.52E01

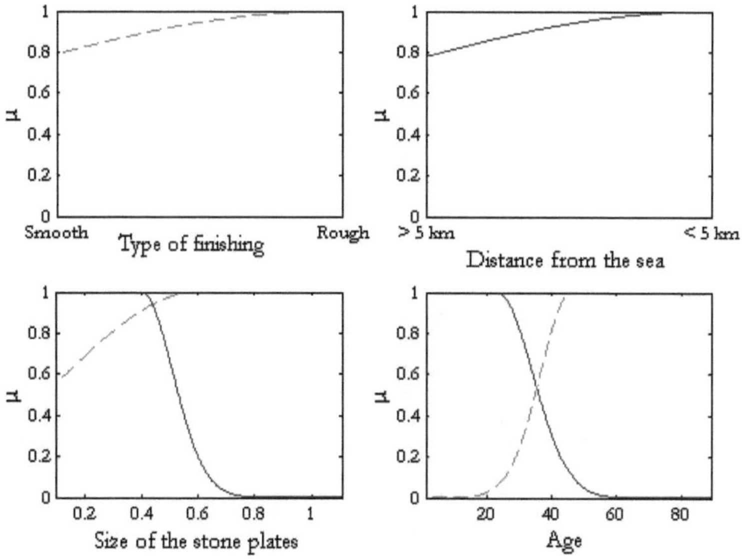

Fig. 5.12 Membership functions for the fuzzy model proposed (data sourced from Silva et al. 2016) (colour figure online)

sea and with rough finish. The degradation severity observed in field work is equal to 0.061 (6.1 %), similar to the value estimated by first order TS fuzzy model (0.069). It is possible to use the same procedure for the remaining cases within the test sample.

To analyze the applicability of the model, the statistical parameters presented in Eqs. (5.3) to (5.7) were used. In these statistical parameters, the predicted values were compared with the values obtained through the field survey. The errors obtained were relatively low; the maximum absolute percentage error ($\varepsilon_{max.}$) were about 24 % and the mean absolute percentage error ($\overline{\varepsilon}$) were equal to 7.7 %. The percentage of patterns with ε greater than $x = 5$ %, $x = 10$ %, $x = 20$ % and $x = 30$ % ($PP_{\varepsilon > x}$) were respectively 48, 32, 8 and 0 %. Furthermore, it was obtained a correlation coefficient (r) equal to 0.88, revealing a high correlation between the predicted and observed values.

Figure 5.13 present the histogram of the estimated service life for first order Takagi–Sugeno model and a normal distribution was adjusted to the histogram. The model proposed leads to an average estimated service life of 90 years for stone claddings. This value is acceptable, however, seems relatively high, in comparison with the values obtained by the remaining methodologies adopted in the present book. The average result is slightly skewed by the high values of expected service life obtained for some case studies (is important to notice that the maximum expected service life obtained is equal to 142 years). The median estimated service life is 71 years, which is in the order of magnitude of the values proposed in the literature and the results obtained in this book (using other models).

Table 5.7 Example of the calculation of the degradation severity using the fuzzy logic model for a case study

Case study	Calculus
	$\beta_1 = 0.1579$; $\beta_2 = 0.8688$ (obtained through "Fuzzy Identification Toolbox for Matlab") $Rule\,1: y_1(k) = (4.25 \cdot 10^{-4} \times 1) + (6.00 \cdot 10^{-4} \times -1) - (2.78 \cdot 10^{-2} \times 0.45) + (1.38 \cdot 10^{-3} \times 42) + 1.56 \cdot 10^{-2}$ $Rule\,1: y_1(k) = 6.09 \cdot 10^{-2}$ $Rule\,2: y_2(k) = (1.04 \cdot 10^{-2} \times 1) + (1.29 \cdot 10^{-2} \times -1) + (1.09 \cdot 10^{-1} \times 0.45) + (5.24 \cdot 10^{-3} \times 42) - 1.96 \cdot 10^{-1}$ $Rule\,2: y_2(k) = 7.06 \cdot 10^{-2}$ $S_w = \frac{\sum_{i=1}^{C} \beta_i f_i(x)}{\sum_{i=1}^{C} \beta_i} = \frac{0.1579 \times 6.09 \cdot 10^{-2} + 0.8688 \times 7.06 \cdot 10^{-2}}{0.1579 + 0.8688} = 0.069$

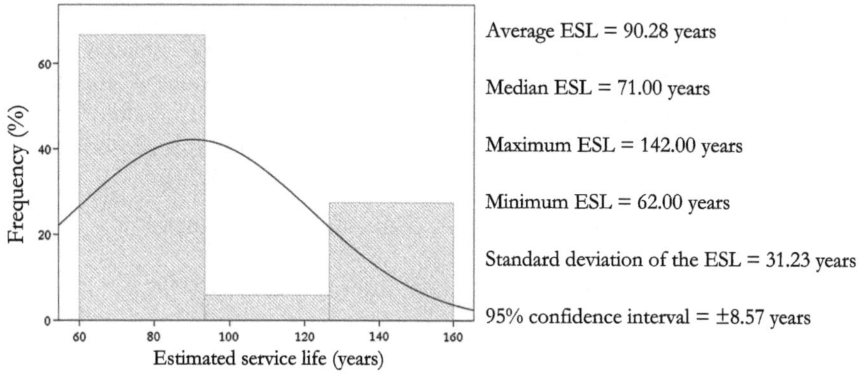

Fig. 5.13 Histogram of the estimated service life for the fuzzy model applied to stone claddings

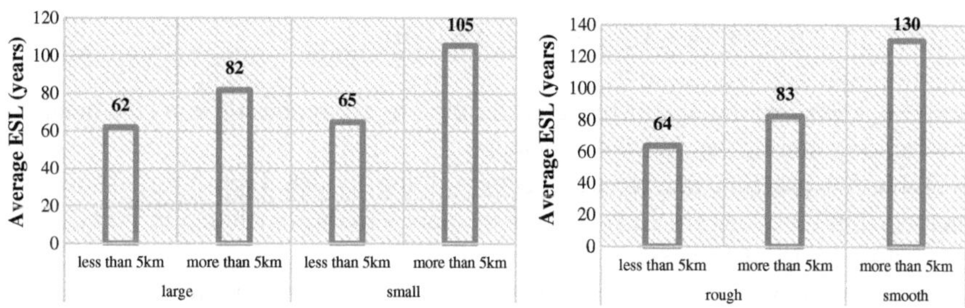

Fig. 5.14 Analysis of the mean estimated service life according to the variables included in the fuzzy model applied to stone claddings (data sourced from Silva et al. 2016)

Figure 5.14 shows an analysis of the average estimated service life according to the variables included in the model. The results obtained are coherent and credible from a physical point of view.

The following conclusions can be taken: (i) at same distance from the sea, stone claddings with larger stone plates have lower values of estimated service lives than claddings with medium size plates; (ii) claddings with smooth finish are more durable than claddings with rough finish; (iii) claddings located at more than 5 km are more durable, with higher service lives than claddings in coastal areas. These results are also in accordance with the results obtained using ANNs-based models.

5.2.2.2 Rendered Façades

In rendered façades, the sample was divided in a training sample composed of 75 case studies (75 % of the sample) and a test sample with 25 case studies (25 %). The fuzzy model proposed include four explanatory variables: age; render type; height of the building; and façades orientation. To applied the model, qualitative variables are codified in the following way: (i) render type—current cement renderings; value 1; cement and lime mortar, value 2; single-layer rendering, value 3; marble agglomerate mortar, value 4; (ii) height of the building—small buildings, value 1; medium height buildings, value 2; high buildings, value 3; (iii) façade orientation—SW, value −4; SE, value −3; South, value −2; West, value −1; East, value 1; North, value 2; NE, value 3; NW, value 4. The fuzzy rules describing the local input–output relation are presented in Eqs. (5.28) and (5.29).

Rule 1: If u_1 is A_{11} and u_2 is A_{12} and u_3 is A_{13} and u_4 is A_{14} then

$$\begin{aligned} y_1(k) &= 1.29 \cdot 10^{-2} u_1 - 1.44 \cdot 10^{-3} u_2 - 2.55 \cdot 10^{-2} u_3 \\ &\quad + 3.40 \cdot 10^{-3} u_4 + 4.18 \cdot 10^{-2} \end{aligned} \tag{5.28}$$

Rule 2: If u_1 is A_{21} and u_2 is A_{22} and u_3 is A_{23} and u_4 is A_{24} then

$$\begin{aligned} y_2(k) &= 1.82 \cdot 10^{-2} u_1 + 1.02 \cdot 10^{-1} u_2 - 7.85 \cdot 10^{-2} u_3 \\ &\quad + 1.31 \cdot 10^{-2} u_4 - 1.67 \cdot 10^{-1} \end{aligned} \tag{5.29}$$

where u_1 represent the age of rendered façades, u_2 the render type, u_3 the height of the building and u_4 the façades orientation.

Table 5.8 presents the clusters centres of the model. Figure 5.15 shows the membership functions for each of the four input variables included in fuzzy model applied to rendered façades.

As in the previous model applied to stone claddings, in this model, continuous lines (blue) represent the membership functions obtained from cluster 1, and the dashed lines (red) represent the membership functions obtained from cluster 2. In this case, age is the most relevant variable in the explanation of façades' degradation, and Fig. 5.15 shows that the clustering clearly divides the data between newer and older buildings. Buildings with less than 30 years strongly belong to cluster one, influencing the output value of rule number one of the fuzzy model. On the other hand, buildings with more than 25 years strongly belong to cluster two (membership degree over 0.5), influencing the output value of rule number two. Concerning the variable render type, only cluster number two can discriminate the different types of render. All the types of render give a contribution to the output of the rule number one; in rule number two, lime–cement renderings has a higher positive contribution to the output of the fuzzy model. In relation to building height, rule number one has high values of membership for all the three categories; the maximum value of membership function correspond to "tall buildings", which indicates that the higher the building the higher the influence on the degradation of rendered façades. The

Table 5.8 Cluster centres of the model proposed (rendered façades)

Cluster	u_1	u_2	u_3	u_4
1	7.82E0	1.37E0	2.14E0	7.51E−02
2	4.09E01	1.70E0	1.88E0	−2.33E−01

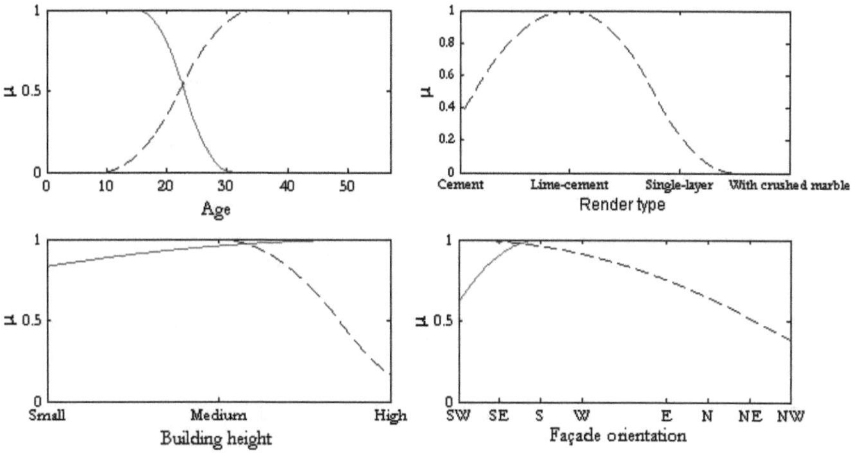

Fig. 5.15 Membership functions for the fuzzy model proposed (data sourced from Vieira et al. 2015) (colour figure online)

consequent parameter associated with building height is negative, revealing a decrease in degradation severity with building height. This result can be explained by the test sample characteristics, since the number of older high buildings within the sample is very low and fuzzy model are unable to learn conveniently the influence of the building height for older buildings in the degradation severity. Concerning façades orientation, it is possible to identify the division between the two membership functions. Orientations N, NE and NW, will increase the rule number one output, thus increasing the degradation severity of rendered façades. Orientations S, SE and SW have a higher membership degree for membership function number two, and as the values are negative the rule output will decrease.

For rendered façades, the fuzzy model proposed leads to a maximum absolute percentage error ($\varepsilon_{max.}$) around 10 % and to a mean absolute percentage error ($\bar{\varepsilon}$) equal to 3.5 %. The percentage of patterns with ε greater than $x = 5$ %, $x = 10$ %, $x = 20$ % and $x = 30$ % ($PP_{\varepsilon > x}$) were, respectively, 24, 4, 0 and 0 %. The Pearson correlation coefficient (r) obtained is equal to 0.92, revealing a very strong correlation between the predicted values and the observed values of the façades' degradation.

The histogram of the estimated service life for first order Takagi–Sugeno model applied to rendered façades is presented in Fig. 5.16. This model leads to an average estimated service life of 18.4 years with a standard deviation of 1.86 years, a consistent value with the results obtained through the methods proposed in this book and the existing literature. Figure 5.17 presents a comparative analysis of the average estimated service life according to the variables considered.

The results obtained are conditioned by the characteristics of the test sample (composed of 25 studies). As mentioned before, fuzzy model are adjusted to the sample used in training process and the characteristics of the sample, in which high buildings with older ages is a very small part of the sample, do not allowing the

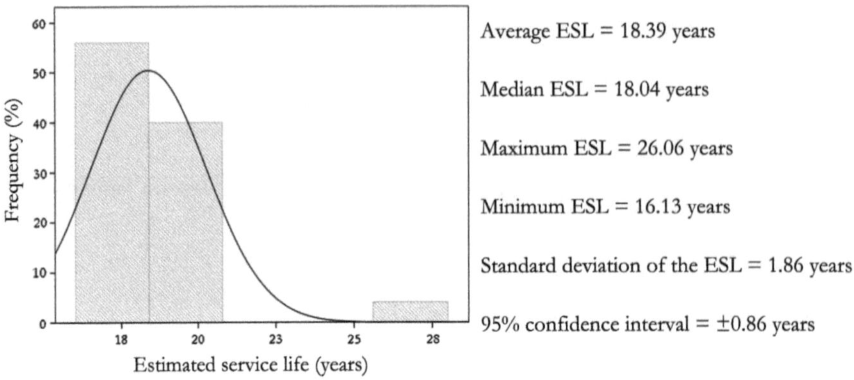

Fig. 5.16 Histogram of the estimated service life for the fuzzy model applied to rendered façades

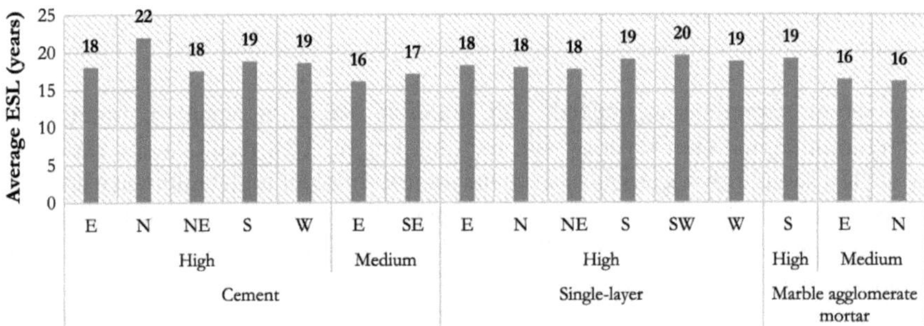

Fig. 5.17 Analysis of the mean estimated service life according to the variables included in the fuzzy model applied to rendered façades (data sourced from Vieira et al. 2015)

model learn the influence of the building height for older buildings in the degradation of rendered façades. In the sample analyzed, higher buildings present higher expected service lives. Concerning the type or render, single-layer and cementitious renderings are the most durable, with higher values of service life (in the sample test none of the case studies correspond to cement–lime renderings).

5.2.2.3 Painted Surfaces

With respect to painted surfaces, it is used a sample of 220 cases of study where 85 % of the sample (187 studies) are used as a training sample and 15 % as test sample. In this case, the test sample is the same sample used in the ANNs model. The proposed model includes as explanatory variables age, the façades orientation and exposure to wind-rain action. The fuzzy rules describing the local input–output relation are presented in Eqs. (5.30) and (5.31).

Rule 1: If u_1 is A_{11} and u_2 is A_{12} and u_3 is A_{13} and u_4 is A_{14} then

$$y_1(k) = -1.43 \cdot 10^{-2} u_1 - 2.63 \cdot 10^{-2} u_2 + 4.25 \cdot 10^{-2} u_3 - 8.54 \cdot 10^{-2} \quad (5.30)$$

Rule 2: If u_1 is A_{21} and u_2 is A_{22} and u_3 is A_{23} and u_4 is A_{24} then

$$y_2(k) = -4.31 \cdot 10^{-3} u_1 - 3.15 \cdot 10^{-3} u_2 + 2.30 \cdot 10^{-2} u_3 - 2.35 \cdot 10^{-2} \quad (5.31)$$

where u_1 represent the façades orientation, u_2 the distance from the sea and u_3 the age of the buildings.

In Table 5.9 is shows the clusters centres of the model and Fig. 5.18 shows the membership functions for each of the four input variables included in fuzzy model applied to painted surfaces.

The analysis of the membership functions reveals that the sample is clearly into two clusters for the variable age of the painted surfaces. In the remaining variables, distance from the sea and façades orientation, all the categories give a contribution to the output of the rules number one and two.

The fuzzy logic model proposed for painted surfaces leads to a relatively small errors, with a mean absolute percentage error ($\bar{\varepsilon}$) equal to 5.1 % and a maximum absolute percentage error ($\varepsilon_{max.}$) equal to 22.1 %. Furthermore, the model shows a Pearson correlation coefficient (r) of 0.98, showing a very strong correlation between the observed values and the values predicted by the model.

Table 5.9 Cluster centres of the model proposed (painted surfaces)

Cluster	u_1	u_2	u_3
1	2.28	2.69	10.90
2	2.55	2.04	3.84

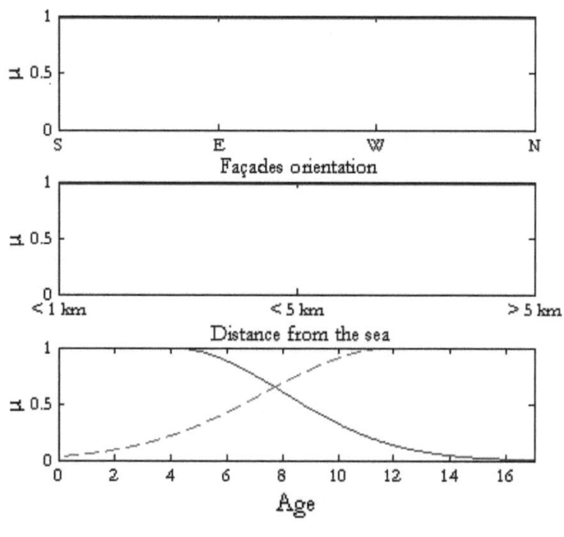

Fig. 5.18 Membership functions for the fuzzy model proposed for painted surfaces

Figure 5.19 presents the histogram of the estimated service life for first order Takagi–Sugeno model applied to painted surfaces. Using this model it was obtained an average estimated service life of 9.7 years with a standard deviation of 0.4 years. This value is closer to the value obtain by ANNs models (around 9.5 years) and similar to the results obtained using the other methods proposed in this book.

In Fig. 5.20 it is shown a comparative analysis of the average estimated service life according to the variables considered. The results obtained do not allow drawing unequivocal conclusions. In fact, painted surfaces present a durability and longevity far below to the durability of stone claddings, which implies that variations in façades' exposure conditions does not lead to significant variations in the expected service lives of painted surfaces. In an overall analysis, the results reveal that under similar conditions, painted surfaces in coastal areas (<1 km) present lower estimated service lives. Regarding the façades orientation, the distinction between different orientations is not clear; however, façades faced north are those with higher estimated service lives. This result is apparently contradictory since in

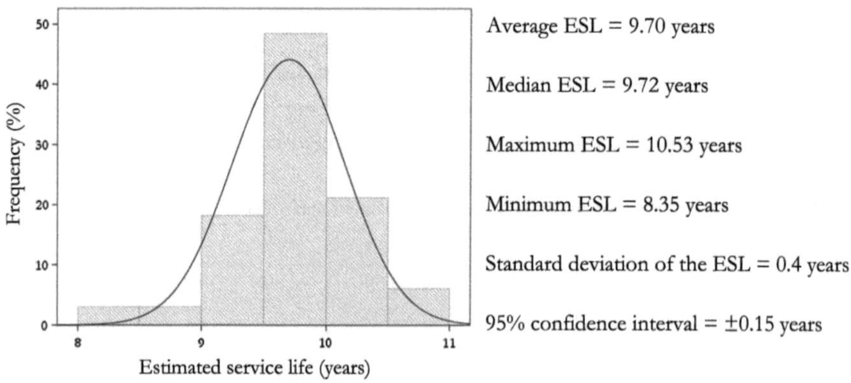

Fig. 5.19 Histogram of the estimated service life for the fuzzy model applied to painted surfaces

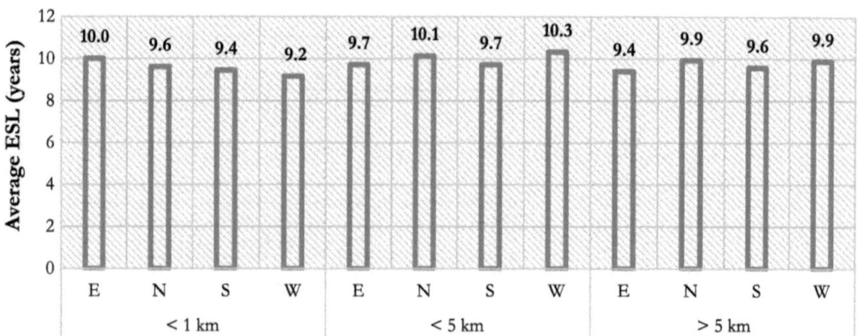

Fig. 5.20 Analysis of the mean estimated service life according to the variables included in the fuzzy model applied to painted surfaces

Portugal north and west are the most aggressive orientations; north is wetter and colder and façades faced west have a greater probability of the simultaneous occurrence of wind and rain. However, in the case of painted surfaces, the exposure to solar radiation contributes significantly to the evolution of degradation, which can, in some way, justify the results obtained in this study.

5.2.2.4 Ceramic Tiling Systems

Concerning ceramic tiling systems, the sample used is composed of 195 case studies; 85 % (166 cases) of the sample is used in the model training and 15 % (29 cases) is used as a test sample. The proposed fuzzy logic model includes as explanatory variables age, the size of ceramic tiles and the distance from the sea. The fuzzy rules describing the local input–output relation are presented in Eqs. (5.34) and (5.35).

Rule 1: If u_1 is A_{11} and u_2 is A_{12} and u_3 is A_{13} and u_4 is A_{14} then:

$$y_1(k) = -2.59 \cdot 10^{-2} u_1 - 2.61 \cdot 10^{-2} u_2 + 6.34 \cdot 10^{-3} u_3 - 8.80 \cdot 10^{-2} \quad (5.34)$$

Rule 2: If u_1 is A_{21} and u_2 is A_{22} and u_3 is A_{23} and u_4 is A_{24} then:

$$y_2(k) = -8.05 \cdot 10^{-3} u_1 - 2.64 \cdot 10^{-3} u_2 + 2.43 \cdot 10^{-3} u_3 - 2.87 \cdot 10^{-3} \quad (5.35)$$

where u_1 represent the distance from the sea, u_2 the size of ceramic tiles and u_3 the age of the buildings.

Table 5.10 presents the clusters centres of the model. Figure 5.21 shows the membership functions for each of the four input variables included in fuzzy model applied to ceramic tiling systems. The analysis of the variables included in the model reveals that the clustering clearly divides the data between newer and older buildings. In the remaining variables, distance from the sea and size of ceramic tiles, all the categories give a contribution to the output of the rules number one and two.

The analysis of the performance of the fuzzy model applied to ceramic tiling systems reveals a strong correlation between the values predicted by the model and the observed values ($r = 0.88$). The percentage of patterns with ε greater than $x = 20$ % and $x = 30$ % are null. Besides that, the fuzzy model proposed leads to a maximum absolute percentage error ($\varepsilon_{max.}$) equal to 19.8 % and to a mean absolute percentage error ($\bar{\varepsilon}$) equal to 7.4 %.

Figure 5.22 shows the histogram of the estimated service life for first order Takagi–Sugeno model applied to ceramic tiling systems. This model leads to an average estimated service life of 50.8 years, with a standard deviation of 6.8 years.

Table 5.10 Cluster centres of the model proposed (ceramic tiling systems)

Cluster	u_1	u_2	u_3
1	−6.11E−01	1.00	38.50
2	4.29E−01	−1.00	18.00

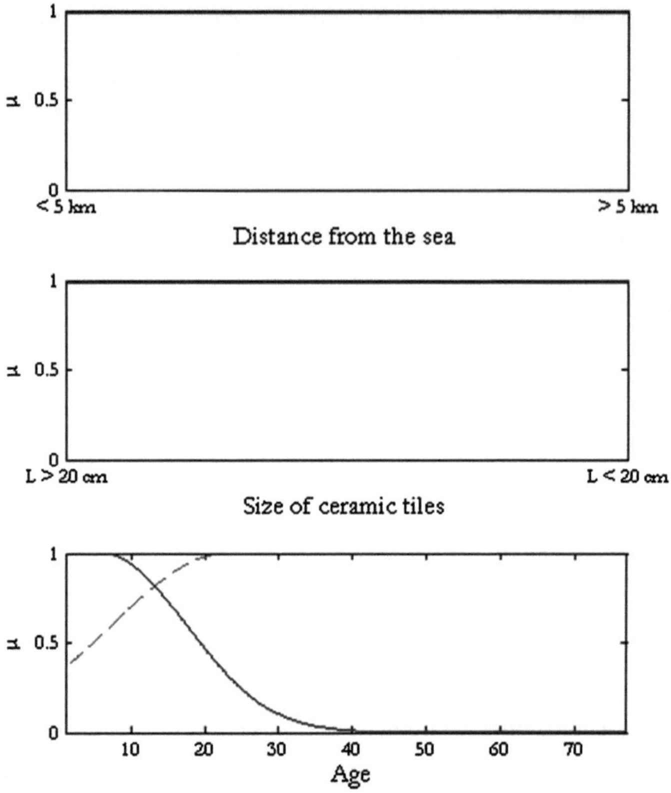

Fig. 5.21 Membership functions for the fuzzy model proposed for ceramic tiling systems

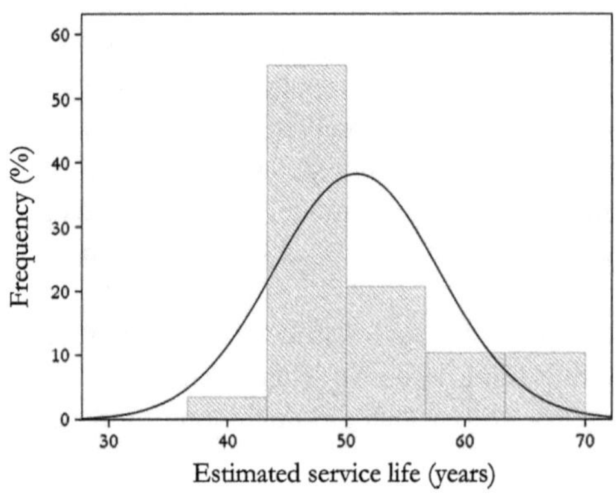

Average ESL = 50.83 years

Median ESL = 48.51 years

Maximum ESL = 67.14 years

Minimum ESL = 37.81 years

Standard deviation of the ESL = 6.84 years

95% confidence interval = ±2.49 years

Fig. 5.22 Histogram of the estimated service life for the fuzzy model applied to ceramic tiling systems

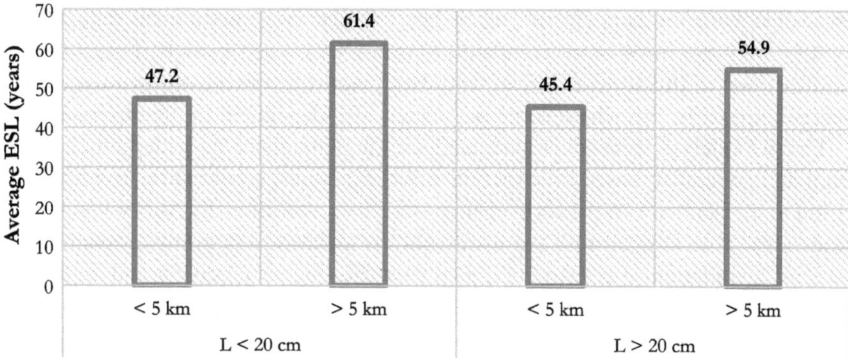

Fig. 5.23 Analysis of the mean estimated service life according to the variables included in the fuzzy model applied to ceramic tiling systems

In the definition of the fuzzy model it was used the same training and test samples used in ANNs-based model; the results obtained through fuzzy model is similar to the value obtained by ANNs model (50.2 years) proposed in Sect. 5.1 of this chapter. Furthermore, this result is similar to the results obtained through other models proposed in the present book.

Figure 5.23 shows an analysis of the mean estimated service life according to the variables included in the fuzzy model applied to ceramic tiling systems. The analysis of the results reveal that: (i) façades in coastal areas (<5 km from the sea) have lower estimated service lives; (ii) ceramic tiles with smaller sized tiles ($L < 20$ cm) has higher estimated service lives. These results are consistent and coherent with the empirical perception about the degradation phenomenon of ceramic tiling systems.

5.2.3 *Conclusions*

In Sect. 5.2 of this chapter is discussed the application of a first order Takagi–Sugeno model in service life prediction of façade coatings. These models are created based on a learning process, in which the model is able to predict the behaviour of the coatings based on a set of input–output patterns. The proposed models are able to predict the degradation and, consequently, the service life of façades based on their age and their characteristics. The fuzzy models proposed in this chapter are able to describe conveniently the degradation of the coatings analyzed. In addition, the estimated service lives obtained for the coatings analyzed are realistic and coherent with the other models proposed in this book and with the values in the literature.

ANNs and fuzzy models are the most popular models used to solve non-linear problems. In fact, these computational methods are flexible mathematical tools and are able to describe reality with the accuracy required. While neural networks are associated with the concept of "black box" (based on experimental data), fuzzy models are more transparent compared to the ANNs based models. Fuzzy models are described in literature as being "grey box" models, which is known the models operating process, but where some parameters still dependent to the observed data.

Fuzzy models present several advantages; the main concerns with their ability to learn and generalize based on experience and examples, combining numerical precision to transparency in the form of linguistic rules. Additionally, fuzzy logic based models are able to deal with the uncertainty associated with complex phenomena such as degradation of construction elements with higher precision and better performance than conventional linear models. The logical structure of the fuzzy logic (in form of logical rules) allows understanding the model in a semi-qualitative way, close to the way humans rationalize reality. In fact, fuzzy logic models are known to be able to effectively deal with inaccurate data; besides that, fuzzy models can deal with samples with outliers, ensuring that outlier's do not biassed the results obtained. However, the complexity associated with these models may condition its use in modelling the service life prediction of façade coatings, because they can become illegible for users who are not familiar with this type of models.

References

Abraham TH (2002) (Physio)logical circuits: the intellectual origins of the McCulloch-Pitts neural networks. J Hist Behav Sci 38(1):3–25

Amirikian B (1999) Neural networks and modeling of neuronal networks. In: Windhorst U, Johansson H (eds) Modern techniques in neuroscience research, 1st edn. Springer, Heidelberg, pp 689–704

Athanasopoulos G, Ribab CR, Athanasopoulouc C (2009) A decision support system for coating selection based on fuzzy logic and multi-criteria decision making. Expert Syst Appl 36 (8):10848–10853

Atkinson PM, Tatnall ARL (1997) Introduction neural networks in remote sensing. Int J Remote Sens 18(4):699–709

Awodele O, Jegede O (2009) Neural networks and its application in engineering. In: Proceedings of informing science & IT education conference (InSITE), Georgia, USA, pp 83–95

Ayat S, Farahani HA, Aghamohamadi M, Alian M, Aghamohamadi S, Kazemi Z (2013) A comparison of artificial neural networks learning algorithms in predicting tendency for suicide. Neural Comput Appl 23(5):1381–1386

Backer E (1995) Computer-assisted reasoning in cluster analysis. Prentice Hall, New Jersey

Basak SC, Grunwald GD, Gute BD, Balasubramanian K, Opitz D (2000) Use of statistical and neural net approaches in predicting toxicity of chemicals. J Chem Inf Comput Sci 40 (4):885–890

Basheer IA, Hajmeer M (2000) Artificial neural networks: fundamentals, computing, design and application. J Microbiol Methods 43(1):3–31

Bebis G, Georgiopoulos M (1994) Feed-forward neural networks: why network size is so important. IEEE Potentials 13(4):27–31

Bebis G, Georgiopoulos M, Kasparis T (1997) Coupling weight elimination with genetic algorithms to reduce network size and preserve generalization. Neurocomputing 17(3–4):167–194

Beck B, Breindl A, Clark T (2000) QM/NN models with error estimation: vapor pressure and logP. J Chem Inf Comput Sci 40(4):1046–1051

Bezdek JC (1981) Pattern recognition with fuzzy objective function algorithms. Plenum Press, New York

Bezdek JC, Pal SK (1992) Fuzzy models for pattern recognition. IEEE Press, New York

Bezdek JC, Ehrlich R, Full W (1984) FCM: the fuzzy c-means clustering algorithm. Comput Geosci 10(2–3):191–203

Boussabaine AH (1996) The use of artificial neural networks in construction management: a review. Constr Manage Econ 14(5):427–436

Carvalho FAT (2007) Fuzzy c-means clustering methods for symbolic interval data. Pattern Recogn Lett 28(4):423–437

Castro JL (1995) Fuzzy logic controllers are universal approximators. IEEE Trans Syst Man Cybern 25(4):629–635

Chai C, de Brito J, Gaspar PL, Silva A (2014) Service life of painted walls: economic analysis of various maintenance strategies. J Constr Eng Manage 140(3):04013057

Chai C, de Brito J, Gaspar PL, Silva A (2015) Statistical modelling of service life prediction of painted surfaces. Int J Strateg Property Manage 19(2):173–185

Chang CC, Chang TYP, Xu YG, Wang ML (2000) Structural damage detection using an iterative neural network. J Intell Mater Syst Struct 11(1):32–42

Chao LC, Skibniewski MJ (1998) Neural networks for evaluating construction technology. In: Flood I, Kartam N (eds) Artificial neural networks for civil engineers: advanced features and applications. ASCE, USA

Cigizoglu HK, Kişi O (2005) Flow prediction by three back propagation techniques using k-fold partitioning of neural network training data. Nord Hydrol 36(1):49–64

Costa MA, Braga AP, Menezes BR (2003) Improving neural networks generalization with new constructive and pruning methods. J Intell Fuzzy Syst 13(2–4):75–83

Cox E (1994) The fuzzy systems handbook: a practitioner's guide to building, using, and maintaining fuzzy systems. AP Professional, Academic Press, San Diego

Cybenco G (1989) Approximation by superposition of a sigmoidal function. Math Control Signals Syst 2(4):303–314

Dias JLR, Silvestre N (2011) A neural network based closed-form solution for the distortional buckling of elliptical tubes. Eng Struct 33(6):2015–2024

Dias JL, Silva A, Chai C, Gaspar PL, de Brito J (2014) Neural networks applied to service prediction of exterior painted surfaces. Build Res Inf 42(3):371–380

Dunn JC (1973) A fuzzy relative of the ISODATA process and its use in detecting compact well-separated clusters. J Cybern 3(3):32–57

Frege G (1879) Begriffsschrift, eine der arithmetischen nachgebildete Formelsprache des reinen Denkens. Halle a/S: Verlag von Louis Nerbert (trans: van Heijenoort J (ed) 1967, From Frege to Gödel: a source book in mathematical logic). Harvard University Press, Cambridge, pp 1879–1931

Frege G (1882) Ueber die wissenschaftliche Berichtigung einer Begriffsschrift. Zeitschrift für Philosophie und philosophische Kritik 81:48–56

Freitag S, Beer M, Graf W, Kaliske M (2009) Lifetime prediction using accelerated test data and neural networks. Comput Struct 87(19–20):1187–1194

Friswell MI, Penny JET, Garvey SD (1998) A combined genetic and eigensensitivity algorithm for the location of damage in structures. Comput Struct 69(5):547–556

Galbusera MM, de Brito J, Silva A (2014) Application of the factor method to the prediction of the service life of ceramic external wall cladding. J Perform Constr Facil (accepted for publication)

Gardner MW, Dorling SR (1998) Artificial neural networks (the multilayer perceptron)—a review of applications in the atmospheric sciences. Atmos Environ 32(14–15):2627–2636

Garrett JH (1994) Where and why artificial neural networks are applicable in civil engineering. J Comput Civ Eng (ASCE) [special issue] 8(2):129–30

Gaspar P (2009) Service life of constructions: development of a methodology to estimate the durability of constructionelements, Application to renders in current buildings (in Portuguese), PhD Thesis in Engineering Sciences atInstituto Superior Técnico, Lisbon, Portugal

Gödel K (1992) On formally undecidable propositions of principia mathematica and related systems, 1st edn. Basic Books Inc., New York

González-Arjona D, López-Pérez G, González AG (2002) Non-linear QSAR modeling by using multilayer perceptron feedforward neural networks trained by back-propagation. Talanta 56 (1):79–90

Halkidi M, Batistakis Y, Vazirgiannis M (2001) On clustering validation techniques. J Intell Inf Syst 17(2–3):107–145

Hartigan JA, Wong MA (1979) A k-means clustering algorithm. Appl Stat 8(1):100–108

Haykin S (1999) Neural networks: a comprehensive foundation, 2nd edn. Prentice Hall Inc., New Jersey

Hebb DO (1949) The organization of behavior: a neuropsychological theory, 1st edn. Wiley, New York

Hepner GF, Logan T, Ritter N, Bryant N (1990) Artificial neural network classification using a minimal training set: comparison to conventional supervised classification. Photogram Eng Remote Sens 56(4):469–473

Hill T, O'Connor M, Remus W (1996) Neural network models for time series forecasts. Manage Sci 42(7):1082–1092

Hornik K, Stinchcombe M, White H (1989) Multilayer feedforward networks are universal approximators. Neural Netw 2(5):359–366

Jain AK, Murty MN, Flynn PJ (1999) Data clustering: a review. ACM Comput Surv (CSUR) 31 (3):264–323

Jang J, Gulley N (1997) MATLAB—fuzzy logic toolbox—user's guide. The MathWorks Inc., USA

Johannet A, Vayssade B, Bertin D (2007) Neural networks: from black box towards transparent box. Application to evapotranspiration modeling. Int J Comput Intell 30(6):162–169 (World Academy of Science, Engineering and Technology)

Kamarthi SV, Pittner S (1999) Accelerating neural network training using weight extrapolations. Neural Netw 12(9):1285–1299

Kazanasmaz T, Günaydin M, Binol S (2009) Artificial neural networks to predict daylight illuminance in office buildings. Build Environ 44(8):1751–1757

Klir GJ, Yuan B (1995) Fuzzy sets and fuzzy logic: theory and applications. Prentice Hall, New Jersey

Kosko B (1994) Fuzzy systems as universal approximators. IEEE Trans Comput 43 (11):1329–1333

Kosko B, Isaka S (1993) Fuzzy logic. Sci Am 269(1):76–83

Kourentzes N, Crone S (2008) Automatic modelling of neural networks for time series prediction—in search of a uniform methodology across varying time frequencies. In: 2nd European symposium on time series prediction (ESTSP'08), Helsinki, Finland, Session Competition 1, 11p

Krishnapuram R, Frigui H, Nasraoui O (1995) Fuzzy and possibilistic shell clustering algorithms and their application to boundary detection and surface approximation—part I. IEEE Trans Fuzzy Syst 3(1):29–43

Kucukvar M, Gumus S, Egilmez G, Tatari O (2014) Ranking the sustainability performance of pavements: an intuitionistic fuzzy decision making method. Autom Constr 40:33–43

Lawrence J (1994) Introduction to neural networks: design, theory and applications, 6th edn. California Scientific Software Publisher, USA

Lefebvre WC, Principe JC, Samson D, Wooten D, Geniesse G, Fancourt C, Gerstenberger J, Euliano N, Lynn G, Allen M (2003) NeuroSolutions 4.42, NeuroDimension, Gainesville, Florida

Leung AWT, Tam CM, Liu DK (2001) Comparative study of artificial neural networks and multiple regression analysis for predicting hoisting times of tower cranes. Build Environ 36 (4):457–467

Lippmann RP (1987) An introduction to computing with neural nets. IEEE ASSP Mag 3(4):4–22

Liu SW, Huang JH, Sung JC, Lee CC (2002) Detection of cracks using neural networks and computational mechanics. Comput Methods Appl Mech Eng 191(25–26):2831–2845

McClelland JL (2009) The place of modeling in cognitive science. Top Cogn Sci 1(1):11–38

McClelland JL, Rumelhart DE, Hinton GE (1986) The appeal of parallel distributed processing. In: Rumelhart DE, McClelland JL (eds) Parallel distributed processing: foundations, vol 1. The PDP Research Group, pp 3–44

McCulloh WS, Pitts W (1943) A logical calculus of the ideas immanent in nervous activity. Bull Math Biophys 5(4):115–133
McNeill FM, Thro E (1994) Fuzzy logic: a practical approach, 1st edn. AP Professional, Boston
Medeiros J (2000) Why does facade tiling fail? In: Proceedings of VI World congress of ceramic tile quality, QUALICER, Castellon, Spain
Mendel JM (1995) Fuzzy logic systems for engineering: a tutorial. Proc IEEE 83(3):345–377
Minsky ML (1954) Theory of neural-analog reinforcement systems and its application to the brain-model problem. PhD thesis, Princeton University, New Jersey
Moselhi O, Hegazy T, Fazio P (1991) Neural networks as tools in construction. J Constr Eng Manage 117(4):606–625
Moselhi O, Hegazy T, Fazio P (1993) DBID: analogy-based DSS for bidding in construction. J Constr Eng Manage 119(3):466–479
Mukaidono M (2001) Fuzzy logic for beginners, 1st edn. World Scientific Publishing Ltd., Singapore
Olden JD, Joy MK, Death RG (2004) An accurate comparison of methods for quantifying variable importance in artificial neural networks using simulated data. Ecol Model 178(3–4):389–397
Pal NR, Bezdek JC (1995) On cluster validity for the fuzzy c-means model. IEEE Trans Fuzzy Syst 3(3):370–379
Pathak V, Dikshit O (2005) Neuro-textural classification of Indian urban environment. Geocarto Int 20(3):65–73
Piaw CY (2006) Basic research statistics (Book 2). McGraw-Hill, Kuala Lumpur (in Malay)
Rafiq MY, Bugmann G, Easterbrook DJ (2001) Neural network design for engineering applications. Comput Struct 79(17):1541–1552
Rezeki S, Sujito B, Subanar, Gurito S (2006) Statistical model selection based on resampling procedure for neural networks classification. In: 1st International conference on mathematics and statistics (ICoMS-1), Bandung Islamic University, Bandung, Indonesia, 8p, 19–21 June 2006
Rivas T, Alvarez E, Mosquera MJ, Alejano L, Taboada J (2010) Crystallization modifiers applied in granite desalination: the role of the stone pore structure. Constr Build Mater 24(5):766–776
Rojas R (1996) Neural networks: a systematic approach, 1st edn. Springer, Berlin
Rosenblatt F (1958) The perceptron: a probabilistic model for information storage and organization in the brain. Psychol Rev 65(6):386–408
Rumelhart DE, Hinton GE, Williams RJ (1986) Learning internal representations by error propagation. In: McClelland JL, Rumelhart DE, The PDP Research Group (eds.) Parallel distributed processing: explorations in the microstructure of cognition: foundations, vol 1. MIT Press, Cambridge, pp 318–362
Ruspini EH (1969) A new approach to clustering. Inf Control 15(1):22–32
Salgado P, Igrejas G (2007) Probabilistic clustering algorithms for fuzzy rules decomposition. Workshop on advanced fuzzy and neural control, Valenciennes, France, TU1-6, 6p
Schalkoff RJ (1997) Artificial neural networks, 1st edn. McGraw-Hill Publishers, New York
Şen Z (1998) Fuzzy algorithm for estimation of solar irradiation from sunshine duration. Sol Energy 63(1):39–49
Shohet IM, Paciuk M (2004) Service life prediction of exterior cladding components under standard conditions. Constr Manag Econ 22(10):1081–1090
Silva A, de Brito J, Vieira SM, Gaspar PL Fuzzy systems in service life prediction of natural stone claddings. Comput Aided Civ Infrastruct Eng (submitted for publication)
Silva A, Dias JL, Gaspar PL, de Brito J (2011) Service life prediction models for exterior stone cladding. Build Res Inf 39(6):637–653
Silva A, de Brito J, Gaspar PL (2012) Application of the factor method to maintenance decision support for stone cladding. Autom Constr 22:165–174
Silva A, Dias JL, Gaspar PL, de Brito J (2013) Statistical models applied to service life prediction of rendered façades. Autom Constr 30:151–160
Silva A, Vieira SM, Brito J, Gaspar PL (2016) Fuzzy systems in service life prediction of natural stone claddings. J Perform Constr Facil 04016005. doi:10.1061/(ASCE)CF.1943-5509.0000860

Sousa JMC, Kaymak U (2002) Fuzzy decision making in modeling and control. In: World scientific series in robotics & intelligent systems. World Scientific Pub. Co., Singapore

Stone M (1974) Cross-validatory choice and assessment of statistical predictions. J R Stat Soc B 36(1):111–147

Suhartono S, Subanar, Guritno S (2006) Model selection in neural networks by using inference of R^2 incremental, PCA and SIC criteria for time series forecasting. J Quant Methods J Devoted Math Stat Appl Var Fields 2(1):41–57

Sung AH (1998) Ranking importance of input parameters of neural network. Expert Syst Appl 15 (3–4):405–411

Szewczyk ZP, Hajela P (1994) Damage detection in structures based on feature-sensitive neural networks. J Comput Civ Eng 8(2):163–178

Tah JHM, Carr V (2000) A proposal for construction project risk assessment using fuzzy logic. Constr Manage Econ 18(4):491–500

Takagi T, Sugeno M (1985) Fuzzy identification of systems and its applications to modelling and control. IEEE Trans Syst Man Cybern 15(1):116–132

Tarski A (1956) On the concept of logical consequence. In: Logic, semantics, metamathematics. The Claredon Press, Oxford, pp 409–420

Tetko IV, Villa AEP, Livingstone DJ (1996) Neural network studies. 2. Variable selection. J Chem Inf Comput Sci 36(4):794–803

Topçu IB, Sarıdemir M (2008) Prediction of compressive strength of concrete containing fly ash using artificial neural networks and fuzzy logic. Comput Mater Sci 41(3):305–311

Usman F, Resdiansyah M (2008) Service life prediction of building components. In: ICCBT—international conference on construction and building technology, Kuala Lumpur, Malaysia, pp 287–296

Vieira SM, Sousa JMC, Durão FO (2005) Fuzzy predictive filters applied to predictive control of a column flotation process. In: Fuzzy logic, soft computing and computational intelligence. 11th international fuzzy systems association World congress, Beijing, China, pp 818–823

Vieira SM, Silva A, Sousa JMC, de Brito J, Gaspar PL (2015) Modelling the service life of rendered façades using fuzzy systems. Autom Constr 51:1–7

Von Neumann J (1958) The computer and the brain, 1st edn. Yale University Press, New Haven

Walczak S (2001) An empirical analysis of data requirements for financial forecasting with neural network. J Manage Inf Syst 17(4):203–222

Wang LX (1992) Fuzzy systems are universal approximators. In: IEEE international conference on fuzzy systems, San Diego, USA, pp 1163–1170

Wang J, Xu Z (2010) New study on neural networks: the essential order of approximation. Neural Netw 23(5):618–624

Whitehead AN, Russell B (1910) Principia mathematica, vol I, 1st edn. Cambridge University Press, London

Whitehead AN, Russell B (1912) Principia mathematica, vol II, 1st edn. Cambridge University Press, London

Whitehead AN, Russell B (1913) Principia mathematica, vol III, 1st edn. Cambridge University Press, London

Worden K (1997) Structural fault detection using a novelty measure. J Sound Vib 201(1):85–101

Xie XL, Beni G (1965) A validity measure for fuzzy clustering. IEEE Trans Pattern Anal Mach Intell 13(8):841–847

Zadeh LA (1965) Fuzzy sets. Inf Control 8(3):338–353

Zadeh L (1973) Outline of a new approach to the analysis of complex systems and decision processes. IEEE Trans Syst Man Cybern (SMC) 3(1):28–44

Zahedi F (1996) A meta-analysis of financial applications of neural networks. Int J Comput Intell Organ 1(3):164–178

Zhang GP, Patuwo BE, Hu MY (1998) Forecasting with artificial neural networks: the state of the art. Int J Forecast 14(1):35–62

Chapter 6
Factorial Models

6.1 Introduction

There are several methodologies available for service life prediction of buildings and their components. Among the existing models, stands out the factorial method, initially proposed by the Architects Institute of Japan (AIJ 1993), which set out the principles for the estimation of the service life of the construction elements, obtained through the multiplication of a reference service life by various durability factors, concerning the characteristics of the elements under analysis. Currently, rather than being a set methodology, the factor method is considered a general framework to estimate the service life of construction elements, allowing different procedures of quantification and expression of the durability factors, leading to distinct results, ranging from absolute values to probabilistic distributions. Furthermore, the factor method combines simplicity and flexibility, becoming one of the main methodologies for service life prediction, established in the international standard for durability already published (ISO 15686-1: 2011). In this chapter, the factorial method is applied to the service life prediction of façade coatings, using two approaches:

(A) A classic or deterministic approach, in which three scenarios are used in the quantification of the durability factors and sub-factors: a neutral scenario (where all the sub-factors are unitary); a scenario where the durability factors are quantified based on ISO 15686: 2011; and an optimized scenario, resulting from an iterative analysis, minimizing the deviations between the observed values and the values predicted by the factor method; in this approach, the sub-factors are quantified as absolute values;
(B) A probabilistic or stochastic approach, where a probability distribution is adjusted to each durability factor; in this approach the estimated service life is also given by a probability distribution.

A. Silva et al., *Methodologies for Service Life Prediction of Buildings*,
Green Energy and Technology, DOI 10.1007/978-3-319-33290-1_6

6.2 Classic Approach

6.2.1 Theoretical Foundations

The factorial method, recommended in the international standard ISO 15686: 2011, allows estimating the service life of buildings elements, covering all the variables that can influence their durability. According to this standard (ISO 15686: 2011), this method can be systematically applied in assessing the service life of the buildings components even if the reference conditions do not fully correspond to the use conditions. Using this method it is possible to estimate the service life of an element or assembly of elements subjected to specific conditions, calculating an estimated service life from a reference service life (usually the expected life for the in-use conditions of an element or assembly of elements) multiplied by a series of "modifying" factors related to the specific conditions of each case (Lair 2003; Silva et al. 2012)—Eq. (6.1). According to Cecconi and Iacono (2005), the number and combination of the durability factors applied in the factorial method depend on the type of building component under analysis.

$$ESL = RSL \cdot A \cdot B \cdot C \cdot D \cdot E \cdot F \cdot G \tag{6.1}$$

where, according to ISO 15686: 2011, *ESL* is the estimated service life, *RSL* the reference service life, *A* the factor related to the quality of the materials, *B* the factor related to the design level, *C* the factor related to the execution level, *D* the factor related to the interior environmental conditions, *E* the factor related to the external environmental conditions, *F* the factor related to the in-use conditions, and *G* the factor related to the level of maintenance.

The factor method is a simple process, relatively easy to apply, but it has been the object of some criticism, because of some limitations, particularly when it is seen as a purely deterministic tool:

- The method does not take into account the variability of the ageing process, i.e. it neglects the fact that the degradation phenomenon is not constant over time; in fact, this phenomenon has periods when degradation occurs more rapidly and others in which it stabilizes (Hovde 2000; Moser and Edvardsen 2002; Mc Duling et al. 2008);
- The service life is obtained based on the characteristics of the construction element, ignoring the degradation condition of the element (Shohet et al. 2002);
- The seven modifying factors cover the essential aspects to service life estimation; however, these factors significantly influence the service life of each specific element and its application must be evaluated for each specific case (Hovde 1998); small variations in the quantification of the durability factors can produce a very large range of estimated service lives (Cusmano et al. 2003); according to Cecconi and Iacono (2005), differences of 10 % in the quantification of the values of the durability factors lead to differences of around 80 %

in the values of the estimated service life; additionally, there is some subjectivity associated with the determination of the reference service life;
- This method leads to an absolute value (that represents the expected limit for the service life of the elements under analysis) and does not provide information concerning the dispersion of the results (Bourke 1999; Emídio et al. 2014).

The shortcomings of this approach led to the investigation for better models that intend to optimize the determination of the service life of the building elements. Aarseth and Hovde (1999) and Moser (1999) proposed a methodology to deal with the uncertainty associated with the factorial method, treating the durability factors as stochastic input parameters. In this methodology, the factorial method is nonetheless a purely deterministic method, since the estimated service life is characterized by a probability distribution, with a confidence interval associated.

In spite of these limitations, factorial methods based on the methodology of the Architects Institute of Japan are the most frequently applied methodologies for service life prediction, being generally accepted in the international community. Bourke and Davis (1999) refer that the factorial method can be used to estimate the expected service life of a construction element when there are no more precise scientific methods, by providing a common matrix for the determination of the service life in the absence of more reliable information. This methodology achieves the desired balance between accuracy, low cost and ease of application.

The durability factors intend to cover the essential aspects to the service life estimation; however, in each case, it is necessary to identify the relevant factors that influence the durability and service life of each element in each specific application (Hovde 1998). Lair (2003) refers that the application of the factorial method can be flexible and various alternative methods based on this approach may be proposed, since the factor method equation corresponds to a general framework that can accommodate different factors, sub-factors and their quantification methods (absolute values, statistical intervals, distribution functions, among others).

The reference service life (*RSL*) is defined according to ISO 15686: 2011 as the period of time that a building or its components are expected to last, when subjected to a set of reference conditions of use and maintenance. However, the reference service life is one of the most difficult variables to quantify. Hovde (1998) states that the reference service life can be determined based on experts opinion and on previous experience and knowledge of buildings and components' behaviour subjected to similar conditions. Other authors (Teplý 1999; Nireki et al. 2002; Aktas and Bilec 2012) suggest alternative methods, including scientific research, regulations and building standards, technical information from producers, laboratory tests and statistical analysis.

The factorial method is seen as a simplified tool (easy to understand and implement), however, the quantification of the durability factors and of the reference service life can become an indefinite calculation process, with numerous possible solutions. To apply this methodology the first step is to identify the durability factors. Hovde (2005) refers that simple procedures for the selection of the durability factors used in the factorial method are needed. The number of durability factors varies according to the element under analysis, although expression (6.1) of ISO 15686: 2011 suggests that the durability sub-factors are grouped into seven main categories:

- Quality of the materials (factor *A*)—this factor can undergo significant changes as apparently similar materials can present different performances and durability in service conditions. The material's characteristics can be evaluated based on laboratory testing, site testing or through the analysis of their behaviour when subjected to aggressive agents (Daniotti et al. 2008);
- Design level (factor *B*)—this factor assesses how the building design influences the service life of the elements under analysis. This factor takes into account variables such as volume and building peculiarities (at a macro level) and the constructive details of the facade cladding (at a micro level, for example, the joints' characteristics, dripping pans, among others) that can confer protection to the construction elements against the degradation agents (for example, the existence of roof eaves) or, on the other hand, can promote the degradation of the elements analysed (Marteinsson 2003);
- Execution level (factor *C*)—this factor analyses during the construction phase the accomplishment of the requirements and the procedures prescribed during the design stage, particularly regarding the quality of labour, storage conditions, the protection of elements during the construction phase, ease of installation, among others (van Nunen 2010). Haapio and Viitaniemi (2008) refer that several factors can affect the service life of buildings and their components, even before being placed into service; in fact, some execution errors inevitably affect the durability of the construction elements;
- Interior environmental conditions (factor *D*)—this factor is related with the exposure to degradation agents in the interior of buildings such as exposure to moisture (in kitchens and sanitary facilities);
- Exterior environmental conditions (factor *E*)—this factor takes into account the conditions of exposure to environmental agents in the exterior of buildings (for example, distance to pollution sources, climatic agents, presence of sea salts, among others);
- In-use conditions (factor *F*)—this factor reflects the type of use of the building; these conditions are relevant in service life prediction, since, for example, elements located in common areas of buildings are more subjected to wear and tear;
- Maintenance conditions (factor *G*)—this factor considers the planning of maintenance actions, including ease of inspection and maintenance (some building components are difficult to access or may require special equipment for that purpose).

ISO 15686 (2011) suggests the use of three standard values: 0.8 for unfavourable situations, 1.0 for current situations; and 1.2 for favourable situations. This proposal, although highly expedient, is over-simplistic, since standard values are unable to represent the inherent variability related to the environmental exposure conditions, such as use and maintenance conditions of the building elements. Marteinsson (2003) states that in reality the sub-factors of durability may have any value, depending on the difference between the actual conditions of use and the reference conditions. Indeed, the values of the sub-factors can vary continuously, and should not be limited to discrete values. Therefore, it seems necessary that methodologies are defined in order to quantify and adjust the sub-factors to each specific case.

Presently, there are several studies that propose more comprehensive but also more complex approaches to the factor method. Shohet et al. (1999) suggest that each durability factor can be achieved through the application of an influence coefficient, which reflects the effect that a given degradation mechanism has on the deterioration process of the overall system. Cecconi and Iacono (2005) propose the establishment of evaluation grids that provide the value of the various durability factors; these grids are built by information collected through expert surveys, in current standards and based on the commonly accepted construction procedures. However, there are no objective methods for quantifying the values that should be included in these grids, and there is even no consensus on which criteria to consider in the preparation of them.

To overcome the difficulties in the estimation of the durability factors, methodologies that combine the factor method with probabilistic models may be a well balanced solution. Cecconi (2004) suggests that durability factors can be treated as random variables; the author adopted a Monte Carlo method and used triangular probability distributions in the definition of durability factors. To a number of authors, the use of random variables seems more appropriate to describe the complexity of the degradation phenomenon, allowing more reliable results of the estimated service of the building elements (Daniotti and Spagnolo 2008).

6.2.2 Methodology

The aim of this chapter is to define models for predicting the service life of façade coatings based on the factorial method. For this, the first step in the development of these models involves the characterization of façade claddings, identifying the factors that influence its durability. Once the different claddings under analysis are characterized, it is possible to quantify the reference service life and the durability factors that enable the application of the factor method. The model validation is done by comparing the estimated service lives (*ESL*) obtained by the factorial method (*FM*) and the estimated service lives obtained by the graphical method (*GM*)—that represents the actual reality observed in the field work—, for each point within the sample.

6.2.2.1 Determination of the Estimated Service Life for Each Point Within the Sample

In practice the estimated service life can be graphically determined for all the case studies. For this, a curve correlated with the average degradation curve is made to pass through each of the case study points (Emídio et al. 2014). Whenever the average curve is conveniently described by a straight line (i.e. when the determination coefficient (R^2) has a sufficiently high value to allow concluding that there is a suitable adjustment of the line to the data), it is relatively straightforward to determine the estimated service life for each point within the sample. For this, it is only necessary to draw straight lines passing through each of the cases, determining the age (x) in which the maximum allowable degradation level (y) is reached. However, when a straight line is not a sufficiently good regression model, it is necessary to choose more complex functions such as polynomials of second or third degrees. In these cases, one has to select the mathematical operation that converts the function f (average degradation curve) into another function f', related to f, which passes through each of the sample's points. In order to obtain the estimated service life for each point, the factor k that relates the ordinates of any two points should be determined. Given a point B, with coordinates (x'; y'), and another A, with coordinates (x'; y), the ratio between the ordinates ($k = y'/y$) is determined and applied to the function f, to determine the function f' that passes through point B, according to expression (6.3) for a third degree polynomial function.

$$f' = k(f) \Leftrightarrow f' = k \cdot a \cdot x^n + k \cdot b \cdot x^{n-1} + \cdots + k \cdot c \cdot x \tag{6.2}$$

where: f represents the function of the average degradation curve of the material, k the conversion factor equal to the ratio between the ordinates of points A (any) and B (belonging to f) with the same abscissa as A.

This method is mathematically very simple and its application guarantees that the criterion is the same for all points. Furthermore, this operation has a very clear physical meaning and a direct application in linear, exponential, logarithmic or square root functions. However, the results in polynomial functions whose second derivative changes sign may suffer some changes, more specifically there may be an unrealistic increase of the predicted service life of the element.

6.2.2.2 Identification and Quantification of the Durability Factors

The four types of façade claddings analysed in this book have different characteristics; therefore and as would be expected, for each type of cladding different durability factors are considered. The sub-factors related to environmental exposure conditions are relatively similar for all types of claddings; in turn, the sub-factors related to material characteristics vary significantly depending on the type of cladding analysed. Given the characteristics of this study (that evaluates façades in service conditions based on visual inspections only), the following assumptions are initially made:

- Since in this book external claddings are being analysed, factor *D* was not considered, because the indoor environmental conditions do not present significant effects on exterior finishes;
- The claddings analysed in this book are evaluated in situ and in some cases many years after their execution, making it difficult or even impossible to determine the sub-factors regarding the execution level (factor *C*). The information concerning the quality of construction processes and the existence of control during the claddings' execution is not available in Municipalities; therefore, this factor is considered neutral in this study;
- During fieldwork, the information concerning the maintenance level (factor *G*) is also difficult to obtain: the contact with the owners can lead to erroneous conclusions and this information is not always available in Municipalities. Therefore, only two sub-factors are considered: (i) the first related to the existence of maintenance; (ii) the second concerning the ease of inspection. Since the first sub-factor is not easy to obtain, as an alternative, it is considered that the existence of maintenance actions can be indirectly evaluated through factor *F*, since the type of the buildings and of their users influences the maintenance actions taken during the buildings' life-cycle. As for ease of inspection, it was assumed that maintenance was easily carried out on buildings up to 5 storey high (current conditions); on the other hand, buildings with more than 5 storey and with irregular volume correspond to the unfavourable situation;
- In addition to the durability factors considered in this study, other factors such as the preparation of the substrate, the characteristics of bonding material (in the case of stone and ceramic claddings, the material for filling the joints, among others) that affect the service life of the claddings analysed may also be considered. However, due to the fact that this study is only based on field survey, they could not be ascertained.

Two different methods are employed for the quantification of the factors that influence the claddings' durability: (i) degradation curves (DCs)—this method is based on plotting the average degradation curve according to the claddings' characteristics and determining an average predicted service life for each characteristic considered (using a maximum degradation level of 20 %); (ii) based on this plotting, graphical models (GMs) express the predicted service life for each cladding as a function of its characteristics. These approaches are used to identify distribution patters separately for each factor and compare the average service life (obtained analytically) corresponding to each set of variables. A preliminary quantification of the durability factors, in order to calculate the reference service life, is required. This preliminary quantification is based on the values suggested by Mc Duling (2006): 0.8 (very aggressive level); 0.9 (aggressive); 0.95 (averagely aggressive); 1.0 (slightly aggressive); 1.1 (slightly favourable); 1.2 (clearly favourable).

In the classic approach to the factor method (as a deterministic model), three different scenarios are adopted: (i) a neutral scenario—all k values are assumed to be 1.00 in order to study the performance of a neutral model, even if without a real counterpart; this model serves only as comparison with other models, allowing ascertaining to what extent the factorial method is useful in service life prediction; (ii) the model based on ISO 15686: 2011—the k values are assigned in accordance with standard ISO 15686, i.e. 1.20 for favourable sub-factors, 1.00 for sub-factors that portray current situations and 0.80 for unfavourable sub-factors; and (iii) an optimized model—a combination of k values is obtained by solving equations or by manual iteration to optimize the solution; in other words, based on the previous scenarios, the k values are changed to obtain the best possible fit to sample, but taking into account the physical reality—favourable sub-factors always equal to or higher than average conditions factors and these always equal to or higher than unfavourable sub-factors.

6.2.2.3 Reference Service Life

In this study three methodologies to determine the reference service life of the claddings analysed are applied:

(i) Based on the degradation curves—in this method, the service life of a building element is obtained through the knowledge of its degradation evolution over time; for that, a comprehensive survey performed on a sufficiently representative sample is required; the average service life (obtained for the overall sample) is then determined graphically by the intersection of the degradation curve with the maximum acceptable degradation level, corresponding to the end of service life of façade claddings; however, generally, this theoretical limit is not easy to specify (Moser 2004), depending on numerous factors, some of them subjective, as mentioned in Chap. 2; in a simplified way, the reference service life can be assumed as equal to the average service life of all case studies obtained by this method;

(ii) The method of the average exposure conditions for a single example—in this method it is assumed that the sub-factors k of the factor method take the values preconized in standard ISO 15686 (1.2 for unfavourable situations, 1.0 for current situations and 0.8 for unfavourable situations). Therefore, based on the sample analysed during the fieldwork, the claddings that are subjected to a current average situation, i.e. those for which all the sub-factors take the value 1.0, are considered. In this case, the reference service life would be equal to the average of the predicted service life values, obtained through the factor method of conversion to the ordinates, for the case studies corresponding to the situation described. For that, the factor method was inversely applied to each element of the sample to evaluate the reference service life (*RSL*)—Eq. (6.3).

$$
\begin{aligned}
ESL &= RSL \cdot A \cdot B \cdot C \cdot D \cdot E \cdot F \cdot G \\
\Leftrightarrow RSL &= \frac{ESL}{A \cdot B \cdot C \cdot D \cdot E \cdot F \cdot G} \\
\Leftrightarrow RSL &= \frac{ESL}{(0.8^{v} \cdot 0.9^{w} \cdot 0.95^{x} \cdot 1.1^{y} \cdot 1.2^{z})}
\end{aligned} \tag{6.3}
$$

in which exponents v, w, x, y and z correspond to the number of times that each k value is present in the case study in question.

(iii) The method of the average conditions for the whole sample—in this method, as in the previous one, it was considered that all sub-factors k present the values specified in the ISO standard. The ratio between the predicted service life and the service life determined by the factor method was determined. As a selection criterion, only those claddings that had values of this ratio with deviations from the overall age less than 2 % were considered. The reference service life is then given by the average of values of the service life estimated that accomplished these criteria.

Finally, the reference service life of façade claddings analysed corresponds to an average of the three values obtained through the three methods listed above.

6.2.2.4 Model's Validation

The results of the various scenarios are evaluated by comparing the results obtained by the factorial method with the results obtained by the graphical method (Gaspar and de Brito 2008). The validation of the model is made assuming that the graphical method (based on the actual reality observed in the fieldwork) better portrays the degradation phenomena, providing a more accurate indicator of the service life of the claddings under analysis. The quality of the results of the simulations was interpreted with statistical markers (average, identification of extremes and cumulative frequency), by analysing the ratio between the service life obtained by applying the factor method (*FM*) and the values estimated through the conversion to the average degradation curve (*GM*). After the simulations, the following acceptance criteria were defined (Emídio et al. 2014; Galbusera et al. 2014):

- The average of the *FM*/*GM* ratios should not differ from 1 by more than 5 %, i.e. its maximum range is [0.95; 1.05];
- The difference between the highest and the lowest values of *FM*, i.e. the range of the values obtained, should be lower than the range of the *GM* values;
- The results of *FM* must fall within physically credible limits;
- The cumulative number of *FM*/*GM* ratios of 0.85 or greater (that is, conservative estimations) should be at least 50 % of the sample;
- The cumulative number of *FM*/*GM* ratios of 1.50 or greater (that is, failed estimations) should not be higher than 10 % of the sample;

- The iterations come to an end when the maximum number of *FM*/*GM* ratios falling in the range [0.85; 1.15] is reached.

6.2.3 Application to Façade Claddings

As mentioned before, the factorial method is currently seen as a general framework for service life prediction that can be applied to different construction elements, allowing taking into account the specific characteristics of each element. This methodology has been successfully applied in service life prediction of façade claddings, namely: (i) rendered façades (Gaspar and de Brito 2008); (ii) stone claddings (Silva et al. 2012); (iii) ceramic tiles (Galbusera et al. 2014); (iv) painted surfaces (Aktas and Bilec 2012); (v) external thermal insulation systems (ETICS) (Daniotti and Paolini 2005). In this study, the classic factorial method (deterministic approach) and the improved method (stochastic approach) are applied to predict the service life of the four types of exterior claddings analysed.

6.2.3.1 Natural Stone Claddings

Based on the degradation evolution of natural stone claddings, it is possible to determine the average estimated service life for the overall sample. In this case, a third degree polynomial curve is the degradation pattern that better adjusts to the sample analysed. The estimated service life can be obtained through the intersection between the degradation curve and the maximum acceptable degradation level, corresponding in this study, to a severity of degradation of 20 %. Using this approach, an estimated service life of 68.6 years is obtained.

To determine the estimated service life for each case study, the ordinates' conversion factor method (described in Sect. 6.2.2.1) is applied. Using this method, a third degree polynomial curve passes through each point in the sample, leading to an average estimated service life of 70.5 years. The maximum value obtained is 157.4 years, corresponding to a case study with 5 years at the inspection time, with a very low value of severity of degradation, around 0.07 %; on the other hand, the minimum value obtained is around 32 years, from a case study with 2 years at the inspection time, with a severity of degradation of 2 %, a relatively high value considering the sample analysed. Figure 6.1 shows the mean degradation curve and dashed curves represent the upper and lower limits of the sample. The results obtained seem realistic and adjusted to the reality and therefore none of the case studies was excluded of the following analysis. The existence of a case study with an expected service life of 157 years (without having been subjected to maintenance operations), when subjected to a very favourable combination of conditions, seems acceptable. Several studies attesting the durability of natural stone as a

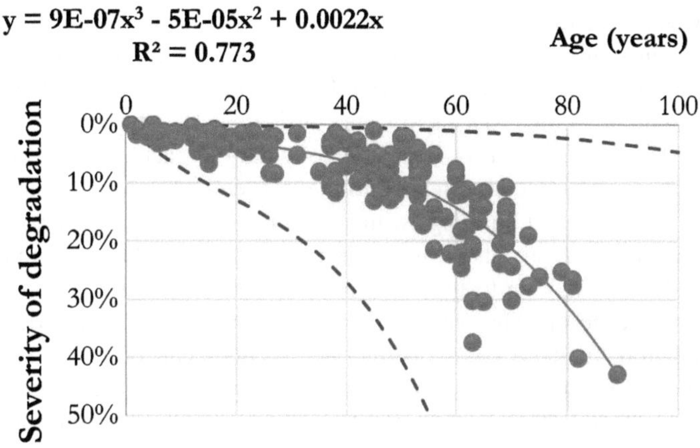

Fig. 6.1 Average degradation curve and upper and lower degradation patterns (stone claddings)

building material can be found in literature (Winkler 1997). The English standard BSI 7543 (1992) recommends that the service life of stone claddings exceeds 60 years for housing and Bell (1993) states that natural stone will be able to meet the demands imposed on it for at least this period. According to Brand (1997), in some countries, natural stone has demonstrated the ability to achieve expected service lives greater than 180 years. In fact, Schafer (2002) mentions that the natural stone claddings have long service lives and are replaced in most situations for aesthetic reasons only.

Identification of the Durability Factors of Stone Claddings

In reality, the durability of stone claddings is influenced by several parameters; however, this study only addresses some of these factors, whose determination can be carried out using visual inspections only. In the first durability factor, related with the quality of the materials used—factor *A*—the type of stone is considered, assuming three categories: granite; marble; and limestone (the most common types of stone used as external coatings). In factor *B*, related to the design level, five durability factors are analysed: *B*1—colour of the natural stone, with two categories: light and dark colours; *B*2—type of finishing (smooth or rough); *B*3—area of stone plates, assuming two categories: large-sized plates, with an area higher than 0.4 m^2; and medium-sized plates for the opposite situation; *B*4—thickness of the natural stone plate, with two groups: natural stone plates with thicknesses of less than 2.5 cm and plates with thicknesses greater than this value; *B*5—location of the cladding, considering two categories: elevated cladding (integral or partial); and in accessible areas (bottom wall cladding). Factor *C* (regarding the execution level) is not considered in this study, since the claddings were assessed based on visual inspections only and many years after execution; which make it very difficult or even impossible to determine the original execution level. Factor *D* is also not considered, since internal environment conditions do not have a significant effect on the external cladding's durability. Regarding factor *E*, related with the external environmental conditions, four factors are considered: *E*1—façades orientation; *E*2—distance from the sea; *E*3—exposure to wind-rain action; *E*4—exposure to damp. Concerning the factor *F* related to the in-use conditions, two categories are considered: housing; and buildings with commerce or services. In factor *G*, related to the maintenance level, only ease of inspection is considered, comprising two situations: (i) buildings up to three stories where the inspection actions are easily achievable from the outside, without the need of scaffolding that correspond to a current situation (in these cases, maintenance actions can be easily carried out); (ii) on the other hand, buildings over three stories with irregular shape that correspond to an unfavourable situation, hindering inspection and maintenance actions (Emídio et al. 2014).

Since the factors affecting the stone claddings' durability have been identified and classified, it is necessary to attribute a weighting value to each sub-factor. For that, two methods are used, the degradation curves (DCs) and plotting graphical models (GMs) that express the predicted service life for each cladding as a function of its characteristics, as described in Sect. 6.2.2.2. Table 6.1 shows the estimated service lives (*ESL*) for the durability sub-factors of stone claddings, as well as the weighting coefficients assigned to each durability factor (*k*). The estimated service lives obtained according to the different characteristics show consistent results, adjusted to the observed reality, allowing obtaining the following conclusions:

- Concerning the type of stone, granite is the more durable one, with the highest service lives, followed by limestone and marble; Schouenborg et al. (2007) tested the mechanical strength of 200 natural stone samples, based on laboratory tests and in situ evaluation, achieving similar results;
- Regarding the claddings' colour, light colours have higher estimated service lives when compared with dark colours; however, there is no indication in the literature that suggests a clear pattern of degradation associated with the claddings' colour;
- Claddings with rough finishing presents lower estimated service lives, probably due to the greater exposed area, thus being more susceptible to the environmental degradation agents;
- Regarding the area of stone claddings, the results reveal that claddings with large-sized plates reach the end of their service life sooner; in fact, as mentioned by Silva et al. (2012), the larger the dimensions of the stone plates, the smaller the area occupied by joints, which can lead to greater concentration of stresses. On the other hand, larger plates correspond to heavier elements also requiring greater thicknesses in order to prevent premature fracture during transport and positioning. Finally, the larger the contact area of each plate the more difficult it is to guarantee that the adhesive material is evenly spread on its back. Therefore, larger stone plates directly adhered to the substrate are more prone to cracking and detachment from the substrate;

Table 6.1 Estimated service lives and weighting coefficients for each durability factor (stone claddings)

Durability factors			Average ESL (DCs)	Average ESL (GMs)	Weighting value (*k*)
*A*1	Type of stone	Limestone	71.4	69.8	1.10
		Granite	83.1	88.0	1.20
		Marble	66.0	68.5	1.00
*B*1	Colour	Light colours	69.2	70.9	1.00
		Dark colours	60.3	64.2	0.80
*B*2	Type of finishing	Smooth	68.5	71.2	1.00
		Rough	68.4	70.1	0.95
*B*3	Size of stone plates	Medium	71.5	74.6	1.20
		Large	66.0	67.7	1.00
*B*4	Thickness of stone plates	Less than 2.5 cm	66.9	68.9	0.80
		≥2.5 cm	69.2	72.7	1.00
*B*5	Location of the cladding	Partial and integral elevated cladding	65.4	66.2	0.90
		Bottom wall cladding	69.0	71.9	1.00
*E*1	Façade orientation	North	64.9	66.4	0.90
		NE/E/SE	69.4	72.1	1.00
		West/NW	63.9	66.0	0.90
		South/SW	71.5	74.0	1.20
*E*2	Distance from the sea	Less than 5 km	66.2	64.8	1.00
		More than 5 km	70.2	73.3	1.20
*E*3	Exposure to wind-rain action	Moderate	68.3	70.5	1.00
		Severe	67.5	69.4	0.90
*E*4	Exposure to damp	Low	70.6	74.6	1.00
		High	62.7	63.5	0.80
*F*1	Type of property	Housing	68.1	71.0	1.10
		Commerce and services	68.6	69.8	1.00
*G*1	Ease of inspection	Current	68.9	71.9	1.20
		Unfavourable	59.7	62.5	1.00

- Donaldson (1990) states that the minimum thickness of stone plates is directly related with the quality of the natural stone, the type of bonding to the substrate, the stresses to which the cladding will be subjected and the applicable standards. In this study, claddings with thicker stone plates present higher expected service lives; in fact, claddings with thinner stone plates are more susceptible to various defects, such as: moisture stains; loss of integrity; and cracking;
- Concerning the location of the cladding, bottom wall claddings present higher service lives, which can be explained by various factors: (i) the better accessibility of this type of claddings, whose maintenance operations are facilitated

(although these type of claddings are more prone to vandalism)—even though these factors are not being considered in this study; (ii) higher protection levels against the environmental degradation agents; and (iii) the type of material applied, since it has a protective function, is usually more durable and thicker when applied in the bottom wall;

- Regarding the façades orientation, stone claddings facing North and West are less durable; as mentioned before, these are the most aggressive quadrants in Portugal, with the highest prevalence of occurrence of moisture and combined occurrence of wind and rain; on the contrary, claddings facing South, are those with higher estimated service lives, as would be expected;
- Concerning distance from the sea, claddings in coastal areas have shorter service lives, due to the exposure to winds containing harmful marine salts;
- Claddings with severe exposure to wind-rain action are less durable, with estimated service lives lower than those with moderate exposure;
- Claddings with high exposure to damp present lower service lives when compared with claddings with low exposure;
- In the sample analysed, buildings with services and commerce present higher estimated service lives, which can be justified by the detailing level of this type of façades (and even though that has not been analysed in this study, it is considered that this type of buildings can be subjected to maintenance operations more often than housing buildings);
- The ease of inspection is directly related to the height of the building; some researchers show that the higher the building, the greater their susceptibility to weathering effects (Westberg et al. 2001); in this study, buildings whose monitoring is more difficult (higher) present lower estimated service lives.

Reference Service Life of Stone Claddings

As mentioned in Sect. 6.2.2.3, three methods are used to determine the reference service life of stone claddings. The first method, in which the reference service life is obtained graphically, leads to a reference service life of 68.6 years. According to Emídio et al. (2014), if the sample were sufficiently representative of the stone claddings in a region or country, then the reference service life could be obtained by this method. Using the second method, in which the average exposure conditions for a single case study are used, a reference service life of 68 years is obtained. For the method of the average conditions for the whole sample, a reference service life of 68.4 years is achieved. Finally, for the application of the factor method to the service life estimation of natural stone claddings, a reference service life of 68.3 years is adopted, corresponding to the average values obtained through the three proposed methods.

Determination of the Estimated Service Life of Stone Claddings

Once the durability factors to include in factor method are defined—and after the determination of the reference service life—the mathematical expression that enables the practical application of the factor method to the service life of stone claddings can be defined—Eq. (6.4).

$$ESL = RSL \cdot A1 \cdot B1 \cdot B2 \cdot B3 \cdot B4 \cdot B5 \cdot E1 \cdot E2 \cdot E3 \cdot E4 \cdot F1 \cdot G1 \quad (6.4)$$

where *ESL* represents the estimated service life, *RSL* the reference service life (equal to 68.3 years), *A*1 the type of stone, *B*1 the colour, *B*2 the type of finishing, *B*3 the size of stone plates, *B*4 the thickness of stone plates, B5 the location of the cladding, *E*1 the façades orientation, *E*2 the distance from the sea, *E*3 the exposure to wind-rain action, *E*4 the exposure to damp, *F*1 the type of property, *G*1 the ease of inspection.

In the determination of the estimated service life using the factor method, three scenarios are analysed: (i) neutral scenario, where all sub-factors *k* assume the value 1; (ii) ISO 15686 (2011) scenario; (iii) optimized scenario, which aims to minimize the differences between the values of estimated service life for each case study, obtained by using the factor method and the corresponding ones estimated by the graphical method. In the optimized scenario, some restrictions are imposed in order to calibrate the durability sub-factors, thus ensuring the physical meaning of these values:

- $f_{\text{granite}} \geq f_{\text{marble}}$;
- $f_{\text{granite}} \geq f_{\text{limestone}}$;
- $f_{\text{medium-sized}} \geq f_{\text{large-sized}}$;
- $f_{\text{more than 2.5cm}} \geq f_{\text{less than 2.5cm}}$;
- $f_{\text{South}} \geq f_{\text{North}}$;
- $f_{\text{South}} \geq f_{\text{West}}$;
- $f_{\text{South}} \geq f_{\text{East}}$;
- $f_{\text{more than 5km from the sea}} \geq f_{\text{less than 5 km from the sea}}$;
- $f_{\text{moderate exposure to wind-rain action}} \geq f_{\text{severe exposure to wind-rain action}}$;
- $f_{\text{low exposure to damp}} \geq f_{\text{high exposure to damp}}$;
- $f_{\text{current inspection conditions}} \geq f_{\text{unfavourable inspection conditions}}$.

In the optimized scenario a non-negative restriction rule is also included, implying that all the durability factors are greater than zero. Additionally, it is ensured that all the durability factors are between 0.6 and 1.5 (assuming a deviation of 25 % from the values prescribed in ISO 15686: 2011). The optimization of the durability factors is performed by varying (in a positive and negative way) their values at intervals of 0.025. Table 6.2 shows the weighting of the durability sub-factors according to the three scenarios analysed. Tables 6.3 and 6.4 present the statistical indicators applied in the results discussion, according to the criteria defined in Sect. 6.2.2.4.

Table 6.2 Weighting of the durability sub-factors according to the three scenarios analysed (stone claddings)

Durability factors			% of case studies (%)	Neutral	ISO 15686: 2011	*A*1
*A*1	Type of stone	Limestone	35	1.0	1.00	1.00
		Granite	27	1.0	1.20	1.10
		Marble	38	1.0	0.80	0.95
*B*1	Colour	Light colours	66	1.0	1.00	1.025
		Dark colours	34	1.0	0.80	1.00
*B*2	Type of finishing	Smooth	47	1.0	1.00	1.10
		Rough	53	1.0	0.80	1.00
*B*3	Size of stone plates	Medium	64	1.0	1.00	1.00
		Large	36	1.0	0.80	0.90
*B*4	Thickness of stone plates	Less than 2.5 cm	76	1.0	1.00	1.00
		≥2.5 cm	24	1.0	1.20	1.10
*B*5	Location of the cladding	Partial and integral elevated cladding	30	1.0	0.80	0.95
		Bottom wall cladding	70	1.0	1.00	1.00
*E*1	Façade orientation	North	14	1.0	0.80	0.90
		NE/E/SE	43	1.0	1.00	0.95
		West/NW	25	1.0	0.80	0.90
		South/SW	17	1.0	1.20	1.00
*E*2	Distance from the sea	Less than 5 km	38	1.0	1.00	1.00
		More than 5 km	62	1.0	1.20	1.15
*E*3	Exposure to wind-rain action	Moderate	84	1.0	1.00	1.00
		Severe	16	1.0	1.20	1.00
*E*4	Exposure to damp	Low	59	1.0	1.20	1.00
		High	41	1.0	1.00	0.90
*F*1	Type of property	Housing	49	1.0	1.20	1.00
		Commerce and services	51	1.0	1.00	1.00
*G*1	Ease of inspection	Current	54	1.0	1.00	1.00
		Unfavourable	46	1.0	0.80	0.90

The optimized scenario presents the best results, in which the durability sub-factors are obtained through successive iterations of the *k* values considering their physical meaning. In this scenario, the percentage of failed estimations (considered as the cumulative number of *FM*/*GM* ratios of 1.50 or greater, which correspond to a deviation higher than 50 % between the values predicted by the *FM* and the *GM*) is very small, around 4.4 %. Using the optimized scenario, a cumulative frequency of *FM*/*GM* greater than or equal to 0.85 of approximately 76.4 % is obtained, also the cumulative frequency of results of *FM*/*GM* ranged between 0.85 and 1.15 is approximately 60.1 %, thus revealing small deviations between the

Table 6.3 Statistical indicators for the scenarios analysed for stone claddings

		Neutral	ISO 15686: 2011	Optimized
Average of *FM/GM*		1.02	0.87	0.99
Standard deviation *FM/GM*		0.24	0.41	0.25
Range of results	*FM* (years)	0	148.5	52.6
	GM (years)	125.4	125.4	125.4
Extreme values of *FM*	Maximum (years)	68.3	170.0	97.4
	Minimum (years)	68.3	21.5	44.8
FM/GM ≥ 0.85 (%)		77.3	45.3	76.4
FM/GM ≥ 0.70 (%)		94.1	61.1	90.6
FM/GM ≥ 1.50 (%)		3.0	7.9	4.4
0.85 ≤ *FM/GM* ≤ 1.15 (%)		50.7	30.5	60.1
Ratio between the *ESL* obtained by *FM* and *GM*		0.97	0.84	0.95

Table 6.4 Statistical indicators regarding the ESL obtained by the different scenarios for stone claddings

Statistical indicator	Scenarios			Graphical method
	Neutral	ISO 15686: 2011	Optimized	
Average ESL	68.3	59.2	66.9	70.5
Median ESL	68.3	50.4	66.3	68.1
Standard deviation ESL	–	25.2	10.2	17.1
Confidence interval of 95 %	–	±4.94	±2.0	±3.4

ESL predicted by the *FM* and the values obtained by *GM*. The scenario based on the ISO 15686: 2011 presents the worst global results, with the highest number of failed estimations, around 7.9 % and with the highest deviations between the values predicted by the two models (with a cumulative number of *FM*/*GM* in the range [0.85; 1.15] of around 30.5 %). Therefore, the sub-factors proposed in the optimized scenario are the most appropriate to estimate the service life of stone claddings and are the values with the best adjustment to the sample analysed. In this scenario, assuming the most optimistic situation, where all the durability factors were favourable, an *ESL* of 107.2 years is obtained; on the contrary, the worst case scenario leads to an *ESL* of 40.4 years; these values fall within the range of estimated service lives obtained by the graphical method.

6.2.3.2 Rendered Façades

Regarding rendered façades, a linear pattern is adopted to express the average degradation curve. This assumption allows simplifying the factor method, applying the principle of parsimony, i.e. when there are two degradation curves that conveniently describe the loss of performance, the curve whose mathematical manipulation is simpler is chosen; in fact, the curve obtained by a third polynomial degree only provides an increase of precision of 3 % when compared with a linear trend line, which does not justify the increase of complexity that the application of this curve implies.

In order to rule out possible outliers, i.e. excessively low or high values, it is considered necessary to establish a lower and an upper limit for the estimated service life values. A lower limit of 5 years is considered, featuring the warranty period of the buildings in Portugal, considering that all the defects that appear during this period will be corrected at the expense of the contractor.

The definition of an upper limit is more complex because involves the question of "how much a current render can last under normal conditions of use and exposure, without maintenance?" The service life values of rendered façades are not consensual and vary significantly in the literature:

- The Japanese guide (AIJ 1993) suggests that the service life of current renderings should exceed 10 years;
- The EOTA (European Organisation for Technical Approvals) relates the durability of construction with the durability of its components; in the case of current buildings (with an expected service life of 50 years), rendered façades should present an estimated service life of 10 years (since this element is easily repairable or replaced);
- Flores-Colen and de Brito (2010) associate the durability of renderings with the maintenance strategies adopted during their life-cycle, defining for single-layer renders the following values for service life: 13.5 years, without maintenance; 16 years, with periodic cleaning; 40 years, with periodic minor repairs; 22 years, subject to a major repair;
- Shohet et al. (1999) obtained an estimated service life for rendered façades of 20 years;
- Shohet and Paciuk (2004) define two minimum performance limits: the first corresponds to situations where the cladding must have a high level of performance, achieving an expected service life of 15 years (with estimated service lives ranging from 12 to 19 years); the second corresponds to a lower level of performance, related to situations where building owners wish to minimize maintenance actions, obtaining an expected service life of 23 years (with estimated service lives ranging from 19 to 27 years);
- ISO 15686: 2000 defines for rendered façades an estimated service life of 25 years;
- A study performed by CSTB (Centre Scientifique et Technique du Bâtiment) reveals that renderings should meet the performance requirements for 50 years, when subjected to periodic maintenance (e.g., cleaning, localized repair and, in some cases, painting);
- Hed (1999) relates the service life of renderings with performance criteria, referring that a rendered façade lasts about 60 years, considering structural

performance criteria, and lasts around 20 years, considering aesthetic criteria; however, these values are associated with maintenance policies;
- The British Standard 7543 (BSI 1992), which proposes that the service life of buildings should be determined based on the type of use, states that renderings must have, with adequate periodic maintenance, a service life similar to the building, i.e. a service life higher than 60 years;
- Mithraratne and Vale (2004) established for renderings an estimated service life of 60 years.

According to the cited, the service life of rendered façades varies between five years (subjected to severe environmental conditions, without maintenance) and 60 years (assuming they are subjected to periodic maintenance). Based on these values, an upper limit for the service life of rendered façades of 60 years is adopted. The adopted range ([5: 60] years) is considered to be physically credible. The case studies with an estimated service life outside these limits are removed from the sample. None of the case studies within the sample has an estimated service life (*ESL*) below 5 years (the minimum estimated service life obtained is around 7 years). On the other hand, there are 12 case studies with an estimated service life that exceeds 60 years and 3 case studies whose degradation is equal to 0 % at inspection time (the model is incapable of predicting the service life of these situations).

Based on these assumptions, the service life is estimated for each case study; an average ESL around 19 years is obtained with a standard deviation around 9 years. Through the intersection of the overall degradation curve (with a sample without outliers—85 study cases) with the maximum acceptable degradation level (considered in this study as a degradation index—S_w of 20 %), an average estimated service life of 14 years is obtained.

Identification of the Durability Factors of Rendered Façades

In the application of the factor method to rendered façades and due to the impossibility of collecting all the information concerning the characteristics of renderings, the following factors are considered:

- Regarding factor *A*, there are various characteristics that could be taken into consideration (e.g. those related to the number of layers of the render, its modulus of elasticity, the mortar's composition, etc.); however, since this study is based on a visual survey, it is not possible to obtain reliable data related to the all of these characteristics. Therefore, only the type of mortar is considered in factor *A*;
- In factor *B*, seven sub-factors are considered: (*B*1) façade colour; (*B*2) building geometry; (*B*3) eaves' protection; (*B*4) protection on parapets in roofs and terraces (flashings or copings); (*B*5) protection on balcony parapets; (*B*6) ground floor protection (socle); (*B*7) detailing/design level;

- Regarding the environmental factors, with serious implications on the renderings' durability, five sub-factors are analysed: (*E*1) façade orientation; (*E*2) distance to the sea; (*E*3) exposure to damp; (*E*4) distance to pollution sources; (*E*5) façade protection level, due to the presence of other buildings, trees or other barriers to prevalent winds;
- For factor *F*, related to the in-use conditions, only the property type was taken into account;
- Regarding factor *G*, related to the level of maintenance, only the ease of inspection of the façade was considered;
- Factor *C* (execution level) was not considered due to the reasons mentioned for stone claddings, likewise, factor *D* was not considered.

Table 6.5 shows the estimated service lives according to the durability factors considered as their preliminary quantification. The results obtained are coherent and make sense from an empirical point of view, leading to the following conclusions:

Table 6.5 Estimated service lives and weighting coefficients for each durability factor (rendered façades)

Durability factors			Average ESL (DCs)	Average ESL (GMs)	Weighting value (*k*)
*A*1	Render type	Lime-cement renderings	13	15	0.90
		Current cement renderings	17	19	1.00
		Renderings with crushed marble	19	23	1.10
		Single-layer renderings	17	24	1.10
*B*1	Colour	White	12	18	0.95
		Light colours	14	19	1.00
		Dark colours	16	19	1.00
*B*2	Building geometry	Compact	13	19	1.00
		Irregular	17	18	0.95
*B*3	Eaves' protection	Without protection	12	20	1.00
		With protection	15	18	1.00
*B*4	Platbands copings	Without copings	14	17	1.00
		With copings	16	24	1.00
*B*5	Balcony copings	Without copings	13	18	1.00
		With copings	18	20	1.10
*B*6	Ground floor protection (socle)	Without protection	14	20	1.00
		With protection	14	19	1.00

(continued)

Table 6.5 (continued)

Durability factors			Average ESL (DCs)	Average ESL (GMs)	Weighting value (k)
*B*7	Detailing/design level	Inferior	14	16	0.90
		Medium	14	20	1.00
		Superior	19	22	1.10
*E*1	Façade orientation	East/SE	15	19	1.00
		North/NE	13	7	0.90
		West/NW	14	20	1.00
		South/SW	17	21	1.10
*E*2	Distance to the sea	Less than 5 km	13	17	0.80
		More than 5 km	14	19	1.00
*E*3	Exposure to damp	Unfavourable	12	17	0.90
		Normal	12	17	0.90
		Favourable	16	20	1.00
*E*4	Distance to pollution sources	Unfavourable	14	19	0.90
		Normal	14	19	0.90
		Favourable	16	22	1.00
*E*5	Façade protection level	Without protection	13	18	1.00
		Normal situation	15	18	1.00
		With protection	16	20	1.10
*F*1	Type of property	Private	14	19	1.00
		Public sector	14	15	0.90
		Commerce and services	22	24	1.10
*G*1	Ease of inspection	Normal	14	18	1.00
		Unfavourable	15	19	1.00

- Renderings with crushed marble (similar to current cement renderings but containing coarser aggregates that confer a rough finishing) and single-layer renderings (rich in cement and usually premixed) have higher estimated service live values, indicating a better performance of these types of mortars, corresponding to the favourable situations. Renderings with mixed binders (mix of cement and lime) are the most prone to degradation, i.e. with the lowest estimated service life, corresponding to the most unfavourable situations. Current cement renderings correspond to an intermediate situation;
- White renderings present the lowest estimated service lives, corresponding to the unfavourable situation. There is no clear distinction between the three categories considered. For light and dark colours the results did not show a clear

degradation trend, therefore, these two categories are considered the current situation. These results can be justified due to the fact that it is easier to visually perceive defects in lighter than in darker façades, which tend to mask the defects;

- Buildings with a more complex shape (irregular), i.e. with protruding elements and complex construction details, present lower estimated service lives, suffering faster degradation, consequently corresponding to the unfavourable situation;
- Concerning protections on eaves, ground floor, parapets and balconies, façades without protection usually correspond to the unfavourable situation, with lower estimated service lives;
- In the sample analysed, as the level of detailing in design increases, the performance of rendered façades improves;
- Regarding the façades orientation, South corresponds to the most favourable situation, with higher estimated service lives; renderings facing East/SE and W/SW correspond to the intermediate or current situations. Renderings facing North reveal the lowest estimated service lives, corresponding to the unfavourable situation;
- Renderings located less than 5 km from the sea show lower estimated service lives, corresponding to the unfavourable situation; on the opposite, rendered façades located more than 5 km from the sea show lower levels of degradation, corresponding to the current situation;
- Concerning exposure to damp and pollution sources, renderings with favourable exposure conditions are more durable, showing higher service lives;
- For the protection level from wind-rain action, protected façades correspond to the favourable situation and are more durable than those without protection;
- In the factor related to the type of use, façades located in buildings with services and commerce, such as hotels or private owned office buildings that regularly subjected to maintenance actions correspond to the favourable situation; this can be explained both by the funds available for maintenance actions and the fact that a well-tended building is part of the corporate image of such activities (Brand 1997; Silva et al. 2016). Based on the sample analysed, it is also considered that low rent residential buildings correspond to the unfavourable situation as opposed to private condominiums;
- For ease of inspection, it is not possible to draw unequivocal conclusions; maintenance actions and the characterization of the degradation condition of renderings were more easily performed in buildings up to three stories; in the sample analysed, renderings located in higher buildings have higher service lives, which can be explained by the difficulty of correctly assessing the degradation condition of renderings located in higher buildings; therefore, this durability factor is considered neutral.

Reference Service Life of Rendered Façades

According to the three methods applied to the estimation of the reference service life of rendered façades (described in Sect. 6.2.2.3), the following results are obtained: (i) in the first method, which allows obtaining the reference service life by the intersection of the average degradation curve and the maximum acceptable degradation level, a reference service life of 14.0 years is obtained; (ii) using the method of average exposure conditions for a single case study, a reference service life of 16.0 years is obtained; (iii) for the method of the average conditions for the whole sample, a reference service life of 18.3 years is achieved. In this study, the reference service life of rendered façades adopted corresponds to an average of the three values obtained, in this case 16.1 years.

Determination of the Estimated Service Life of Rendered Façades

Based on the generic model defined in ISO 15686-1, it is possible to define the mathematical expression to estimate the service life of rendered façades, taking into account the factors that contribute to their degradation—Eq. (6.5).

$$ESL = RSL \cdot A1 \cdot B1 \cdot B2 \cdot B3 \cdot B4 \cdot B5 \cdot E1 \cdot E2 \cdot E3 \cdot E4 \cdot E5 \cdot F1 \cdot G1 \quad (6.5)$$

where *ESL* represents the estimated service life, *RSL* the reference service life (equal to 16.1 years), *A*1 the render type, *B*1 façade colour, *B*2 building geometry, *B*3 eaves' protection, *B*4 protection of parapets in roofs and terraces, *B*5 protection of balcony parapets, *B*6 ground floor protection (socle), *B*7 detailing/design level, *E*1 façade orientation, *E*2 distance from the sea, *E*3 exposure to damp, *E*4 distance from pollution sources, *E*5 façade protection level, *F*1 in-use conditions, and *G*1 ease of inspection of the façade.

As mentioned in Sect. 6.2.2.2, the application of the factor method implies the analysis of three scenarios in the quantification of the durability factors (Table 6.6). In the first scenario, all the values of *k* are equal to 1.00. In the second scenario, the quantification of the durability factors is made based on the values proposed by ISO 15686 (2011). In the third scenario, a set of restrictions is imposed in order to adjust the durability sub-factors, taking into account the influence of the different characteristics in the renderings' degradation:

- $f_{\text{compact}} \geq f_{\text{irregular}}$;
- $f_{\text{with platbands copings}} \geq f_{\text{without platbands copings}}$;
- $f_{\text{with balcony copings}} \geq f_{\text{without balcony copings}}$;
- $f_{\text{with ground floor protection}} \geq f_{\text{without ground floor protection}}$;
- $f_{\text{superior design level}} \geq f_{\text{current design level}} \geq f_{\text{inferior design level}}$;
- $f_{\text{South}} \geq f_{\text{North}}$;
- $f_{\text{South}} \geq f_{\text{West}}$;
- $f_{\text{South}} \geq f_{\text{East}}$;

Table 6.6 Weighting of the durability sub-factors according to the three scenarios analysed (rendered façades)

Durability factors			% of case studies (%)	Neutral	ISO 15686: 2011	Optimized
*A*1	Render type	Lime-cement renderings	23.5	1.0	0.8	0.875
		Current cement renderings	58.8	1.0	1.0	1.025
		Renderings with crushed marble	5.9	1.0	1.2	1.325
		Single-layer renderings	11.8	1.0	1.2	1.425
*B*1	Colour	White	17.6	1.0	0.8	0.700
		Light colours	51.8	1.0	1.0	1.025
		Dark colours	30.6	1.0	1.0	1.125
*B*2	Building geometry	Compact	58.8	1.0	1.2	1.025
		Irregular	41.2	1.0	1.0	0.975
*B*3	Eaves' protection	Without protection	43.5	1.0	1.0	1.000
		With protection	56.5	1.0	1.0	1.125
*B*4	Platbands copings	Without copings	78.8	1.0	1.0	0.850
		With copings	21.2	1.0	1.2	1.200
*B*5	Balcony copings	Without copings	64.7	1.0	1.0	1.000
		With copings	35.3	1.0	1.2	1.200
*B*6	Ground floor protection (socle)	Without protection	16.5	1.0	1.0	1.000
		With protection	83.5	1.0	1.0	1.050
*B*7	Detailing/design level	Inferior	29.4	1.0	0.8	0.800
		Medium	65.9	1.0	1.0	1.000
		Superior	4.7	1.0	1.2	1.150
*E*1	Façade orientation	East/SE	23.5	1.0	1.0	1.050
		North/NE	37.6	1.0	0.8	0.825
		West/NW	17.6	1.0	1.0	1.050
		South/SW	21.2	1.0	1.2	1.250
*E*2	Distance from the sea	Less than 5 km	25.9	1.0	0.8	0.950
		More than 5 km	74.1	1.0	1.0	1.050
*E*3	Exposure to damp	Unfavourable	12.9	1.0	0.8	1.000
		Normal	23.5	1.0	1.0	1.100
		Favourable	63.5	1.0	1.2	1.125
*E*4	Distance from pollution sources	Unfavourable	24.7	1.0	0.8	0.800
		Normal	69.4	1.0	1.0	0.800
		Favourable	5.9	1.0	1.2	1.500
*E*5	Façade protection level	Without protection	47.1	1.0	0.8	1.000

(continued)

Table 6.6 (continued)

Durability factors			% of case studies (%)	Neutral	ISO 15686: 2011	Optimized
		Normal situation	21.2	1.0	1.0	1.000
		With protection	31.8	1.0	1.2	1.000
*F*1	Type of property	Private	80.0	1.0	1.0	0.800
		Public sector	16.5	1.0	0.8	1.075
		Commerce and services	3.5	1.0	1.2	1.075
*G*1	Ease of inspection	Normal	61.2	1.0	1.0	1.225
		Unfavourable	38.8	1.0	1.0	1.000

- $f_{\text{more than 5 km from the sea}} \geq f_{\text{less than 5 km from the sea}}$;
- $f_{\text{favourable exposure to damp}} \geq f_{\text{normal exposure to damp}} \geq f_{\text{unfavourable exposure to damp}}$;
- $f_{\text{favourable exposure to pollution}} \geq f_{\text{normal exposure to pollution}} \geq f_{\text{unfavourable exposure to pollution}}$;
- $f_{\text{with protection against wind-rain action}} \geq f_{\text{current protection against wind-rain action}} \geq f_{\text{without protection against wind-rain action}}$;
- $f_{\text{current inspection conditions}} \geq f_{\text{unfavourable inspection conditions}}$.

Table 6.7 shows some statistical indicators used to evaluate the results produced by the factor method according to the various scenarios analysed. The optimized scenario is the one that provides more consistent results from a physical point of view, leading to the closest results between the values obtained by the factor method and the graphical method. Table 6.8 summarizes the statistical indicators used in the evaluation of the scenarios studied. For the three scenarios, the average ratio between the results of *FM* and *GM* is close to 1, thus revealing low deviations between the

Table 6.7 Statistical indicators for the scenarios analysed for rendered façades

		Neutral	ISO 15686: 2011	Optimized
Average of *FM/GM*		1.04	1.03	0.97
Standard deviation *FM/GM*		0.44	0.56	0.40
Range of results	*FM* (years)	0	44.8	26.2
	GM (years)	39.1	39.1	39.1
Extreme values of *FM*	Maximum (years)	16.1	48.1	33.1
	Minimum (years)	16.1	3.2	7.0
FM/GM ≥ 0.85 (%)		63.5	54.1	71.8
FM/GM ≥ 0.70 (%)		75.3	72.9	77.6
FM/GM ≥ 1.50 (%)		17.6	11.8	8.2
0.85 ≤ *FM/GM* ≤ 1.15 (%)		32.9	24.7	50.6
Ratio between the *ESL* obtained by *FM* and *GM*		0.86	0.91	0.86

Table 6.8 Statistical indicators regarding the ESL obtained by the different scenarios for rendered façades

Statistical indicator	Scenarios			Graphical method
	Neutral	ISO 15686: 2011	Optimized	
Average *ESL*	16.1	17.2	16.2	18.8
Median *ESL*	16.1	17.1	15.2	15.7
Standard deviation *ESL*	–	7.8	5.9	9.0
Confidence interval of 95 %	–	±1.5	±1.2	±1.8

predicted and the observed values. The model's validity is ensured when the cumulative frequency of *FM*/*GM* is greater than or equal to 0.85 in at least 50 % of the sample; for this coating, the three scenarios analysed fulfil this requirement. In a global analysis, although the neutral scenario presents acceptable results, with a cumulative frequency of *FM*/*GM* greater than 0.85 in 64 % of the case studies, it also presents the highest percentage of failed estimations, with a cumulative number of *FM*/*GM* ratios greater than 1.50 in 17.6 % of the sample. The second scenario (based on ISO 15686: 2011) leads to the lowest percentage of cases with a cumulative frequency of *FM*/*GM* greater than or equal to 0.85. Finally, the optimized scenario presents the best results, with a cumulative frequency of *FM*/*GM* greater than or equal to 0.85 of approximately 71.8 %, with a cumulative frequency of results of *FM*/*GM* ranged between 0.85 and 1.15 of approximately 50.6 % and the lowest percentage of failed estimations (8.2 %).

Assuming the best case scenario, i.e. that all the durability factors are favourable, an *ESL* of 124.8 years would be obtained; for the opposite situation, the worst case scenario, an *ESL* of 3.3 years would be achieved. These values are outside the range considered reasonable (between 5 and 60 years), and although these may correspond to extreme cases, whose probability of occurrence is practically nil, such results suggest point out to the need of additional research on the identification and quantification of factors. But generally speaking and given the variability of the factors affecting the durability of renderings, the majority of results obtained seem coherent, with small deviations only. The quantification of the durability factors proposed by the optimized scenario, although appropriate to the description of the degradation phenomena of the sample studied, should not be taken as a standard model applicable to all rendered façades. This scenario provides some indications for further investigation, allowing a continuous updating when additional data becomes available. Naturally, the higher the number of cases studies used in the application of the factor method, the greater its reliability and applicability.

6.2.3.3 Painted Surfaces

The sample analysed regarding the painted surfaces comprises 220 case studies; to determine the estimated service life of each point within the sample, the ordinates' conversion factor method is applied (described in Sect. 6.2.2.1), assuming that a third-degree polynomial curve is the most appropriate model to describe the degradation phenomena of painted surfaces. In the sample analysed, there are 27 case studies whose severity of degradation at the time of inspection is equal to 0 %. These values cannot be modelled by the factor method, since in these situations the estimated service life tends to infinite; in fact, the absence of visible degradation does not allow a reliable prediction of the behaviour of these coatings over time. Therefore, these case studies are excluded for the application of this methodology. A lower and an upper limit for the estimated service life values should be established, ensuring that the values obtained are consistent and adjusted to reality. In the literature, there are several studies that address the life cycle of painted surfaces, with the following results:

- According to the NYC Housing Maintenance Code (2013), the BS 7543 (1992) and Surgelas et al. (2010), painted surfaces located in public buildings should be replaced every three years;
- Lippiatt (2008) proposes a service life of four years for painted surfaces;
- Kelly (2007) and Scheuer et al. (2003) refer a service life of five years for painted surfaces;
- A study performed by Pullen (2000) states that painted surfaces should be replaced every 7 years;
- Aktas and Bilec (2012) estimate an average service life of 7.1 years for painted surfaces (considering indoor coatings), with a variation range between 3.3 and 12 years;
- Mithraratne and Vale (2004) propose a service life of eight years for painted surfaces;
- Several authors (Adalberth 1997; Keoleian et al. 2001; Hed 1999; Fay et al. 2000) suggest an average service life of ten years;
- The Japanese guide (AIJ 1993) recommend an estimated service life equal or greater than 10 years;
- Flores and de Brito (2010) suggest that painted coatings should be subjected to periodic cleaning every two years, should be repainted every five years and subjected to localized repairs every 10–15 years;
- Finally, ISO 15686: 2011 proposes that exterior painted surfaces should have a service life of 25 years for current buildings.

Based on the literature, a minimum service life of three years (in harsh environments) and a maximum service life of 25 years (for a coating subjected to a set of favourable conditions) for painted surfaces are established. The case studies that exceed these limits are excluded from the analysis. The final sample comprises 177 case studies, leading to a maximum estimated service life of 25 years and a minimum service life of 4 years. Through the intersection of the average degradation curve with the maximum allowable degradation level (corresponding to a severity of degradation of 20 %), an average estimated service life of 9.7 years is obtained. The determination of the estimated service life for each case study, using the ordinates' conversion factor method (described in Sect. 6.2.2.1) leads to an average estimated service life of 10.4 years.

Identification of the Durability Factors of Painted Surfaces

As in the remaining claddings under analysis, also for painted surfaces the execution level (factor *C*) and the indoor environmental exposure conditions (factor *D*) are not included in the model. Regarding factor *A*—quality of the materials—only the type of paint is considered, with four categories: (i) plain paints; (ii) elastic membranes; (iii) silicate and silicone paints; textured paint. For factor *B* (design level), three sub-factors are considered: *B*1—colour; *B*2—type of finishing; *B*3—building geometry. Regarding factor *E*, five sub-factors are analysed: *E*1—façade orientation; *E*2—wind-rain action; *E*3—distance to the sea; *E*4—exposure to damp; *E*5—distance to pollution sources. As for the factor *F*, related to the in-use conditions, two categories are considered: commerce and services; housing. For factor *G*, concerning maintenance conditions, the ease of inspection is analysed, considering two categories: buildings with unfavourable inspection conditions (tall and difficult to access); and buildings under normal conditions. Table 6.9 presents the estimated service lives values for the durability sub-factors analysed and the weight of each factor in the coatings' estimated service life (k). These values are obtained using the two methods described in Sect. 6.2.2.2: (i) degradation curves (DCs) (ii) based on plotting graphical models (GMs).

The values obtained seem appropriate in the description of the degradation phenomena of painted surfaces, leading to the following conclusions:

- The results reveal that plain (traditional and non-traditional) and textured paints are those with higher estimated service lives; plain paints or water-based emulsions may have high resistance to weathering, with great flexibility and hardness; in turn, textured paints with sand in their formulation present, when properly applied, high strength and durability (Chai et al. 2011); in this study, these two types of paint correspond to the favourable situation. Elastic membranes correspond to the current situation since they present some disadvantages when compared with the previous types of paint analysed, such as their waterproofing characteristics that restrain the outward flow of water from the substrate, which can cause a set of premature defects. Silicate and silicone paints are those with lower estimated service lives. However on must take into account that the sample comprises current buildings only, relatively recent, as opposed to old buildings with lime-based mortars, for which silicate paints have excellent performances when compared to other types of paint considered in this study;
- Regarding the colour of the façade, white and light coloured painted surfaces present higher estimated service lives; in fact, Teo et al. (2005) refer that coatings with light colours have a higher reflective power (with a lower coefficient of solar absorption), avoiding the façade' overheating, which may lead to surface irregularities; therefore, white colours correspond to the favourable

Table 6.9 Estimated service lives and weighting coefficients for each durability factor (painted surfaces)

Durability factors			Average ESL (DCs)	Average ESL (GMs)	Weighting value (*k*)
*A*1	Type of paint	Plain paints	10.0	10.8	1.2
		Elastic membranes	9.4	10.1	1.0
		Silicate and silicone paints	9.0	9.5	0.9
		Textured paint	9.9	10.6	1.2
*B*1	Colour	White	9.7	10.5	1.1
		Light colours	9.7	10.4	1.0
		Dark colours	5.8	9.1	0.8
*B*2	Type of finishing	Smooth	9.8	10.8	1.0
		Rough	9.9	10.5	0.95
*B*3	Building geometry	Compact	9.7	10.5	1.0
		Irregular	9.0	9.4	0.95
*E*1	Façade orientation	East/SE	10.0	10.1	1.0
		North/NE	10.0	11.3	1.1
		West/NW	9.5	10.8	1.0
		South/SW	9.0	9.6	0.9
*E*2	Wind-rain action	Severe	9.5	10.7	0.9
		Moderate	9.8	10.8	1.0
		Low	10.0	10.3	1.1
*E*3	Distance from the sea	Less than 5 km	9.4	10.2	0.9
		More than 5 km	9.9	10.4	1.0
*E*4	Exposure to damp	High	9.4	10.2	0.9
		Low	9.9	10.4	1.0
*E*5	Distance from pollution sources	Unfavourable	9.8	10.7	1.0
		Normal	9.7	10.4	1.0
*F*1	Type of use	Commerce and services	9.8	10.2	1.0
		Housing	9.5	10.4	1.0
*G*1	Ease of inspection	Unfavourable	9.6	10.5	0.95
		Normal	9.8	10.6	1.0

situation, light colours to the current situation and dark colours to the unfavourable conditions;

- In the sample analysed, the distinction between painted surfaces with smooth and rough finishing is practically nil; the literature is not consensual about this matter, since textured coatings can have higher strength than smooth paints, but its surface roughness promotes the accumulation of dirt. In Table 6.9, the durability sub-factor related with rough finishing is slightly penalized, according to the results obtained; however, this is a preliminary quantification (necessary for the determination of the reference service life) and the quantification of the durability factors will be optimized in the next sections;

- Regarding the building volume, painted surfaces in compact buildings have higher estimated service lives than those in buildings with an irregular shape; buildings with compact volume correspond to 86 % of the sample, so this characteristic is assumed to be the current situation;
- Concerning the façades orientation, Garrido et al. (2012) suggest that painted surfaces facing South and West are the ones with the lowest service lives, with the highest rates of cracking, detachments, colour changes and presence of dust. Painted surfaces facing North have better performance, which can be justified by lower exposure to UV radiation, responsible for accelerating the degradation process of this type of coatings;
- For wind-rain action, according to the estimated service life values obtained, painted surfaces with severe exposure to this action correspond to the unfavourable situation, painted surfaces with moderate exposure correspond to the current situation and painted surfaces with low exposure correspond to the favourable situation;
- As for distance to the sea, painted surfaces in buildings located less than 5 km from the sea correspond to the unfavourable situation (having lower estimated service lives when compared to painted surfaces away from coastal areas);
- In relation to the exposure to damp, claddings with high exposure are considered as the unfavourable situation;
- Regarding distance from pollution sources, the results obtained contradict what would be expected, with small differences between the two categories; therefore, this sub-factor is considered neutral and its quantification is further evaluated in the next sections;
- For type of use, both methods lead to contradictory results and it is impossible to draw clear conclusions regarding this factor; consequently, this factor assumed as neutral;
- Concerning ease of inspection, as expected, higher buildings correspond to the unfavourable situation, since they require additional—often expensive—means to carry out inspections as opposed to low buildings that can easily be surveyed.

Reference Service Life of Painted Surfaces

To determine the reference service life of painted surfaces, the three methods described in Sect. 6.2.2.3 are used. For painted surfaces, the following results are obtained: (i) in the first method, based on the average degradation curve, a reference service life of 9.7 years is obtained; (ii) in the second method, of the average exposure conditions for a single case study, a reference service life of 9.6 years is achieved; (iii) for the third method, of the average conditions for the whole sample, a reference service life of 9.8 years is reached. Therefore, a reference service life of

9.7 years is adopted for painted surfaces, corresponding to an average of the values obtained by the three methods proposed. This result is in accordance with the results obtained by different methods presented in this book and consistent with existing literature, which suggest 10 years as the average service life of painted surfaces (Adalberth 1997; Keoleian et al. 2001; Hed 1999; Fay et al. 2000; Chai et al. 2014, 2015).

Determination of the Estimated Service Life of Painted Surfaces

The expression (6.6) shows the mathematical equation for the application of factor method to obtain the estimated service life of painted surfaces.

$$ESL = RSL \cdot A1 \cdot B1 \cdot B2 \cdot B3 \cdot E1 \cdot E2 \cdot E3 \cdot E4 \cdot E5 \cdot F1 \cdot G1 \qquad (6.6)$$

where *ESL* represents the estimated service life, *RSL* the reference service life (equal to 9.7 years), *A* the type of paint, *B*1 façade colour, *B*2 type of finishing, *B*3 building geometry, *E*1 façade orientation, *E*2 wind-rain action, *E*3 distance to the sea, *E*4 exposure to damp, *E*5 distance to pollution sources, *F*1 in-use conditions, and *G*1 ease of inspection of the façade.

Table 6.10 shows the weighting of the durability sub-factors according to the three scenarios analysed for painted surfaces. In the optimized scenario, a set of restrictions is imposed in order to adjust the durability sub-factors, regarding the influence of the different characteristics on the painted surfaces' degradation:

- $f_{\text{compact}} \geq f_{\text{irregular}}$;
- $f_{\text{North}} \geq f_{\text{South}}$;
- $f_{\text{more than 5 km from the sea}} \geq f_{\text{less than 5 km from the sea}}$;
- $f_{\text{low exposure to wind-rain action}} \geq f_{\text{moderate exposure to wind-rain action}} \geq f_{\text{severe exposure to wind-rain action}}$;
- $f_{\text{low exposure to damp}} \geq f_{\text{high exposure to damp}}$;
- $f_{\text{current inspection conditions}} \geq f_{\text{unfavourable inspection conditions}}$.

Tables 6.11 and 6.12 show the statistical indicators used to evaluate the results obtained by the factor method according to the various scenarios analysed. The optimized scenario leads to the best results, with a cumulative frequency of results of *FM*/*GM* ranged between 0.85 and 1.15 of approximately 59.9 %, revealing small deviations between the predicted and the observed values, with only 5.1 % of failed estimations. The scenario based on the values prescribed in ISO 15686: 2011 presents the worst performance, with the highest percentage of failed estimations, around 24.3 %, and with higher deviation between the values predicted by the factor method and the values obtained by the graphical method. In the best case scenario, with all the durability factors with the most favourable conditions, a service life of 15.4 years is obtained; in the opposite situation, with all the

Table 6.10 Weighting of the durability sub-factors according to the three scenarios analysed (painted surfaces)

Durability factors			% of case studies (%)	Neutral	ISO 15686: 2011	Optimized
*A*1	Type of paint	Plain paints	34	1.0	1.0	1.025
		Elastic membranes	36	1.0	1.0	1.015
		Silicate and silicone paints	5	1.0	0.8	1.000
		Textured paint	25	1.0	1.2	1.025
*B*1	Colour	White	32	1.0	1.2	1.075
		Light colours	63	1.0	1.0	1.000
		Dark colours	5	1.0	0.8	0.975
*B*2	Type of finishing	Smooth	56	1.0	1.2	1.000
		Rough	44	1.0	1.0	1.150
*B*3	Building geometry	Compact	86	1.0	1.0	1.000
		Irregular	14	1.0	0.8	0.975
*E*1	Façade orientation	East/SE	32	1.0	1.0	1.000
		North/NE	18	1.0	1.2	1.050
		West/NW	29	1.0	1.0	0.950
		South/SW	21	1.0	0.8	0.925
*E*2	Wind-rain action	Severe	33	1.0	0.8	0.950
		Moderate	40	1.0	1.0	0.975
		Low	28	1.0	1.2	1.000
*E*3	Distance from the sea	Less than 5 km	45	1.0	1.0	0.975
		More than 5 km	55	1.0	1.0	1.000
*E*4	Exposure to damp	High	45	1.0	1.0	1.000
		Low	55	1.0	1.2	1.050
*E*5	Distance from pollution sources	Unfavourable	23	1.0	0.8	0.950
		Normal	77	1.0	1.0	1.000
*F*1	Type of use	Commerce and services	66	1.0	1.0	1.000
		Housing	34	1.0	1.0	1.000
*G*1	Ease of inspection	Unfavourable	27	1.0	1.0	0.985
		Normal	73	1.0	1.0	1.000

durability factors quantified as unfavourable, an average estimated service life of 3.9 years is obtained. The results obtained in this scenario are consistent with the values obtained by the graphical method (considered as representative of the observed reality) and with the empirical perception regarding the durability of painted surfaces.

Table 6.11 Statistical indicators for the scenarios analysed for painted surfaces

		Neutral	ISO 15686: 2011	Optimized
Average of *FM/GM*		1.01	1.22	1.03
Standard deviation *FM/GM*		0.30	0.43	0.27
Range of results	*FM* (years)	0	15.1	4.7
	GM (years)	21.5	21.5	21.5
Extreme values of *FM*	Maximum (years)	9.7	20.1	7.9
	Minimum (years)	9.7	5.0	4.7
FM/GM ≥ 0.85 (%)		74.0	81.4	83.6
FM/GM ≥ 0.70 (%)		89.8	91.0	89.3
FM/GM ≥ 1.50 (%)		5.6	24.3	5.1
0.85 ≤ *FM/GM* ≤ 1.15 (%)		53.7	33.9	59.9
Ratio between the *ESL* obtained by *FM* and *GM*		0.93	1.15	0.96

Table 6.12 Statistical indicators regarding the ESL obtained by the different scenarios for painted surfaces

Statistical indicator	Scenarios			Graphical method
	Neutral	ISO 15686: 2011	Optimized	
Average *ESL*	9.7	12.0	10.4	10.0
Median *ESL*	9.7	11.2	9.7	9.9
Standard deviation *ESL*	–	3.6	3.3	1.1
Confidence interval of 95 %	–	±0.5	±0.5	±0.2

6.2.3.4 Ceramic Tiling Systems

Concerning ceramic tiling systems, the initial sample comprises 195 case studies. The methodology previously applied to the other types of coatings is also applied to ceramic tiling systems. The first step is the determination of the estimated service life of each case study, applying the ordinates' conversion factor method is applied (described in Sect. 6.2.2.1), assuming a third-degree polynomial curve to describe the degradation pattern of ceramic claddings over time. Some case studies lead to abnormally high values and it is necessary to define an upper limit for the estimated service life for ceramic claddings. Adherent ceramic tiling systems comprise three main durability layers: the substrate; the material that promotes adhesion between the ceramic tile and the substrate; and the ceramic tile. In general, ceramic tiles are highly durable elements, being compact materials, generally glazed, with low permeability to water and excellent mechanical and chemical properties (Sánchez et al. 2010). Portugal has a long tradition concerning the production and marketing of ceramic tiles, which are considered one of the most important and noblest decorative elements. In the Portuguese building stock, some evidences of the durability of these claddings can be found, such as buildings more than a hundred years old with their ceramic cladding intact (Bordalo et al. 2011) and only traces of the natural aging process. These claddings are an important asset within the cultural heritage and should be preserved. In fact, there are some indications that ceramic tiles have an estimated durability of more than 100 years, but the lack of studies in this area does not allow obtaining unequivocal conclusions regarding the expected service life of this type of coatings. Nine of the case studies analysed present very high estimated service lives (one case study with an estimated service life of 160 years and the remaining cases with an estimated service life greater than 200 years). These case studies show a severity of degradation of 4 %, which means that at the inspection time these claddings present very low degradation indexes, not allowing the mathematical model to accurately predict their behaviour over time. These values seem too optimistic, bearing in mind that these claddings are not subjected to major maintenance actions, notwithstanding the existence of testimonies of very old ceramic claddings in Lisbon with an acceptable state of conservation. In this study, an upper limit of 120 years is thus established, assuming that ultimately, an adherent ceramic cladding subject to a set of favourable conditions can fulfil the minimum performance requirements during this period of time. Therefore, the nine cases with an estimated service life above the upper limit are excluded from further analysis.

Regarding the lower limit, the minimum estimated service life value is equal to 23 years and was obtained for a case study with 6 years, with a severity of degradation of 4.6 % at the inspection time. This value seems acceptable considering the worst case scenario and agrees with the values in the literature: (i) Shohet and Laufer (1996) established an expected service life for ceramic claddings between 10 and 15 years (subject to unfavourable environmental conditions); (ii) Shohet et al. (2002) set a range for the service life of ceramic claddings between 18 and 27 years; (iii) Shohet and Paciuk (2004) for a stricter level of demand propose an expected service life of 18 years (ranging between 15 and 21 years) and for lower level of requirement, an expected service life of 27 years (ranging between 24 and 29 years); (iv) BCIS (2001) suggests an average service life of 35 years with a range between 20 and 50 years; (v) Shohet et al. (1999) suggest an expected service life of 25 years; (vi) data from BMI (Building Maintenance Information) suggests an expected service life of 45 years for indirect ceramic claddings (with a range between 25 and 55 years); (vii) and Carani et al. (1997) assume an average service life for ceramic claddings around 20 years, based on users preferences (i.e. assuming that the replacement of ceramic cladding is conditioned by aesthetic reasons).

Identification of the Durability Factors of Ceramic Tiling Systems

There are several factors that affect the durability of adhered ceramic claddings, but this study is based on the quantification of the degradation condition of the claddings analysed through visual inspections, not allowing the accurate identification of some characteristics. Regarding the factor related with the quality of the materials (factor *A*), only the type of surface is analysed, according to two categories: glazed and not glazed. In factor *B* (design level), five sub-factors are considered: *B*1—colour, with two categories: light and dark colours; *B*2—tiles size, considering two situations: ceramic tiles larger than 20 cm in one dimension; and tiles where the two sizes are less than 20 cm; *B*3—type of substrate (concrete or masonry); *B*4—the presence of peripheral joints (yes or no); *B*5—the presence of peripheral protection (yes or no). Also in this case, factors *C* (execution level) and *D* (interior environmental conditions) are not considered. Concerning the exterior environmental conditions (factor *E*), four sub-factors are analysed: *E*1—façade orientation; *E*2—distance from the sea; *E*3—wind-rain action; *E*4—exposure to damp. For ceramic tiling systems, the use conditions are not evaluated. Regarding factor *G* (maintenance level), the ease of inspection is analysed, considering two categories: buildings with unfavourable inspection conditions; and buildings under normal conditions.

Once the sub-factors that affect the durability of ceramic tiling systems have been identified, two methodologies are applied in order to evaluate the service life of these claddings according to their characteristics (as mentioned in Sect. 6.2.2.2): (i) degradation curves (DCs) (ii) based on plotting graphical models (GMs). Table 6.13 shows the estimated service lives values for the durability sub-factors analysed and the weight of each feature on the coatings' estimated service life (*k*). From the analysis of the estimated service lives of ceramic tiling systems according to their characteristics, the following conclusions can be drawn:

- Regarding the type of surface, supposedly glazed ceramic tiles present higher estimated service lives due to their lower permeability and consequently to their higher resistance to mechanical and chemical degradation agents; however, in the sample analysed (with 186 case studies), ceramic tiles without glazed show higher estimated service lives, corresponding to only 19 case studies (around 10 % of the sample); therefore, this sub-factor is considered neutral in this preliminary quantification;
- For the colour of ceramic tiles, dark colours present higher estimated service lives than light colours, which can be explained by the fact that pathological situations are easily detected in claddings with lighter colours;
- Concerning the tiles size, as expected, the sample reveals that claddings with larger tiles present higher degradation indexes and, consequently, lower estimated service lives;
- Considering the type of substrate, in the sample analysed masonry substrates present lower estimated service lives as these claddings correspond to older buildings (with higher degradation levels at inspection date), with distinctive building typology when compared with more recent buildings, and thus more prone to differential movements that may affect the durability of the ceramic claddings system;
- As for peripheral joints, the two methods applied lead to inconclusive results; therefore, this sub-factor is assigned as neutral in this preliminary assessment;
- Regarding the existence of peripheral protection, 52 % of the sample has protection, with higher estimated service lives, therefore corresponding to the favourable situation;

Table 6.13 Estimated service lives and weighting coefficients for each durability factor (ceramic tiling systems)

Durability factors			Average ESL (DCs)	Average ESL (GMs)	Weighting value (k)
*A*1	Type of surface	Glazed	49.7	53.7	1.0
		Not glazed	61.0	65.9	1.0
*B*1	Colour	Light colours	48.6	53.1	1.0
		Dark colours	53.3	59.8	1.2
*B*2	Size of Tiles	L ≤ 20 cm	50.1	54.3	1.0
		L > 20 cm	44.6	51.6	0.9
*B*3	Substrate	Masonry	48.6	51.6	1.0
		Concrete	51.6	57.2	1.2
*B*4	Peripheral joints	Without	49.5	54.4	1.0
		With	52.1	53.4	1.0
*B*5	Peripheral protection	Without	49.5	53.1	0.9
		With	50.3	54.4	1.0
*E*1	Façade orientation	East/SE	52.1	57.1	1.0
		North/NE	49.3	52.9	0.9
		West/NW	51.1	57.1	1.0
		South/SW	47.4	49.7	0.8
*E*2	Distance to the sea	Less than 5 km	49.2	52.8	1.0
		More than 5 km	49.5	59.1	1.2
*E*3	Wind-rain action	Severe	48.6	53.6	0.9
		Moderate	50.4	54.2	1.0
		Low	51.0	56.9	1.2
*E*4	Exposure to damp	High	49.5	51.3	0.8
		Low	52.3	56.4	1.0
*G*1	Ease of inspection	Unfavourable	49.8	54.2	1.0
		Normal	50.5	56.7	1.1

- Claddings facing South are those with the lowest estimated service lives, followed by claddings facing North (thus corresponding to the most unfavourable situations); claddings facing East and West present higher estimated service lives and are therefore considered the current situation;
- Concerning distance to the sea, ceramic claddings located less than 5 km from the sea present higher degradation indexes, with lower estimated service lives;
- Claddings with severe exposure to wind-rain action correspond to the unfavourable situation (quantified with a value of k equal to 0.9, since the values obtained are close to those for claddings with moderate exposure); ceramic claddings with moderate exposure correspond to the current situation and claddings with low exposure correspond to the favourable situation (with higher estimated service lives);

- For exposure to damp, claddings with low exposure are defined as the current situation (corresponding to 55 % of the sample analysed); claddings with high exposure to damp correspond to the unfavourable situation;
- Concerning ease of inspection, current low rise buildings have higher estimated service lives and are therefore considered the favourable situation.

Reference Service Life of Ceramic Tiling Systems

The reference service life of ceramic tiling systems is obtained based on the three methods described in Sect. 6.2.2.3: (i) with the method based on the average degradation curve, a reference service life of 50 years is obtained; (ii) with the method of the average exposure conditions for a single case study, a reference service life of 50.7 years is obtained; (iii) for the method of the average conditions for the whole sample, a reference service life of 52.3 years is reached. Therefore, a reference service life of 51 years is adopted for ceramic claddings, corresponding to an average of the values obtained by the three methods proposed. The adopted reference service life is consistent with the average service life values obtained by the different methods proposed in this book and agrees with the values in the literature that suggest 50 years as the average estimated service life of adherent ceramic tiling systems (Tam et al. 1993; BCIS 2001; Galbusera et al. 2014).

Determination of the Estimated Service Life of Ceramic Tiling Systems

Once the durability factors that affect the durability of ceramic claddings are identified and quantified, the practical application of the factor method to obtain the estimated service life of ceramic claddings can be proposed based on Eq. (6.7).

$$ESL = RSL \cdot A1 \cdot B1 \cdot B2 \cdot B3 \cdot B4 \cdot B5 \cdot E1 \cdot E2 \cdot E3 \cdot E4 \cdot G1 \qquad (6.7)$$

where *ESL* represents the estimated service life, *RSL* the reference service life (equal to 51 years), *A* the type of surface, *B*1 façade colour, *B*2 tiles size, *B*3 the substrate, *B*4 presence of peripheral joints, *B*5 presence of peripheral protection, *E*1 façade orientation, *E*2 distance to the sea, *E*3 wind-rain action, *E*4 exposure to damp and *G*1 ease of inspection of the façade.

According to the procedures adopted for the remaining types of claddings, also in this case, three scenarios are analysed—neutral, based on ISO 15686: 2011 and optimized—for the application of the factor method to the service life prediction of ceramic tiling systems. In the optimized scenario, the following restrictions are imposed in order to calibrate the durability sub-factors:

- $f_{\text{glazed}} \geq f_{\text{not glazed}}$;
- $f_{\text{tiles with } L \leq 20\text{ cm}} \geq f_{\text{tiles with } L > 20\text{ cm}}$;
- $f_{\text{concrete}} \geq f_{\text{masonry}}$;
- $f_{\text{with joints}} \geq f_{\text{without joints}}$;

- $f_{\text{with peripheral protection}} \geq f_{\text{without peripheral protection}}$;
- $f_{\text{more than 5 km from the sea}} \geq f_{\text{less than 5 km from the sea}}$;
- $f_{\text{low exposure to wind-rain action}} \geq f_{\text{moderate exposure to wind-rain action}} \geq f_{\text{severe exposure to wind-rain action}}$;
- $f_{\text{low exposure to damp}} \geq f_{\text{high exposure to damp}}$;
- $f_{\text{current inspection conditions}} \geq f_{\text{unfavourable inspection conditions}}$.

Table 6.14 presents the weighting of the durability sub-factors according to the three scenarios analysed for ceramic tiling systems.

Table 6.14 Weighting of the durability sub-factors according to the three scenarios analysed (ceramic tiling systems)

Durability factors			% of case studies (%)	Neutral	ISO 15686: 2011	Optimized
*A*1	Type of surface	Glazed	90	1.0	1.2	1.000
		Not glazed	10	1.0	1.0	0.950
*B*1	Colour	Light colours	58	1.0	0.8	1.000
		Dark colours	42	1.0	1.0	1.050
*B*2	Size of tiles	L ≤ 20 cm	84	1.0	1.2	1.000
		L > 20 cm	16	1.0	1.0	0.825
*B*3	Substrate	Masonry	55	1.0	0.8	1.000
		Concrete	45	1.0	1.0	1.050
*B*4	Peripheral joints	Without	85	1.0	1.0	1.000
		With	15	1.0	1.0	1.150
*B*5	Peripheral protection	Without	48	1.0	1.0	1.000
		With	52	1.0	1.2	1.125
*E*1	Façade orientation	East/SE	24	1.0	1.0	1.125
		North/NE	31	1.0	1.0	1.125
		West/NW	24	1.0	0.8	0.945
		South/SW	22	1.0	1.0	1.045
*E*2	Distance to the sea	Less than 5 km	72	1.0	0.8	0.850
		More than 5 km	28	1.0	1.0	1.000
*E*3	Wind-rain action	Severe	28	1.0	0.8	0.900
		Moderate	50	1.0	1.0	0.995
		Low	22	1.0	1.2	1.000
*E*4	Exposure to damp	High	45	1.0	0.8	0.900
		Low	55	1.0	1.0	1.000
*G*1	Ease of inspection	Unfavourable	34	1.0	1.0	0.925
		Normal	66	1.0	1.0	1.050

Tables 6.15 and 6.16 show the statistical indicators used to evaluate the results obtained by the factor method according to the various scenarios analysed. As expected, the optimized scenario leads to the best results, with the best overall performance, with a cumulative frequency of results of *FM/GM* ranged between 0.85 and 1.15 of approximately 50 %, therefore revealing that in half of the sample analysed, the deviation between the observed and the prediction values is relatively low (below 15 %). Additionally, the average of the ratio between the *FM* and the *GM* is closer to 1 for the optimized model, and in this scenario, only 5.9 % of the estimations are considered as failed estimations. If all the durability sub-factors were favourable (best case scenario) an average estimated service life of 85.9 years would be obtained. If all the durability sub-factors were unfavourable, an average estimated service life of 24.1 years would be obtained. These values fall within the range of the service life values estimated by the graphical method, considered acceptable and consistent with the empirical perception regarding the durability of adherent ceramic claddings.

Table 6.15 Statistical indicators for the scenarios analysed for ceramic tiling systems

		Neutral	ISO 15686: 2011	Optimized
Average of *FM/GM*		1.03	0.90	1.00
Standard deviation *FM/GM*		0.32	0.37	0.33
Range of results	*FM* (years)	0	68.9	42.9
	GM (years)	95.4	95.4	95.4
Extreme values of *FM*	Maximum (years)	51.0	88.1	70.8
	Minimum (years)	51.0	19.3	27.9
$FM/GM \geq 0.85$ (%)		69.4	52.2	69.9
$FM/GM \geq 0.70$ (%)		85.5	67.7	82.3
$FM/GM \geq 1.50$ (%)		5.4	7.5	5.9
$0.85 \leq FM/GM \leq 1.15$ (%)		39.2	34.4	50.0
Ratio between the *ESL* obtained by *FM* and *GM*		0.94	0.84	0.91

Table 6.16 Statistical indicators regarding the ESL obtained by the different scenarios for ceramic tiling systems

Statistical indicator	Scenarios			Graphical method
	Neutral	ISO 15686: 2011	Optimized	
Average *ESL*	51.0	45.5	49.8	54.5
Median *ESL*	51.0	45.1	48.9	50.1
Standard deviation *ESL*	–	15.8	9.0	18.1
Confidence interval of 95 %	–	±2.3	±1.3	±2.6

6.3 Stochastic Approach

6.3.1 Theoretical Foundations

The shortcomings of the classic approach led to the search for more precise models that aim at optimizing the evaluation of the service life of construction components. Various authors (Aarseth and Hovde 1999; Moser 1999) proposed different methodologies to deal with the uncertainty associated with the factor method, proposing the application of the durability factors as stochastic input parameters. Using this methodology, the factor method is no longer a purely deterministic model, since the estimated service life is characterized by a probability distribution, with a confidence interval associated. Moser (2004) proposes a hybrid approach for the application of the factor method in which a deterministic value for the reference service life is used, along with distinct probability distribution to each durability factor. Van Nunen (2010) suggests a similar methodology, proposing some statistical indicators to select the probability distribution which best fits each durability factor.

In this study, a stochastic approach to the factor method is proposed applied to the service life prediction of façade claddings. In this approach, the factors affecting the durability of the various types of claddings analysed are identified and are translated into probability distributions. Therefore, the estimated service life of each case study within the sample is given by a probability distribution. For this purpose, the following steps are performed:

- The procedure begins by calculating the first durability factor (factor A), as described in Eq. (6.8):

$$A1 = \frac{ESL_{GM}}{RSL \times A2 \times \cdots \times AN \times B1 \times \cdots \times BN \times E1 \times \cdots \times EN \times F1 \times \cdots \times FN \times \cdots \times G1 \times \cdots \times GN} \tag{6.8}$$

  where ESL_{GM} represents the estimated service life obtained by the graphical method; the values of the product of all other sub-factors in the denominator ($B1 \times B2 \times B3 \times B4 \times B5 \times B6 \times B7 \times E1 \times E2 \times E3 \times E4 \times E5 \times F1 \times G1$) correspond to the values obtained by the optimized scenario; for example, in a case study with an estimated service life calculated by the graphical method equal to 17 years, it is possible to calculate the value of A (1.11, in this example), knowing that the product of all other factors (excluding A) is 0.95, for a reference service life of 16.1 years. This methodology ensures that the estimated service life obtained by both methods—graphical method and Eq. (6.8)—is similar. This process is then repeated for all case studies within the sample so that an average value for each category of the factor A can be obtained;
- Then, the values obtained are normalized to ensure that the average value of the probability distribution for each category is equal to the value given by the optimized scenario (Tables 6.2, 6.6, 6.10 and 6.14 for stone claddings,

renderings, painted surfaces and ceramic claddings, respectively) and the average value (obtained in the previous step) for each category within the sub factor Eq. (6.9); notice that the values of the optimized scenario are those from the factor method that optimize the quality of the results, i.e. that minimize the errors between the values obtained by the factor method and the values obtained by the graphical method (that supposedly portray reality);

$$A_{ic} = \frac{A_i \cdot \bar{A}_{\text{Optimized scenario}}}{\bar{A}_i} \tag{6.9}$$

where A_{ic} represents the value of each case sample i for category c of factor A, A_i represents the value obtained for factor A in the previous step of calculation, $\bar{A}_{\text{Optimized scenario}}$. The average value for factor A, obtained in the optimized scenario (Tables 6.2, 6.6, 6.10 and 6.14 for stone claddings, renderings, painted surfaces and ceramic claddings, respectively), and $\bar{A}_i$ the average value of factor A_i;

- Using the same procedure used to calculate the sub-factor A, the remaining durability factors are sequentially determined.

To associate a probability distribution to each durability factor, the software EasyFit 5.4 was used. This software allows adjusting 65 theoretical probability distributions to the data. The selection of the best distribution is made based on goodness-of-fit tests. In this study, three statistical tests were used to determine the distribution that is more appropriate to represent the data, by minimizing the deviation or error (van Nunen 2010):

(i) **The Anderson-Darling test (*A-D* test)**—which is suitable for continuous distributions and whose soundness and reliability for more than 50 data samples have been demonstrated by Wetherill (1986);
(ii) **The Kolmogorov-Smirnov test (*K-S* test)**—which can be applied to continuous distributions, even with samples of small size;
(iii) **The Chi-square test**—the most common of the three, which can be applied to continuous and discrete distributions; in this test, the values of the abscissa are grouped in several subsets and there are no predefined rules for this division; due to this feature, this test loses some objectivity; however, for larger samples this test has better results than the previous ones.

Each of these tests provides a ranking of the fit of the data to each distribution. In this study, the three methods are taken into account and are combined into one—the overall result—obtained by the sum of the results of each test. However, to address situations of "tie" (i.e. when the best overall result are obtained in more than one distribution) two additional tests are performed (van Nunen 2010): (i) the first considers the effect of small samples on the test results, thus not taking into account the chi-square test, and using as indicator the sum of *A-D* and *K-S* tests; (ii) the second indicator puts special emphasis on the tail of the distribution, applying an index that expresses the sum of twice the result of the *A-D* test and the result of the

K-S test. As the purpose of this study is to define practical and easily understandable methods, only six distributions are analysed. The distributions selected are the ones most common for continuous distributions and the easiest to apply by non-specialists in statistics (since the methodology proposed will be used by stakeholders in the construction industry): (i) Normal; (ii) Lognormal; (iii) Gamma; (iv) Weibull; (v) Gumbel; (vi) Logistic.

6.3.2 Application to Façade Claddings

Using the procedure described in the previous section, the durability factors for the application of the factor method to the service life prediction of façade claddings are measured using a stochastic approach. In the next sections, the application of this methodology for the four types of claddings analysed is described. Once the probability distributions that best fit each durability sub-factor is known, the estimate of the service life for each case study is obtained. For that, the reference service life is multiplied by probability distributions related with the durability factors that characterize the case study. This operation is performed using an algorithm created in Matlab that generates a set of random numbers (in this study, 100,000 random numbers are generated). The algorithm must be run for each case in the sample, thereby calculating the probability distribution function of its estimated service life based on their characteristics. An example of this estimation is given for each one of the claddings under analysis.

6.3.2.1 Natural Stone Claddings

As mentioned before, the durability sub-factors are determined by the probabilistic approach to ensure that the average (assuming a normal distribution) of each category is equal to the value established in the deterministic approach. Table 6.17 presents the probability distributions associated with the durability factors and some statistical indicators that define the corresponding distributions.

The translation of the durability factors into probabilistic distributions allows predicting the variability associated with real life situations. These probability distributions consider the possibility of a façade showing a good performance, even when subjected to some unfavourable conditions. Likewise, a case study subjected to a set of unfavourable conditions may present higher degradation indexes, thus reaching the end of its service life before the expected average value for the whole sample.

Based on the probability distributions that best fit each of the factors that affect the durability of natural stone claddings, the estimated service life of each case study can be achieved by multiplying the reference service life by the probability distributions related with the durability factors that characterize the case study. In this study, the stochastic approach of the factor method is defined for a case study,

Table 6.17 Probability distributions associated with the durability factors related with natural stone claddings

Durability factors			Sample size	Distribution	Average	Standard deviation	Variance	Value of the factor with a probability higher than 95 % to be exceeded	Value of the factor with a probability higher than 5 % to be exceeded
*A*1	Type of stone	Limestone	72	Lognormal	1.000	0.219	0.048	0.658	1.376
		Granite	54	Gumbel	1.100	0.327	0.107	0.673	1.711
		Marble	77	Lognormal	0.946	0.216	0.047	0.697	1.355
*B*1	Colour	Light colours	134	Lognormal	1.024	0.240	0.058	0.682	1.460
		Dark colours	69	Gumbel	1.000	0.281	0.079	0.633	1.525
*B*2	Type of finishing	Smooth	96	Gumbel	1.100	0.295	0.087	0.715	1.651
		Rough	107	Lognormal	0.999	0.233	0.054	0.668	1.421
*B*3	Size of stone plates	Medium	129	Lognormal	0.999	0.258	0.066	0.647	1.473
		Large	74	Gumbel	0.900	0.203	0.041	0.635	1.280
*B*4	Thickness of stone plates	Less than 2.5 cm	49	Lognormal	1.000	0.241	0.058	0.638	1.427
		≥2.5 cm	154	Lognormal	1.093	0.284	0.081	0.794	1.630
*B*5	Location of the cladding	Partial and integral elevated cladding	61	Lognormal	0.949	0.295	0.087	0.557	1.497
		Bottom wall cladding	142	Lognormal	1.000	0.213	0.045	0.684	1.379
*E*1	Façade orientation	North	29	Lognormal	0.902	0.197	0.039	0.681	1.275
		NE/E/SE	88	Lognormal	0.949	0.239	0.057	0.613	1.384
		West/NW	51	Lognormal	0.900	0.256	0.065	0.598	1.384
		South/SW	35	Normal	1.000	0.213	0.045	0.650	1.351

(continued)

Table 6.17 (continued)

Durability factors			Sample size	Distribution	Average	Standard deviation	Variance	Value of the factor with a probability higher than 95 % to be exceeded	Value of the factor with a probability higher than 5 % to be exceeded
*E*2	Distance from the sea	Less than 5 km	77	Lognormal	0.998	0.267	0.071	0.702	1.504
		More than 5 km	126	Lognormal	1.150	0.263	0.069	0.744	1.606
*E*3	Exposure to wind-rain action	Moderate	170	Lognormal	0.999	0.254	0.065	0.641	1.463
		Severe	33	Gumbel	1.000	0.212	0.045	0.723	1.397
*E*4	Exposure to damp	Low	119	Gumbel	1.000	0.225	0.050	0.707	1.420
		High	84	Lognormal	0.899	0.245	0.060	0.562	1.350
*F*1	Type of property	Housing	104	Gumbel	1.000	0.203	0.041	0.735	1.380
		Commerce and services	99	Lognormal	0.999	0.296	0.088	0.620	1.553
*G*1	Ease of inspection	Normal	110	Lognormal	0.999	0.248	0.062	0.648	1.450
		Unfavourable	93	Gumbel	0.900	0.223	0.050	0.610	1.316

whose characteristics are described in Table 6.18. Figure 6.2 shows the probability distribution of the estimated service life of the case study analysed, obtained by applying the stochastic approach to the factor method. This approach leads to an average estimated service life of 71.5 years, with a standard deviation of 8.8 years. The probability distribution with better adjustment is the Gamma distribution.

From the analysis of the probability distribution function associated with the estimated service life it is possible to conclude that an average service life of 58 years has a probability higher than 95 % of being exceeded; on the other hand, the case study analysed has a probability lower than 5 % of having a service life higher than 86.9 years. Replacing the values of the sub-factors in Eq. (6.4) for the average values of the distributions (in a semi-probabilistic approach) an estimated service life of 73.2 years is obtained for this case study. Using the graphical method, the estimated service life of the case study is 74.2 years.

The graphical method is used to validate the results, since it is based on the observed degradation analysed during the field work, assuming that it expresses the reality. The values obtained by the factorial method are close to those obtained by

Table 6.18 Description of the natural stone cladding analysed

Case study	Description
	Stone cladding with 73 years and with a degradation severity index equal to 19.1 %; Sub factor *A*: marble—lognormal distribution (μ = 0.946; σ = 0.216); Sub factor *B*1: light colours—lognormal distribution (μ = 1.024; σ = 0.240); Sub factor *B*2: rough finishing—lognormal distribution (μ = 0.999; σ = 0.233); Sub factor *B*3: medium-sized plate—lognormal distribution (μ = 0.999; σ = 0.258); Sub factor *B*4: less than 2.5 cm—lognormal distribution (μ = 1.093; σ = 0.284); Sub factor *B*5: bottom wall cladding—lognormal distribution (μ = 1.000; σ = 0.213); Sub factor *E*1: cladding faced southwest—normal distribution (μ = 1.000; σ = 0.213); Sub factor *E*2: less than 5 km from the sea—lognormal distribution (μ = 0.998; σ = 0.267); Sub factor *E*3: moderate exposure to wind-rain action—lognormal distribution (μ = 0.999; σ = 0.254); Sub factor *E*4: low exposure to damp—Gumbel distribution (μ = 1.000; σ = 0.225); Sub factor *F*: commerce—lognormal distribution (μ = 0.999; σ = 0.296); Sub factor *G*: current inspection conditions—lognormal distribution (μ = 0.999; σ = 0.248)

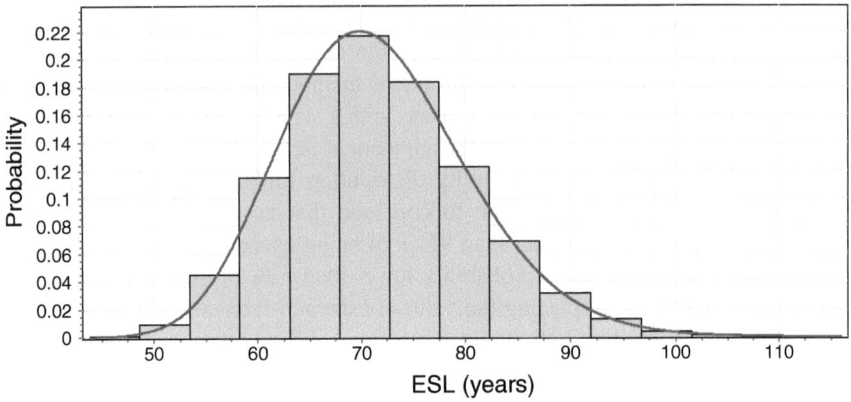

Fig. 6.2 Probability density distribution of the estimated service life of the stone cladding analysed

the graphical method and are in agreement with the values obtained by the other methods proposed in this book. The case study analysed illustrates the applicability of the probabilistic approach to the factor method, which produce reliable and consistent results with the information obtained in the fieldwork. Therefore this approach seems worth to be applied to a wider range of cases in order to define a probability distribution that characterizes all the case studies analysed.

6.3.2.2 Rendered Façades

Table 6.19 presents the probability distributions associated with the durability factors related with rendered façades. As in the previous section, this approach is also illustrated for a case study, whose characteristics are described in Table 6.20. Figure 6.3 presents the probability distribution of the estimated service life of the case study analysed, obtained by applying the stochastic approach to the factor method. Through the application of this approach, an average estimated service life of 16.1 years is obtained (by unintended coincidence, equal to the average of the sample), with a standard deviation of 4.5 years. The probability distribution with better adjustment is the Gamma distribution and the density probability function reaches a maximum at 14.6 years.

Based on the density probability function, it is possible to conclude that an average service life of 9.8 years has a probability higher than 95 % of being exceeded; on the other hand, the case study analysed has a probability lower than 5 % of having a service life higher than 24.3 years. Replacing the values of the sub-factors in Eq. (6.5) for the average values of the distributions (in a semi-probabilistic approach) an estimated service life of 17.5 years is obtained for this case study. Using the graphical method, the estimated service life of the case study is 17.7 years.

Table 6.19 Probability distributions associated with the durability factors related with rendered façades

Durability factors			Sample size	Distribution	Average	Standard deviation	Variance	Value of the factor with a probability higher than 95 % to be exceeded	Value of the factor with a probability higher than 5 % to be exceeded
*A*1	Render type	Lime-cement renderings	20	Lognormal	0.875	0.203	0.041	0.583	1.241
		Current cement renderings	50	Lognormal	1.025	0.498	0.248	0.440	1.964
		Renderings with crushed marble	5	Gumbel	1.325	0.463	0.214	0.720	2.189
		Single-layer renderings	10	Lognormal	1.425	0.726	0.526	0.651	2.801
*B*1	Colour	White	15	Lognormal	0.700	0.378	0.143	0.343	1.387
		Light colours	44	Lognormal	1.025	0.485	0.235	0.442	1.940
		Dark colours	26	Lognormal	1.125	0.390	0.152	0.682	1.860
*B*2	Building volume	Compact	50	Lognormal	1.025	0.434	0.189	0.486	1.838
		Irregular	35	Lognormal	0.975	0.545	0.297	0.487	1.979
*B*3	Eaves' protection	Without protection	37	Lognormal	1.000	0.496	0.246	0.501	1.933
		nWith protection	48	Lognormal	1.125	0.460	0.118	0.554	1.986
*B*4	Platbands copings	Without copings	67	Lognormal	0.850	0.409	0.167	0.378	1.619
		With copings	18	Lognormal	1.200	0.386	0.149	0.727	1.928
*B*5	Balcony copings	Without copings	55	Lognormal	1.000	0.460	0.221	0.451	1.863
		With copings	30	Lognormal	1.200	0.577	0.333	0.651	2.283

(continued)

Table 6.19 (continued)

Durability factors			Sample size	Distribution	Average	Standard deviation	Variance	Value of the factor with a probability higher than 95 % to be exceeded	Value of the factor with a probability higher than 5 % to be exceeded
*B*6	Ground floor protection (socle)	Without protection	14	Lognormal	1.000	0.612	0.381	0.498	2.109
		With protection	71	Lognormal	1.050	0.443	0.197	0.509	1.883
*B*7	Detailing/design level	Inferior	25	Weibull	0.800	0.363	0.132	0.415	1.518
		Medium	56	Lognormal	1.000	0.438	0.192	0.474	1.821
		Superior	4	(*)	1.150	–	–	–	–
*E*1	Façade orientation	East/SE	20	Lognormal	1.050	0.490	0.240	0.452	1.966
		North/NE	32	Lognormal	0.825	0.401	0.161	0.456	1.548
		West/NW	15	Lognormal	1.050	0.337	0.113	0.635	1.684
		South/SW	18	Lognormal	1.250	0.558	0.312	0.567	2.304
*E*2	Distance from the sea	Less than 3 km	22	Lognormal	0.950	0.490	0.240	0.496	1.851
		More than 3 km	63	Lognormal	1.050	0.444	0.197	0.491	1.880
*E*3	Exposure to damp	Unfavourable	11	Lognormal	1.000	0.557	0.310	0.498	2.030
		Normal	20	Lognormal	1.100	0.547	0.299	0.648	2.060
		Favourable	54	Lognormal	1.125	0.490	0.241	0.531	2.047
*E*4	Distance from pollution sources	Unfavourable	21	Lognormal	0.800	0.326	0.106	0.465	1.413
		Normal	59	Lognormal	0.800	0.364	0.133	0.368	1.486
		Favourable	5	Gamma	1.500	0.996	0.992	0.310	3.421

(continued)

Table 6.19 (continued)

Durability factors			Sample size	Distribution	Average	Standard deviation	Variance	Value of the factor with a probability higher than 95 % to be exceeded	Value of the factor with a probability higher than 5 % to be exceeded
E5	Façade protection level	Without protection	40	Lognormal	1.000	0.454	0.206	0.555	1.849
		Normal situation	18	Lognormal	1.000	0.477	0.228	0.575	1.885
		With protection	27	Lognormal	1.000	0.551	0.304	0.401	2.031
F1	Type of property	Private	68	Lognormal	0.800	0.431	0.206	0.398	1.600
		Public sector	14	Normal	1.075	0.294	0.228	0.591	1.559
		Commerce and services	3	(*)	1.075	–	0.304	–	–
G1	Ease of inspection	Normal	33	Lognormal	1.225	0.565	0.185	0.473	2.049
		Unfavourable	52	Lognormal	1.000	0.491	0.086	0.606	2.142

(*) It is impossible to fit a distribution to this category. To adjust a distribution probability to a sample at least 5 cases are necessary

Table 6.20 Description of the rendered façade analysed

Case study	Description
	Rendered façade with 8 years and with a degradation severity index equal to 9 %; Sub factor *A*: current cement rendering—lognormal distribution ($\mu = 1.021$; $\sigma = 0.498$); Sub factor *B*1: light colours—lognormal distribution ($\mu = 0.690$; $\sigma = 0.378$); Sub actor *B*2: compact—lognormal distribution ($\mu = 1.020$; $\sigma = 0.434$); Sub factor *B*3: without eaves' protection—lognormal distribution ($\mu = 1.001$; $\sigma = 0.496$); Sub factor *B*4: without platbands copings—lognormal distribution ($\mu = 0.845$; $\sigma = 0.409$); Sub factor *B*5: without balcony copings—lognormal distribution ($\mu = 0.993$; $\sigma = 0.460$); Sub factor *B*6: without ground floor protection—lognormal distribution ($\mu = 1.007$; $\sigma = 0.612$); Sub actor *B*7: inferior design level—Weibull distribution ($\alpha = 1.153$; $\beta = 0.439$; $\gamma = 0.382$); Sub factor *E*1: east—lognormal distribution ($\mu = 1.042$; $\sigma = 0.490$); Sub factor *E*2: less than 3 km from the sea—lognormal distribution ($\mu = 0.948$; $\sigma = 0.490$); Sub factor *E*3: normal exposure to damp—lognormal distribution ($\mu = 1.090$; $\sigma = 0.547$); Sub factor *E*4: favourable exposure to pollutant sources—gamma distribution ($\mu = 1.500$; $\sigma = 0.996$); Sub factor *E*5: Façade without protection—gamma distribution ($\alpha = 2.269$; $\beta = 0.661$); Sub factor *F*1: commerce and services—lognormal distribution ($\mu = 1.075$; $\sigma = 0.049$); Sub factor *G*1: normal conditions—lognormal distribution ($\mu = 1.221$; $\sigma = 0.491$)

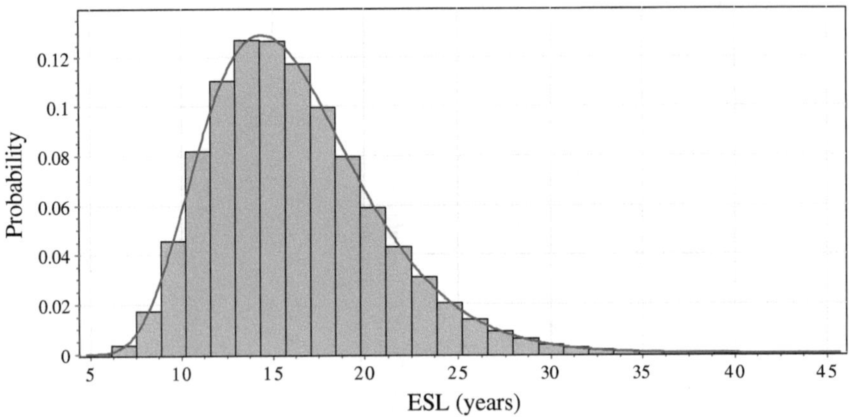

Fig. 6.3 Probability density distribution of the estimated service life of the rendered façade analysed. Data sourced from Silva et al. (2016)

6.3.2.3 Painted Surfaces

Using the stochastic approach of the factor method, the estimated service life of each painted surface within the sample analysed can be obtained, given by a probability distribution. Therefore, the reference service life (9.7 years) is multiplied by the probability distributions that describe the case study. Table 6.21 shows the probability distributions associated with the durability factors related with painted surfaces. To illustrate the application of the stochastic approach of the factor method, a case study is analysed (Table 6.22). This approach leads to an average estimated service life of 11 years, with a standard deviation of 1.4 years (Fig. 6.4). The analysis of the probability distribution associated with the estimated service life by the stochastic approach for this case study reveals that an estimated service life of 8.8 years has a probability higher than 95 % of being exceeded; on the other hand, a case study with these characteristics presents a probability of the less than 5 % of having have a service life higher than 13.3 years. The deterministic model (using the optimized scenario) leads to an average service life for the same case study of 11.4 years, and the graphical method allows obtaining an estimated service life of 10.5 years.

6.3.2.4 Ceramic Tiling Systems

In the stochastic approach to the factor method, the durability sub-factors are defined by probability distributions whose average value of each category is equal to the value assumed in the deterministic approach (optimized scenario). Table 6.23 shows the probability distributions and the respective statistical indicators for the durability sub-factors related with ceramic tiling systems. Once defined the durability distributions that characterize each durability sub-factor, the estimated service life of each case study within the sample can be calculated. Once again, an example for a ceramic cladding—with the characteristics described in Table 6.24—is provided in order to illustrate the application of the stochastic approach of the factor method.

The estimated service life of this case study is obtained by multiplying the reference service life (51 years) by the probability distributions related with its characteristics. This methodology leads to an average estimated service life of 54 years, with a standard deviation of 17 years. The histogram that characterizes the probability distribution of the estimated service life for the ceramic cladding analysed is shown in Fig. 6.5. The analysis of the probability distribution associated with the estimated service life shows that, for the case study analysed, an estimated service life of 30.8 years has a probability higher than 95 % of being exceeded; on the other hand, the probability of having an estimated service life greater than 85 years is less than 5 %. Using the deterministic approach of the factor method

Table 6.21 Probability distributions associated with the durability factors related with painted surfaces

Durability factors			Sample size	Distribution	Average	Standard deviation	Variance	Value of the factor with a probability higher than 95 % to be exceeded	Value of the factor with a probability higher than 5 % to be exceeded
*A*1	Type of paint	Plain paints	61	Lognormal	1.023	0.322	0.103	0.635	1.630
		Elastic membranes	63	Lognormal	1.010	0.297	0.088	0.654	1.571
		Silicate and silicone paints	9	Gumbel	1.000	0.258	0.067	0.519	1.337
		Textured paint	44	Lognormal	1.025	0.200	0.040	0.780	1.383
*B*1	Colour	White	57	Lognormal	1.073	0.324	0.105	0.684	1.685
		Light colours	112	Lognormal	0.996	0.275	0.076	0.639	1.510
		Dark colours	8	Normal	0.975	0.281	0.079	0.607	1.437
*B*2	Type of finishing	Smooth	100	Lognormal	0.997	0.330	0.109	0.580	1.617
		Rough	77	Lognormal	1.149	0.238	0.057	0.826	1.589
*B*3	Building geometry	Compact	152	Lognormal	0.997	0.300	0.090	0.618	1.559
		Irregular	25	Gumbel	0.975	0.216	0.047	0.693	1.379
*E*1	Façade orientation	East/SE	57	Gumbel	1.000	0.207	0.043	0.730	1.387
		North/NE	32	Gama	1.050	0.278	0.077	0.638	1.545
		West/NW	51	Lognormal	0.945	0.369	0.136	0.590	1.632
		South/SW	37	Lognormal	0.925	0.233	0.054	0.585	1.343
*E*2	Wind-rain action	Severe	58	Lognormal	0.946	0.284	0.081	0.596	1.481
		Moderate	70	Lognormal	0.973	0.328	0.107	0.560	1.588
		Low	49	Lognormal	0.999	0.188	0.035	0.754	1.349

(continued)

Table 6.21 (continued)

Durability factors			Sample size	Distribution	Average	Standard deviation	Variance	Value of the factor with a probability higher than 95 % to be exceeded	Value of the factor with a probability higher than 5 % to be exceeded
E3	Distance from the sea	Less than 5 km	79	Lognormal	0.924	0.275	0.076	0.557	1.435
		More than 5 km	98	Lognormal	0.996	0.278	0.077	0.657	1.520
E4	Exposure to damp	High	80	Lognormal	0.998	0.295	0.087	0.605	1.545
		Low	97	Lognormal	1.045	0.295	0.087	0.687	1.602
E5	Distance from pollution sources	Unfavourable	41	Lognormal	0.944	0.253	0.064	0.632	1.420
		Normal	136	Lognormal	0.998	0.294	0.087	0.620	1.548
F1	Type of use	Commerce and services	61	Lognormal	1.000	0.283	0.080	0.584	1.505
		Housing	116	Lognormal	0.994	0.280	0.079	0.660	1.524
G1	Ease of inspection	Unfavourable	48	Lognormal	0.984	0.259	0.067	0.632	1.462
		Normal	129	Lognormal	0.996	0.297	0.088	0.623	1.554

Table 6.22 Description of the painted surface analysed

Case study	Description
	Painted surface with 18 years and with a degradation severity index equal to 57.3 %; Sub factor *A*: textured paint—lognormal distribution (μ = 1.025; σ = 0.200); Sub factor *B*1: light colours—lognormal distribution (μ = 0.996; σ = 0.275); Sub factor *B*2: rough finishing—lognormal distribution (μ = 1.149; σ = 0.238); Sub factor *B*3: compact—lognormal distribution (μ = 0.997; σ = 0.300); Sub factor *E*1: east—Gumbel distribution (μ = 1.000; σ = 0.207); Sub factor *E*2: low exposure to wind-rain action—lognormal distribution (μ = 0.999; σ = 0.188); Sub factor *E*3: more than 5 km from the sea—lognormal distribution (μ = 0.996; σ = 0.278); Sub factor *E*4: low exposure to damp—Gumbel distribution (μ = 1.045; σ = 0.295); Sub factor *E*5: unfavourable exposure to pollutants—lognormal distribution (μ = 0.944; σ = 0.253); Sub factor *F*1: housing—lognormal distribution (μ = 0.994; σ = 0.280); Sub factor *G*1: current inspection conditions—lognormal distribution (μ = 0.996; σ = 0.297)

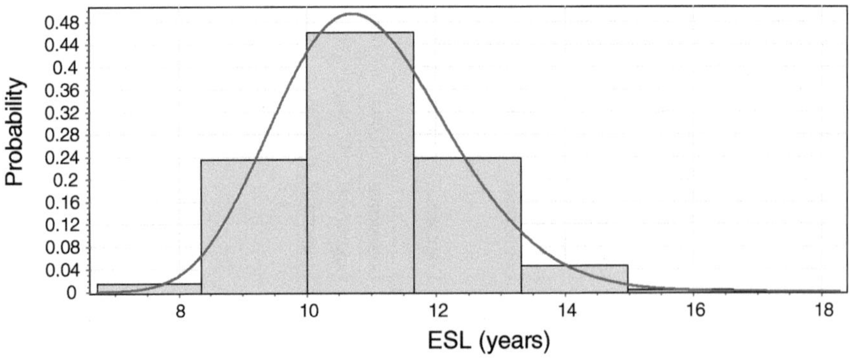

Fig. 6.4 Probability density distribution of the estimated service life of the painted surface analysed

(optimized scenario), an estimated service life of 50.5 years is obtained for this case study, while the graphical method leads to an estimated service life of 50.1 years.

The examples above are used to illustrate the applicability of the stochastic approach of the factor method as the results obtained are similar to those of the other methods and approached; their application may be extended to the whole sample.

Table 6.23 Probability distributions associated with the durability factors related with ceramic tiling systems

Durability factors			Sample size	Distribution	Average	Standard deviation	Variance	Value of the factor with a probability higher than 95 % to be exceeded	Value of the factor with a probability higher than 5 % to be exceeded
A1	Type of surface	Glazed	167	Lognormal	0.950	0.347	0.121	0.558	1.650
		Not glazed	19	Normal	0.980	0.262	0.069	0.518	1.382
B1	Colour	Light colours	108	Lognormal	0.998	0.360	0.130	0.543	1.674
		Dark colours	78	Lognormal	1.049	0.332	0.110	0.610	1.665
B2	Tiles dimension	L ≤ 20 cm	157	Lognormal	0.999	0.316	0.100	0.577	1.586
		L > 20 cm	29	Lognormal	0.823	0.394	0.155	0.392	1.568
B3	Substrate	Masonry	102	Gumbel	1.000	0.284	0.081	0.624	1.530
		Concrete	84	Lognormal	1.048	0.434	0.188	0.519	1.866
B4	Peripheral joints	Without	158	Lognormal	0.998	0.344	0.118	0.575	1.646
		With	28	Weibull	1.124	0.337	0.113	0.561	1.672
B5	Peripheral protection	Without	89	Lognormal	0.997	0.337	0.113	0.586	1.632
		With	97	Lognormal	1.124	0.360	0.130	0.642	1.792
E1	Façade orientation	East/SE	44	Gumbel	1.125	0.317	0.101	0.711	1.718
		North/NE	57	Lognormal	1.124	0.363	0.132	0.628	1.793
		West/NW	44	Lognormal	0.944	0.384	0.148	0.503	1.672
		South/SW	41	Lognormal	1.044	0.341	0.116	0.588	1.675
E2	Distance from the sea	Less than 5 km	134	Lognormal	0.848	0.291	0.085	0.477	1.394
		More than 5 km	52	Lognormal	0.999	0.346	0.119	0.549	1.643

(continued)

Table 6.23 (continued)

Durability factors			Sample size	Distribution	Average	Standard deviation	Variance	Value of the factor with a probability higher than 95 % to be exceeded	Value of the factor with a probability higher than 5 % to be exceeded
E3	Wind-rain action	Severe	53	Lognormal	0.898	0.306	0.094	0.549	1.478
		Moderate	93	Lognormal	0.993	0.348	0.121	0.536	1.640
		Low	40	Lognormal	0.999	0.300	0.090	0.501	1.566
E4	Exposure to damp	High	83	Lognormal	0.899	0.305	0.093	0.495	1.465
		Low	103	Lognormal	0.997	0.336	0.113	0.552	1.620
G1	Ease of inspection	Unfavourable	63	Lognormal	0.921	0.372	0.139	0.510	1.626
		Normal	123	Lognormal	1.050	0.307	0.094	0.603	1.600

Table 6.24 Description of the ceramic cladding analysed

Case study	Description
	Ceramic cladding with 63 years and with a degradation severity index equal to 34.1 %; Sub factor *A*: glazed—lognormal distribution (μ = 0.998; σ = 0.347); Sub factor *B*1: light colours—lognormal distribution (μ = 0.998; σ = 0.360); Sub factor *B*2: L < 20 cm—lognormal distribution (μ = 0.999; σ = 0.316); Sub factor *B*3: masonry—Gumbel distribution (μ = 1.000; σ = 0.284); Sub factor *B*4: without joints—lognormal distribution (μ = 0.998; σ = 0.344); Sub factor *B*5: with peripheral protection—lognormal distribution (μ = 1.124; σ = 0.360); Sub factor *E*1: east—Gumbel distribution (μ = 1.125; σ = 0.317); Sub factor *E*2: less than 5 km from the sea—lognormal distribution (μ = 0.848; σ = 0.291); Sub factor *E*3: moderate exposure to wind-rain action—lognormal distribution (μ = 0.993; σ = 0.348); Sub factor *E*4: low exposure to damp—lognormal distribution (μ = 0.993; σ = 0.348); Sub factor *G*1: current inspection conditions—lognormal distribution (μ = 1.050; σ = 0.307)

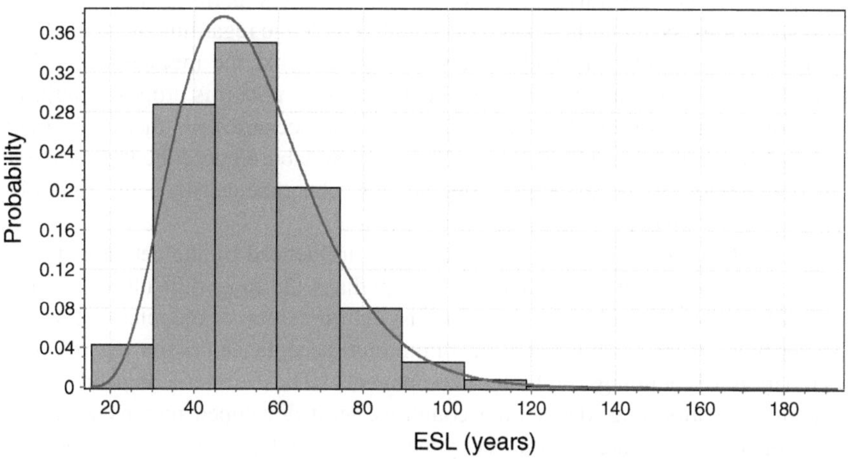

Fig. 6.5 Probability density distribution of the estimated service life of the ceramic cladding analysed

6.4 Conclusions

This chapter addresses the implementation of the factor method for service life prediction of façade claddings. This method has been the object of criticism and has some limitations when used as a purely deterministic method. However, the factor method combines flexibility with a relative ease of application and is currently seen as a general framework for the estimation of the service life, whose application is normalized in the international standard for durability of building components (partially published).

In the last decades, the factor method has been the subject of numerous studies but there are no implemented methodologies to determine the durability sub-factors. The method is extremely sensitive to small changes in the sub-factors quantification (since they are usually multiplied consecutively). Moreover, as more durability sub-factors are adopted, the complexity of the method increases. Therefore, the most important characteristics that affect the durability of the elements under analysis should be identified, since not all the features contribute in the same way to establish the end of the claddings' service life. Some authors suggest that only the essential durability sub-factors must be taken into account. By reducing the number of sub-factors, the applicability of the method increases, leading to a rougher analysis though; this may compromise the fine adjustment between the method and the observed reality.

In this chapter, it is possible to conclude that the quantification of the durability sub-factors is not unequivocal, based on the analysis performed. In fact, the sub-factors must be quantified based on actual knowledge regarding the degradation phenomena of the element under analysis. In this book, the measurement of the durability sub-factors in absolute terms (deterministic model) is proposed, adjusted to the degradation of façades claddings (optimized scenario). Furthermore, a new approach to the factor method is presented, more suitable to describe the variability and complexity of the claddings' degradation phenomena, since the durability sub-factors are quantified in a stochastic way.

The application of the factor method is also influenced by the determination of the reference service life. This parameter is perhaps the most difficult to establish, since it is affected by subjective criteria. The determination of the reference service life depends on the establishment of a minimum acceptance criteria regarding the degradation of buildings and their components. These criteria depend on the building's location (e.g. developing countries, least developed regions within the same country or even areas within a city having a more degraded built environment) and on the social and economic context (the funds available for maintenance and rehabilitation, which affect the performance of the construction elements). Throughout this book, a minimum performance threshold based on the observation of the in situ degradation condition of the façade claddings under analysis is defined. The reference service life is established based on these criteria, using several methodologies for that purpose. The results obtained are supported by the standards and the literature on the subject.

The results obtained using the factor method are similar to the values obtained by the graphical method, which demonstrates the validity of the proposed models. The proposed methodologies are simple to apply, despite some uncertainty associated. This uncertainty does not exclude its application because the method can be used as an approximation to reality, providing some information for further investigation and development. The new stochastic approach to the factor method intends to overcome some of these limitations, by presenting estimations of the service life represented by a probability distribution. Such approach is extremely relevant since it provides (i) an average value for the service life of façade claddings, and (ii) the service life value with the highest probability of being achieved based on the characteristics of the claddings analysed. Likewise it is possible to estimate, for a given confidence interval, the probability of a given value of service life to be exceeded. This information can be applied in the development of maintenance and rehabilitation plans, providing valuable information to help stakeholders in the decision process.

The methodologies proposed in this chapter seem adequate to describe the variability and the complexity of the degradation phenomena of façade claddings. These models easily allow further updating when additional data becomes available and can be applied to other types of claddings and other countries with different exposure conditions provided that the durability factors are adjusted to the specific cases under analysis.

References

Aarseth LI, Hovde PJ (1999) A stochastic approach to the factor method for estimating service life. 8th international conference on durability of building materials and components. Vancouver, Canada, pp 1247–1256

Adalberth K (1997) Energy use during the life cycle of buildings: a method. Build Environ 32 (4):317–320

AIJ (1993) The English edition of the principal guide for service life planning of buildings. Architectural Institute of Japan, Tokyo

Aktas CB, Bilec MM (2012) Service life prediction of residential interior finishes for life cycle assessment. Int J Life Cycle Assess 17(3):362–371

BCIS Life Expectancy of Buildings Components (2001) Surveyor's experiences of building in use—a practical guide. Building Cost Information Service, London

Bell FG (1993) Durability of carbonate rocks as building stone with comments on its preservation. Environ Geol 21:187–200

Bordalo R, de Brito J, Gaspar P, Silva A (2011) Service life prediction modelling of adhesive ceramic tiling systems. Build Res Inf 39(1):66–78

Bourke K, Davis H (1999) Estimating service lives using the factor method for use in whole life costing. 8th international conference on durability of building materials and components. Vancouver, Canada, pp 1518–1526

Brand S (1997) How buildings learn: what happens after they're built?, 1st edn. Phoenix Illustrated, London

BSI (1992) Guide to durability of buildings and building elements, products and components. BS 7543. British Standards Institution, London

Carani G, Palmonari C, Timellini G (1997) The costs of the materials for floorings. Ceram Acta 6 (9):5–16 (in Italian)
Cecconi F (2004) Engineering method for service life planning: the evolved factor method. In: Building the future: 16th CIB world building congress. Toronto, Canada
Chai C (2011) Service life prediction of painted surfaces in external walls, Master Dissertation in Civil Engineering, Instituto Superior Técnico, Technical University of Lisbon, Portugal. (In Portuguese)
Chai C, de Brito J, Gaspar P, Silva A (2014) Predicting the service life of exterior wall painting: techno-economic analysis of alternative maintenance strategies. J Constr Eng Manage 140 (3):04013057
Chai C, de Brito J, Gaspar PL, Silva A (2015) Statistical modelling of the service life of exterior painted surfaces. International Journal of Strategic Property Management 19(2):173–185
Cusmano G, Lucchini A (2003) Implementing the factor method for service life estimation using weighted-factors. International workshop on "management of durability in the building process". Italy, Milan, pp 25–26
Daniotti B, Paolini R (2005) Durability design of external thermal insulation composite systems with rendering. 10th international conference on durability of building and components. Lyon, France, p TT5-164
Daniotti B, Spagnolo SL (2008) Service life prediction tools for buildings' design and management. 11th international conference on durability of building materials and components. Istanbul, Turkey, p T72
Donaldson B (1990) New stone technology, design and construction for exterior wall systems, 2nd edn. American Society for Testing and Materials, Baltimore
Emídio F, de Brito J, Gaspar P, Silva A (2014) Application of the factor method to the estimation of the service life of natural stone cladding. Constr Build Mater 66:484–493
Fay R, Treloar G, Iyer-Raniga U (2000) Life-cycle energy analysis of buildings: a case study. Build Res Inf 28(1):31–41
Flores-Colen I, de Brito J (2010) A systematic approach for maintenance budgeting of buildings façades based on predictive and preventive strategies. Constr Build Mater 24(9):1718–1729
Galbusera MM, de Brito J, Silva A (2014) Application of the factor method to the prediction of the service life of ceramic external wall claddings. J Build Perform Constr Facil pp 19–29. doi:10.1016/j.conbuildmat.2014.05.045
Garrido MA, Paulo PV, Branco FA (2012) Service life prediction of façade paint coatings in old buildings. Constr Build Mater 29:394–402
Gaspar P, de Brito J (2008) Service life estimation of cement-rendered facades. Build Res Inf 36 (1):44–55
Haapio A, Viitaniemi P (2008) How workmanship should be taken into account in service life planning. 11th international conference on durability of building materials and components. Istanbul, Turkey, p T45
Hed G (1999) Service life planning of building components. 8th international conference on durability of building materials and components. Vancouver, Canada, pp 1543–1551
Hovde PJ (1998) Evaluation of the factor method to estimate the service life of building components. CIB World Building Congress, Gaevle, pp 223–232
Hovde PJ (2000) Factor methods for service life prediction: a state-of-the-art. Draft report, Norwegian University of Science and Technology, Trondheim
Hovde PJ (2005) The factor method—a simple tool to service life estimation. 10th international conference on durability of building materials and components. Lyon, France, pp 522–529
ISO 15686-1 (2011) Buildings and constructed assets—service life planning—part 1: general principles and framework. International Organization for Standardization, Switzerland
Kelly DJ (2007) BRE, design life of buildings—a scoping study. Scottish Building Standards Agency, Glasgow
Keoleian GA, Blanchard S, Reppe P (2001) Life-cycle energy, costs, and strategies for improving a single-family house. J Ind Ecol 4(2):135–156

Lair J (2003) Failure modes and effect analysis and service life prediction, intermediary report (D4-C2-jl-01 Draft 2), IEA task 27 (Project C2: Failure Mode Analysis). CSTB, France, pp 166–212

Lippiatt BC (2008) Building for environmental and economic sustainability (BEES). The National Institute of Standards and Technology (NIST), Gaithersburg

Marteinsson B (2003) Assessment of service lives in the design of buildings—development of the factor method. Licentiate thesis, KTH's Research School—HiG, Centre of Built Environment. University of Gävle, Sweden

Mc Duling JJ (2006) Towards the development of transition probability matrices in the Markovian model for the predicted service life of buildings. Ph.D. Thesis in Civil Engineering, Faculty of Engineering, Built Environment and Information Technology, University of Pretoria, Pretoria

Mc Duling J, Horak E, Cloete C (2008) Service life prediction beyond the 'factor method'. 11th international conference on durability of building materials and components. Istanbul, Turkey, p T42

Mithraratne N, Vale B (2004) Life cycle analysis model for New Zealand houses. Build Environ 39(4):483–492

Moser K (1999) Towards the practical evaluation of service life—illustrative application of the probabilistic approach. 8th international conference on durability of building materials and components. Vancouver, Canada, pp 1319–1329

Moser K (2004) Engineering design methods for service life prediction. CIB W080/RILEM 175 SLM: service life methodologies prediction of service life for buildings and components, task group: performance based methods of service life prediction. Trondheim, Norway, pp 52–95

Moser K, Edvardsen C (2002) Engineering design method for service life prediction. In: 9th international conference on the durability of building materials and components. Brisbane, Australia, paper 222

Nireki T, Inukai T, Motohashi K (2002) Toward practical application of factor method for estimating service life of building. In: 9th international conference on the durability of building materials and components. Brisbane, Australia, paper 218

NYC Housing Maintenance Code (2013) Subchapter 2: maintenance, services, and utilities. Section 27-2013: painting of public parts and within dwellings, New York

Pullen S (2000) Energy assessment of institutional buildings. In: 34th annual conference of the Australia and New Zealand Architectural Science Association. Adelaide, Australia

Re Cecconi F, Iacono P (2005) Enhancing the factor method—suggestions to avoid subjectivity. 10th international conference on durability of building materials and components. Lyon, France, p TT4-172

RICS Books, "Building maintenance: Strategy, planning & performance", The Royal Institution of Chartered Surveyors, UK, 2000

Sánchez E, García-Ten J, Sanz V, Moreno A (2010) Porcelain tile: almost 30 years of steady scientific-technological evolution. Ceram Int 36(3):831–845

Schafer BL (2002) Durability of acrylic waterproofing membranes for wet areas. In: 9th international conference on durability of building materials and components. Brisbane, Australia, paper 042

Scheuer C, Keoleian GA, Reppe P (2003) Life cycle energy and environmental performance of a new university building: modeling challenges and design implications. Energy Build 35 (10):1049–1064

Schouenborg B, Grelk B, Malaga K (2007) Testing and assessment of marble and limestone (TEAM)—important results from a large European research project on cladding panels. ASTM Int 4(5):10–22

Shohet I, Laufer A (1996) Exterior cladding methods: A technoeconomic analysis. J Constr Eng Manag 122(3):242–247

Shohet IM, Paciuk M (2004) Service life prediction of exterior cladding components under standard conditions. Constr Manag Econ 22(10):1081–1090

Shohet I, Rosenfeld Y, Puterman M, Gilboa E (1999) Deterioration patterns for maintenance management—a methodological approach. 8th international conference on durability of building materials and components. Vancouver, Canada, pp 1666–1678
Shohet IM, Puterman M, Gilboa E (2002) Deterioration patterns of building cladding components for maintenance management. Constr Manag Econ 20(4):305–314
Silva A, Gaspar PL, de Brito J (2012) Application of the factor method to maintenance decision support for stone cladding. Autom Constr 22:165–174
Silva A, de Brito J, Gaspar PL (2016) Stochastic approach to the factor method applied to service life prediction of rendered façades. J Mater Civ Eng 28(2):04015130
Surgelas FMA, Marques GF, Rodrigues CDS (2010) Life cycle analysis of ceramic versus painting materials applied to external walls. Int J Hous Sci 34(1):37–46
Tam C, Loo Y, Quek S, Saw W (1993) Simulated thermal fatigue testing of wall-to-tile bond. In: Nagataky S, Nireki T, Tomosawa F, Spon FN (eds) Durability of building materials and components
Teo EAL, Chew MYL, Harikrishna N (2005) An assessment of factors affecting the service life of external paint finish on plastered facades. 10th international conference on durability of building materials and components. Lyon, France, p TT4-152
Teplý B (1999) Service life prediction of structures—factor method. Struct Horiz 8:137–141 (in Czech)
Van Nunen H (2010) Assessment of the sustainability of flexible building. The improved factor method: service life prediction of buildings in the Netherlands, applied to life cycle assessment, 1st edn. Aeneas, The Netherlands
Westberg K, Norén J, Kus H (2001) On using available environment data in service life estimations. Build Res Inf 29(6):428–439
Wetherill GB (1986) Regression analysis with applications, 1st edn. Chapman and Hall, London
Winkler EM (1997) Stone in architecture: properties, durability. In: 3rd completely revised and extended edition. Springer, New York

Chapter 7
Comparative Analysis of Service Life Prediction Methods

7.1 Theoretical Foundations

The aim of this Book is to model the service life prediction of façade claddings. For this purpose, the applicability of different approaches for service life prediction have been analysed in the previous chapters. The application of the various methodologies led to different results, with (Lourenço et al. 2007): (i) different levels of complexity (from simple regression techniques to complex formulations involving a wide range of nonlinear equations); (ii) different levels of availability (ranging from models easily applied in professional offices to models only available in R&D institutions, universities or specialized consultants); (iii) different processing times, ranging from the instantaneous output of results to models that may require hours of computer processing; and (iv) different costs. The methodologies described are conceptually distinct and naturally lead to different results, which is not an argument to prefer one model over another (Lourenço 2002). In fact, all the analytical techniques employed in this Book are useful and appropriate in the description of the degradation phenomena of façade claddings, and can be applied in other engineering fields, if properly adapted to the reality that they intend to model.

Happern (1978) refers that there are different criteria that must be taken into account when analysing the applicability of a service life prediction model, such as: (i) the probability of failure of the model, making it necessary to measure the obtained errors; (ii) the performance of the model, which establishes a set of requirements to use the model to perform the task for which it was designed; (iii) the time required by the model to solve the problem; and (iv) its adaptability to new situations. In this chapter, a comparative analysis of the different models is performed. For each service life prediction method proposed, several parameters are analysed to evaluate its predictive capability. The comparison between service life prediction models takes into account

A. Silva et al., *Methodologies for Service Life Prediction of Buildings*,
Green Energy and Technology, DOI 10.1007/978-3-319-33290-1_7

some basic principles (Ratkowsky 1983): (i) the model must be adjusted to the data; (ii) the models must have a significant explanatory power; (iii) the models should be applicable to the entire population and not only to the sample, i.e. it is necessary to avoid the overfitting of the model, ensuring the capability of prediction of the behaviour of other case studies not used in the models' definition; (iv) the parsimony principle—which corresponds to the evaluation of the models according to the number of estimation parameters—should be taken into account: in the presence of two models that equally represent the reality, the stakeholder should choose the model that requires a smaller number of parameters; (v) the model parameters should be easily interpretable; (vi) the model should be straightforward to apply, considering the ease with which the user understands how to use the model, the amount and complexity of data required for the model to work properly, as well as the necessary software and its processing time.

A comparative analysis of the different proposed models is thus performed, first based on the predictive ability of each model, which is evaluated through the comparison between the values obtained by each model and the data collected in field work, corresponding to the real degradation condition of the façades analysed. In the second step of the analysis, the sensitivity and specificity of each model is evaluated based on ROC curves, to analyse the accuracy of each model to correctly classify the case studies that may have reached the end of their service life. Once the predictive accuracy and the classification capabilities of each model are evaluated, a multi-criteria analysis is performed. This multi-criteria analysis encompasses different criteria that should be taken into account by different stakeholders to choose the best model for a specific application. A sensitivity analysis is performed in order to analyse the best model for distinct decision-making profiles. Based on these analyses, some recommendations are provided, referring the advantages and limitations of the models analysed and thus allowing a more rational and informed selection of a service life prediction model as far as the purpose of the model and the planner's profile and the user's perspective are concerned.

7.2 Service Life Prediction Models

Currently, there are several methodologies for service life prediction, as mentioned in Chap. 2 of this Book. According to various authors (Moser 2004; Lacasse and Sjöström 2004), service life prediction methods can be divided into deterministic, stochastic and engineering methods. Using a different approach, Shohet and Paciuk (2004) suggest that the different methodologies to service life prediction of building are grouped into: analytical models; statistical models; empirical methods; and experimental methods. Rudbeck (1999) proposes the use of statistical tools as an improvement to existing methods and Moser (2004), in an extensive analysis of the work carried out by different authors in this area, concludes that more studies are needed in order to identify the parameters that influence the life cycle of the building elements, making it necessary to create viable mathematical relations that allow the application of these methods. Therefore, there are different approaches in the definition of service life prediction methodologies. Based on the literature and taking into account the particularities of the models analysed, the service life prediction methods are classified in this study in four categories: (i) deterministic models (Chap. 3), based on degradation paths that illustrate the façade's loss of performance (regression analysis: simple nonlinear regression; multiple linear regression; and multiple nonlinear regression); (ii) stochastic models (Chap. 4) (logistic regression and Markov chains); (iii) computational methods (Chap. 5) (artificial neural networks and fuzzy systems); (iv) factor methods (Chap. 6) (using a classical and a probabilistic approach). This characterization allows grouping the different methods in methodological families.

In deterministic models, mathematical and/or statistical formulations are used, trying to describe the relationship between the degradation factors and the façade's condition. Setunge and Hasan (2011) refer that these methods are extremely efficient when used in large and representative samples, although, in general, they ignore the random errors associated with the predictions as well as the randomness associated with the degradation phenomenon. All regression techniques are based on the same assumptions, i.e. these methods intend to obtain the function that best fits a set of random data. Simple regression analysis allows obtaining the estimated service life of the façade claddings based on the evolution of the degradation of the overall sample over time. Furthermore, the independent degradation curves for the different characteristics that affect the durability of claddings can be defined, transforming the claddings' features into quantitative variables used in multiple linear regression (see Silva et al. 2012). Generally, purely deterministic models are unable to take into account the uncertainty and variability of the data, neglecting the influence of the variables with explanatory power in the buildings' degradation (Jiang and Sinha 1989; Madanat et al. 1995). Multiple regression analysis (linear or nonlinear) can be used to overcome this limitation, since the evaluation of the claddings' degradation is obtained using an expression that includes all the variables that influence their degradation process.

Stochastic models (logistic regression analysis and Markov chains model) allow analysing the claddings' service life based on probabilistic distribution functions. Stochastic models are usually superior when compared with pure deterministic formulations, showing a considerably better performance dealing with the uncertainty associated to a complex phenomenon, such as the degradation of buildings and their components (Frangopol et al. 1997; Lounis and Mirza 2001). Using these methods it is possible to assess: (i) the probability of each façade having a given degradation condition according to its age, its characteristics and the environmental exposure conditions; (ii) the period of time with maximum probability of transition between a degradation condition to the next one (more severe); (iii) the probability of each case study reaching the end of their service life over a given period of time.

In computational methods, two approaches are used, leading to similar results: artificial neural networks (ANNs) and fuzzy systems (fuzzy logic). These models are able to find the nonlinear function that best fits the dataset to be modelled, using a learning process based on experiences and examples, and to generalize for new samples hitherto unknown. Models based on "artificial intelligence", as in the case of artificial neural networks, use computational techniques that intend to simulate and automate intelligent behaviour. According to Morcous et al. (2002), this method applies previous knowledge related to the reality that one intends to model, transforming raw data into models easy to apply. On the other hand, fuzzy logic models combine mathematical systematization with linguistic rules, being able to handle with vague and inaccurate data. Computational methods allow obtaining the estimated service life of façades according to the variables considered as explanatory and statistical relevant in the degradation phenomena.

Today, factor methods are the most common methodologies for service life prediction of building components, being partially published in the international standard for the durability (ISO 15686: 2011). Various authors (Hovde 2000; Moser and Edvardsen 2002) criticize this methodology, referring that it is too simplistic in the description of the degradation process. Notwithstanding these limitations, this method can be easily learned and implemented, and has become a framework methodology for the service life prediction of building components. In this Book, two approaches to the factor method are suggested: (i) the classic approach, in which the factor method is applied as a deterministic tool, as described in the literature, in which the durability factors are given by absolute values according to the characteristics of the element; (ii) and the stochastic approach (an improved methodology), in which the durability factors are associated to a probabilistic distribution function, providing an estimated service life given by a probabilistic distribution function for each case study within the sample. The stochastic approach allows analysing the uncertainty associated with the degradation phenomenon.

Table 7.1 presents a summary of the different methodologies discussed in this Book, with their characteristics, advantages and limitations, and the information produced by each of them.

Some features are transversal to all methods: (i) in all the models, the results are achieved by minimizing the deviation between the predicted and the observed values; (ii) the models react to modifications of the data, which means that the models are established for a dataset, and some variations in the models parameters can be expected with the acquisition of more data. Furthermore, all the models discussed in this Book consider that the claddings are not subject to any maintenance action during the period of time under consideration. Sanders and Zhang (1994) refer that this assumption is due to the difficulty of estimating the impact of various maintenance strategies in buildings deterioration. This assumption is extremely relevant for the application of stochastic models and, especially, Markov chains, since these models analyse the probability of transition between degradation conditions, and it is not physically understandable that the claddings can change

Table 7.1 Description of the different proposed methodologies

Model	Description	Advantages	Limitations	Model's output
Simple regression	This model relates the severity of degradation of the claddings analysed with their age, allowing the definition of the graphical description of the loss of performance of the claddings	Simple and straightforward method This model can be easily adapted and complemented with new case studies. It does not require specialized knowledge or specific software	It only allows relating the degradation of façades with one variable (x) This model is unable to capture the random nature of the modelled phenomenon (it does not estimate the model deviations) The variable age is unable to explain 100 % of the variability of the severity of degradation, which implies that some of the variability of the severity of degradation may be explained by other factors	$S_w = B_0 + B_1 \cdot x + \varepsilon$ This model leads to a simple expression which gives the severity of degradation (S_w) estimated based on the façades age

(continued)

Table 7.1 (continued)

Model	Description	Advantages	Limitations	Model's output
Multiple linear regression	This model can be seen as an extension of simple regression analysis, encompassing more than one independent variable	Method easy to learn The method is able to handle with an arbitrarily large number of independent variables and allows covering all the relevant variables for the explanation of the degradation phenomenon. The extension of this type of analysis, involving causal relationships between variables, allows a better understanding of the degradation phenomenon of façades, revealing the most significant variables in this process	For a user unfamiliar with this model, the analysis of the results produced can be complex Models that include causal relationships are more complex and the manipulation of mathematical equations is more difficult	$S_w = B_0 + B_1 \cdot x_1 + B_2 \cdot x_2 + \cdots + B_k \cdot x_k + \varepsilon$ This model leads to a linear mathematical equation, which gives the severity of degradation (S_w) according to the age and the claddings' characteristics (considered as explanatory variables)

(continued)

Table 7.1 (continued)

Model	Description	Advantages	Limitations	Model's output
Multiple nonlinear regression	This model is an extension of simple regression analysis, encompassing more than one independent variable and producing a nonlinear model	Nonlinear models are, usually, more effective to model real phenomena, whether physical, chemical or biological This model allows covering all the relevant variables to the explanation of the degradation of façade claddings	In these models, the arithmetic equations that describe the degradation of the façade claddings are unknown, whereby the definition of these models is an iterative process The researcher is unable to identify with certainty the variables that should be included in the model and, therefore, a sensitivity analysis must be performed. These models are more complex than the previous models analysed, requiring more time in the model's	These models lead to a nonlinear expression, which give the severity of degradation (S_w) estimated according to the age and the characteristics of the claddings analysed (explanatory variables). Given as an example of these models, an expression of a multiple polynomial model is presented: $S_w = B_0 + B_1x_1 + B_2x_2 + \cdots + B_ix_i + B_{12}x_1^2 + B_{22}x_2^2 + \cdots + B_{i2}x_i^2 + \cdots + B_{ik}x_i^k + \varepsilon$

(continued)

Table 7.1 (continued)

Model	Description	Advantages	Limitations	Model's output
			definition, with a large number of estimation parameters, whose estimation is performed by an iterative process. In these models, the mathematical equations are more complex and the analytical or numerical solutions are usually more difficult to achieve	
Logistic regression	This methodology is an extension of the previous regression analysis, but in this case this model allows establishing a relationship between the	This model allows a probabilistic analysis of the degradation condition of façades over time, transforming data collected during field work into	This statistical technique leads to more complex models, whose results are difficult to interpret, as well as its statistical validity To obtain these models, statistical	This model provides an expression that gives the probability of a given façade presenting a given degradation condition based on its age. The mathematical expression is given by a logistic function, such as: $P(Y=1\|X)=\frac{e^{B_{10}+B_{11}\cdot X_1+\cdots+B_{1p}\cdot X_p}}{1+e^{B_{10}+B_{11}\cdot X_1+\cdots+B_{1p}\cdot X_p}+e^{B_{20}+B_{21}\cdot X_1+\cdots+B_{2p}\cdot X_p}}$

(continued)

Table 7.1 (continued)

Model	Description	Advantages	Limitations	Model's output
	target or dependent variable and qualitative variables	statistical distributions This methodology allows evaluating the probability of transition between degradation conditions It evaluates the probability of a cladding reaching the end of its service life, depending on its age and its characteristics	software is required; in this Book, SPSS is used to define these models	
Markov chains	Markov chains simulate the evolution of the degradation condition over time, defining the probability of the future performance of a construction element based on	This model allows: (i) analysing the probability of the façade claddings belonging to each of the degradation conditions; (ii) evaluating the probability of transition	In this model, the acquisition of the necessary knowledge for the application of the methodology can be time consuming It requires knowledge regarding	The results are expressed by an intensity matrix $\boldsymbol{Q}$, which can be related to a $\boldsymbol{P}$ Markov transition matrix, given by the generic expression: $P(\Delta t) = \exp(Q \cdot \Delta t)$ In this model, the transition rate between degradation states is given by the ratio between the number of elements that move between consecutive degradation states (i to j) and the sum of intervals between inspections for elements whose initial state is i

(continued)

Table 7.1 (continued)

Model	Description	Advantages	Limitations	Model's output
	its current performance only, ignoring the history of deterioration, the maintenance actions performed during buildings life cycle, its age, as well other parameters	between degradation conditions; and (iii) estimating the mean time of permanence in each degradation condition. These models have the ability to extrapolate the performance of building elements, even if the information on the state of these elements at a given time is omitted. Markov chains avoid regression techniques that have, in some cases, the presence of multicollinearity effects in the explanation of the degradation	programming techniques, i.e. a specific algorithm adjusted to the phenomenon to be modelled must be defined Furthermore, this method requires the application of specific software; in this case, the Matlab is used	

(continued)

Table 7.1 (continued)

Model	Description	Advantages	Limitations	Model's output
		phenomenon, which may jeopardizes the analysis		
Artificial neural network	These models are based on the emulation of biological neural systems The model is created based on a learning process through behaviour patterns This method produces a model to predict the future performance of the façades based on knowledge previously acquired	These models can be trained to solve problems, whose numerical solution is difficult to achieve by more conventional approaches This method has low sensitivity to errors, featuring a massively parallel processing and a distributed storage of information They have high generalization ability and adaptability to new data These models are able to obtain the best nonlinear function to	The determination of the variables to be included in the model can be a time-consuming process when there is no previous knowledge regarding the phenomenon under analysis The model does not provide obvious explanations regarding the modelling process, which is also known as the "black box" concept. The model produced may be a quite complex for users	This model provides a mathematical expression, where the severity of degradation is obtained based on the explanatory variables considered and a set of coefficients estimated for the proposed formulation, as shown below: $S_w = h_0 + \sum_{i=1}^{k} h_i H_i$ $H_i = \tanh\left(c_{i0} + \sum_{n=1}^{j} c_{in} V_n\right)$ where k represents the number of variables included in the model

(continued)

Table 7.1 (continued)

Model	Description	Advantages	Limitations	Model's output
		describe a given phenomenon	with no expertise in this area This method requires specific software to be implemented	
Fuzzy logic systems	Fuzzy logic systems are defined based on a set of independent systems of multiple inputs and a single output The degradation of the façade claddings is evaluated based on a set of fuzzy rules that describe local relationships of input–output	The application of this method is especially interesting when the problem to model is subjected to uncertainties. These models are known to be flexible and able to deal with inaccurate data, modelling with precision naturally vague and difficult to describe systems. This method can be developed based on the experience of experts (accepting as	This model presents a high computational demand, requiring specific software and a specific algorithm in Matlab. The application of the model also requires the use of a toolbox, in order to obtain the values of the membership functions Conceptually, this model is complex, with a time-consuming learning process	The model allows predicting the severity of degradation of façade claddings based on the following expression: $S_w(x) = \frac{\sum_{i=1}^{C} \beta_i f_i(x)}{\sum_{i=1}^{C} \beta_i}$ where $f_i(x)$ represents the consequent function for each rule and βi the membership function associated with the set of conditions that describe each case study

(continued)

Table 7.1 (continued)

Model	Description	Advantages	Limitations	Model's output
		input, information based on human observation)		
Deterministic factor method	This method consists in the determination of an estimated service life by multiplying a reference service life by a set of durability factors regarding the specific conditions of each case	This methodology allows achieving the desired balance between accuracy, speed, low cost and ease of application This method may be easily complemented with additional data at any time. Through the durability sub-factors, this methodology evaluates the influence of the claddings characteristics on their estimated service lives Due to its simplicity, this approach can be	One of the main disadvantages of this method is that it is unable to take into account the variability of the ageing process. The service life is estimated based on the characteristics of the construction element only, ignoring its degradation condition at the time of inspection This method is highly sensitive to small variations in the value of each sub-factor The result obtained is an absolute value	In this case, instead of estimating the severity of degradation of façade claddings, this method directly provides the estimated service life, therefore using the following equation: $\text{ESL} = \text{RSL} \cdot A \cdot B \cdot C \cdot D \cdot E \cdot F \cdot G$ where, according to ISO 15686: 2011, ESL is the estimated service life, RSL the reference service life, *A* the factor related to the quality of the materials, *B* the factor related to the design level, *C* the factor related to the execution level, *D* the factor related to the interior environmental conditions, *E* the factor related to the external environmental conditions, *F* the factor related to the in-use conditions, and *G* the factor related to the level of maintenance

(continued)

Table 7.1 (continued)

Model	Description	Advantages	Limitations	Model's output
		easily applied by any user	(which represents the estimated service life), not revealing any information about the results dispersion	
Stochastic factor method	This approach is similar to the previous method but, in this case, the durability sub-factors are characterized by probability distributions	The model combines the application of a simple methodology (the factor method) with probabilistic models. This solution has a relatively good balance between a higher precision (since random variables allow describing with more accuracy the degradation phenomenon) and a relative simple conceptual approach	Despite being a conceptually simple model to apply and understand, the practical application of this model becomes complex when using probabilistic values for the quantification of the durability sub-factors. In this case, an algorithm in Matlab has been created in order to allow multiplying different probability distributions	Using this methodology, the estimated service life for each case study is given by a probability distribution, as the example given below 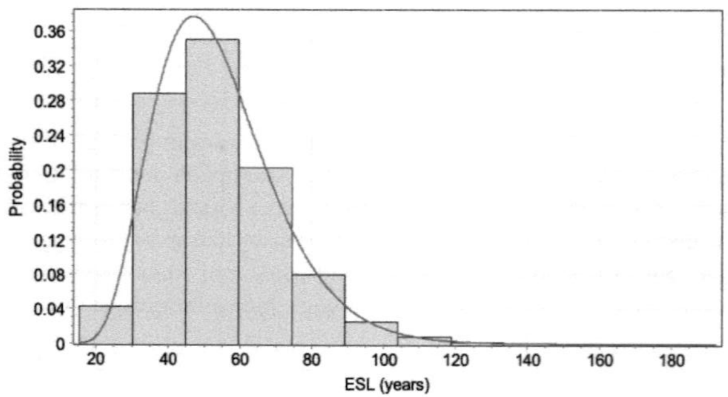

from a more severe to a less serious condition. By not considering maintenance actions, the degradation of façade claddings is considered a continuous process, assuming the current condition of the claddings to predict their future performance.

7.3 Predictive Capability of the Service Life Prediction Models

In the definition of service life prediction models, it is essential to ensure that these models are able to express the reality they intend to model, with adequate levels of accuracy and reliability. Currently, it is possible to find in the literature different models for service life prediction of claddings, some of them based on mathematical tools, rarely used in construction. Mathematical models are not unique for a given application, i.e. depending on the user's perspective, a given reality can be represented in different ways and by different mathematical models.

It should be noted that there are no exact models in the field of service life prediction as in any field of science—to such an extent that Albert Einstein (1921) himself once stated that "as far as the laws of mathematics refer to reality, they are not certain; and as far as they are certain, they don't refer to reality". In fact, all the service life prediction models present some restrictions, essentially due to two main reasons: (i) the accurate prediction of the future is impossible; (ii) measurements of the real-world concepts with the precision of a mathematical model is impossible and no model is able to perfectly translate the natural and frequent changes of the real-world phenomenon. Therefore, the results of prediction models should be analysed as a rough translation of the real world. According to ISO 15686-1: 2011, the service life prediction models are rarely as accurate and reliable as desired, due to the variability related with the degradation processes and the large number of variables that affect the buildings' life cycle, thus leaving some variability of the phenomenon of degradation unexplained. ISO 15686-1: 2011 also refers that an acceptable uncertainty level for the service life prediction model should be defined; furthermore, the standard suggests that a confidence limit of 80 % may be acceptable for non-structural components (easily accessible and maintained), and it is necessary to set a higher confidence level for structural elements (whose failure may jeopardize the safety of the building).

In the definition of mathematical and prediction models it is therefore necessary to deal with the uncertainty inherent to real-world modelling. In general, uncertainty can be classified as epistemic or random (Oberkampf et al. 2004). The distinction is not always clear but it can be said that the random uncertainty (also called as noise) is the result of the variability associated with the physical processes and environmental degradation, caused by the random nature of the data. On the other hand, the epistemic uncertainty [also designated as imprecision by Limbourg and Point (2005)] is due to the lack of knowledge regarding the phenomenon to be modelled. Zimmermann (2000) states that the lack of information is the major source of (epistemic) uncertainty and, naturally, with the acquisition of more and better data, more accurate and reliable models can be achieved (Ross 1996).

The methodologies applied in this Book have some uncertainty associated, the most relevant of which are related with the definition of the maximum acceptable degradation level. In fact, the definition of the instant in which a cladding reaches the end of its service life is a complex issue that depends on subjective criteria (referred to in Chap. 2). This theoretical limit varies, essentially according to the buildings' social and economic context, and is defined based on users' expectations. To overcome this difficulty, an expert survey can be performed, defining different models with different minimum performance levels that could be adjusted depending on the stakeholders requirements, allowing obtaining the remaining service life of buildings according to the level of requirement adopted. In this study, the maximum level of degradation adopted is based on the analysis of the degradation of the inspected façades. Therefore, the limit which establishes the end of the service life is associated with a given degradation condition obtained from a ratio of extent and severity of defects. From the analysis of the samples collected during the fieldwork, a maximum degradation level of 10 % seems too conservative, since for this level of degradation only aesthetic defects are found, which do not compromise the claddings' mechanical safety. On the other hand, a maximum degradation level of 40 % corresponds to a much too advanced deterioration state. In this study, a maximum degradation level of 20 % is therefore adopted, which seems reasonable and coherent to describe the end of service life of inspected claddings.

An uncertainty inherent to the modelling process is also present in the models proposed. In fact, as mentioned before, all the models react to modification in the data. Therefore, a balance between the mathematical function approximation to the data and the model's generalization ability must be achieved; this equilibrium is weaker the greater the number of explanatory variables in the model and the smaller the size of available data. As the complexity of the mathematical model increases, the risk of over fitting the model to the data also increases, compromising its generalization capability, thus increasing the prediction errors (Dieterle 2003). The model's generalization capability is one of the fundamental characteristics that must be guaranteed; scientifically, a prediction model will be more effective the higher its generalization capability is, thereby ensuring that the model is representative of the study's population, being able to predict the behaviour of other case studies, outside the sample used in the model's definition. Besides that, other uncertainty sources related with the modelling of the service life prediction models can be identified:

- The uncertainty associated with the claddings description—some properties are clearly quantifiable and measurable during fieldwork (such as the building's height); however, there are features such as the environmental exposure conditions, which are yet quantified using a qualitative scale such as "favourable", "current" and "unfavourable", which may lead to some errors in the façades' characterization; furthermore, in some prediction models, the qualitative variables must be transformed into quantitative values, leading to some inaccuracies in this transformation;
- The uncertainty inherent to the possibility of design, execution or use errors, which are impossible to identify during fieldwork but compromise the claddings' natural

degradation process, thus altering the performance of the predictive models. Furthermore, defects caused by accidental actions cannot be modelled over time but influence the results of the prediction models.

On top of the uncertainty inherent to some features of the prediction models, it is important to understand that each building is unique, both at the design and execution level (along with the technical capacity of the stakeholders in the construction process, available budget, the construction context, among others) and in terms of its environmental exposure and use conditions; thus it is not easy—not to say conceptually impossible—to make direct comparisons between buildings (even if they are apparently very similar) (Bordass et al. 2000). For this reason, the service life prediction models are not perfect in the description of every case study analysed. The models proposed in this study seek to adapt to a representative sample of the different types of cladding analysed, taking into consideration the different characteristics of the sample.

In the following sections, the predictive capability of each of the proposed models is analysed, proceeding to its validation.

7.3.1 Validation of the Service Life Prediction Models

In theory, the models' validation through testing is the one that gives greater accuracy to the results obtained. In this case, the validation tests: (i) the ability to obtain the same result, when the same research conditions are exactly replicated; and (ii) the model's capability to predict causal relationships between phenomena. In the field of service life prediction, the model's experimental validation is not reasonable, since an excessively long period of time (an incompatible with any research work) is necessary to monitor the behaviour of the claddings; often testing campaigns are not even feasible, given their costs and the number of case studies to be monitored. In practice, in studies involving predictive models, the results validation can be performed in two ways: (i) internal validation—which evaluates the effectiveness of the model, either based on the obtained prediction errors or by statistical indicators that evaluate the model's capability to predict the behaviour of the dataset; (ii) external validation—by using a validation sample (data not used to create the model).

In this study an internal validation is therefore performed; the validation of the predictive capability of the proposed models is done by comparing the values obtained by each model ($S_{w,\text{predicted}}$) and the values observed during the fieldwork ($S_{w,\text{observed}}$). It is important to notice that the performance of a prediction model is usually worse for the external validation data than for the training sample (Vergouwe 2003). The results obtained are evaluated based on the following statistical indicators: (i) the correlation coefficient (r) and the determination coefficient (R^2), which allow evaluating the goodness of fit between the models and the data; (ii) the mean percentage error normalized with respect to the maximum ($\overline{\varepsilon}$) and the maximum normalized percentage error ($\varepsilon_{\max}$); (iii) the percentage of patterns where

ε is greater than $x = 5$ %, $x = 10$ %, $x = 20$ %, and $x = 30$ % ($PP_\varepsilon > x$); (iv) the mean absolute error (MAE), and (v) the root mean square error (RMSE). These indicators were explained in Part 5 of this Book (Computational models).

Indicators such as the mean absolute error and the root mean square error analyse the model's accuracy in relative terms, i.e. these measures are usually applied to compare the predictive capability of various models. Lewis (1997) refers that the mean absolute percentage error (MAPE) measures the mean relative size of the absolute error (produced by the model) as a percentage of the corresponding observed value. The authors propose a scale that relates the mean absolute percentage error values with the model's predictive capability: MAPE values below 10 % reveal a model providing potentially very good predictions; MAPE values lower than 20 % indicate a model providing potentially good predictions; MAPE values lower than 30 % indicate that the model provides potentially reasonable predictions; and MAPE values higher than 30 % indicate that the model is potentially inaccurate.

In this study, other parameters are used to analyse the predictive capability of the proposed models. If a perfect model is achieved, the values predicted by the model should be equal to the observed values, meaning thereby that there are no deviations between the predicted and the real values. Therefore, the ratio between the predicted and the observed values should be equal to one. An indicator was used to express the deviations from 1 of the ratio between S_w predicted by the model and S_w observed. If the ratio between the predicted and the observed values is lower than 1, it means that the real values are higher than the predicted values, indicating thereby that the model is optimistic. An optimistic model can comprise the claddings safety, since it underestimates the values of the severity of degradation, and may assume that a façade is in better condition than the actual degradation levels that may be occurring.

In the following sections, the predictive capability of the models proposed for the different claddings analysed is evaluated. These analysis, concerning the prediction errors, are only performed for deterministic and computational models, since the stochastic and factor models do not provide as output information regarding the severity of degradation of the claddings; the stochastic models provide an indication of the probability of transition between degradation conditions and the probability of the end of façades service life being reached, according to their age and characteristics; the factor models allow obtaining directly the estimated service life of each case study within the sample.

7.3.1.1 Stone Claddings

In the definition of the service life prediction models applied to stone claddings the entire same sample is used in the regression analysis, in the stochastic and factor models (no case study was excluded); due to the inherent properties of the computational models, part of the sample is used to create the model and the other part is used to test the model, being also applied to evaluate the predictive capacity of these models. Tables 7.2 and 7.3 show various statistical indicators that evaluate the

Table 7.2 Statistical indicators used in the analysis of the predictive ability of the proposed models applied to stone claddings

Model	R^2	r	MAPE	ε_{max}	$PP_{\varepsilon>5\%}$	$PP_{\varepsilon>10\%}$	$PP_{\varepsilon>20\%}$	$PP_{\varepsilon>30\%}$	MAE	RMSE
Simple nonlinear regression	0.775	0.880	6.347	49.230	46.31	22.66	3.45	0.99	0.027	0.039
Multiple linear regression	0.711	0.843	7.128	45.546	49.75	24.14	5.42	1.48	0.031	0.044
Interaction effects	0.711	0.843	7.131	45.555	49.75	24.14	5.42	1.48	0.031	0.044
Path analysis	0.660	0.813	7.984	49.494	53.69	31.03	5.42	2.46	0.034	0.048
Polynomial model	0.785	0.886	6.388	40.553	44.83	22.66	3.45	0.49	0.027	0.038
Gompertz curve	0.818	0.904	6.101	37.404	45.81	18.72	2.96	0.49	0.026	0.036
Von Bertalanffy curve	0.801	0.895	6.502	37.220	49.75	21.67	3.94	0.99	0.028	0.038
Richards curve	0.805	0.897	6.295	36.435	45.32	21.67	2.46	0.99	0.027	0.037
Morgan-Mercer-Flodin curve	0.714	0.845	7.117	44.153	51.23	25.62	5.42	1.48	0.031	0.044
Weibull curve	0.775	0.880	6.443	48.736	46.80	22.17	2.96	0.99	0.028	0.039
Brody curve	0.639	0.799	8.198	50.165	53.20	31.03	6.90	2.46	0.035	0.049
Exponential model	0.772	0.879	6.353	50.054	44.33	21.67	3.94	1.48	0.027	0.040
Potential model	0.766	0.875	7.213	47.941	54.19	22.66	3.45	0.99	0.031	0.041
Artificial neural networks	0.704	0.839	7.717	35.633	49.02	29.41	11.76	1.96	0.033	0.047
Fuzzy systems model	0.778	0.882	7.635	24.375	47.06	31.37	7.84	0.00	0.033	0.043

Table 7.3 Statistical indicators of the ratio between the values predicted by the proposed models and the observed values applied to stone claddings

Model	Average deviations from 1 between $S_{w,\text{predicted}}$ and the $S_{w,\text{observed}}$ (%)	Percentage of results $S_{w,\text{predicted}}/S_{w,\text{observed}} > 1$ (%)	Percentage of results $S_{w,\text{predicted}}/S_{w,\text{observed}}$ varying ±25 % (%)	Percentage of results $S_{w,\text{predicted}}/S_{w,\text{observed}} > 1.5$ (%)
Simple nonlinear regression	51.7	43.8	26.6	51.7
Multiple linear regression	49.8	36.9	24.1	49.8
Interaction effects	49.3	36.9	24.1	49.3
Path analysis	51.2	34.5	30.5	51.2
Polynomial model	50.2	39.9	20.2	50.2
Gompertz curve	39.4	36.5	15.3	39.4
Von Bertalanffy curve	36.9	33.5	13.8	36.9
Richards curve	41.4	32.0	15.8	41.4
Morgan-Mercer-Flodin curve	46.8	38.4	23.6	46.8
Weibull curve	54.7	39.9	27.1	54.7
Brody curve	49.3	31.5	25.1	49.3
Exponential model	50.7	43.8	26.6	50.7
Potential model	35.0	43.8	26.6	35.0
Artificial neural networks	64.7	35.3	27.5	64.7
Fuzzy systems model	70.6	39.2	31.4	70.6

predictive ability of the proposed models. From the analysis of the obtained results, the following conclusions can be reached:

- All the proposed models have a MAPE value lower than 10 %, revealing a potentially very good predictive capacity, based on the scale proposed by Lewis (1997); therefore, the proposed models seem able of conveniently describing the degradation phenomena of stone claddings;
- The nonlinear simple regression model leads to acceptable results, with correlation and determination coefficients higher than those obtained for multiple linear regression models. These results indicate that nonlinear models are potentially more suitable in the description of the degradation of stone claddings;
- Multiple linear regression models with the analysis of causal effects between variables lead to more complex models; however, these models do not reveal an increase in their statistical significance when compared with simpler methods, thus not improving the model's explanation power;
- The multiple nonlinear regression models show better overall results compared to the multiple linear regression models. The Gompertz curve is the model with

the best overall performance, with a higher predictive ability; the remaining multiple nonlinear models show similar results, the Brody curve being the model with higher estimation errors;

- In some of the proposed models (multiple linear regression, interaction effects analysis, Gompertz curve, von Bertalanffy (CVB) curve, Richards curve, Morgan-Mercer-Flodin (MMF) curve, Brody and potential model), the majority of the values predicted are lower than the observed values (between 65 % in the potential model, and 50.2 % in multiple linear regression model). As mentioned before, this particularity can compromise the reliability of the results; however, it should be noticed that for all these models, the deviations between the predicted and the observed values is very small (with MAPE values lower than 10 %), revealing that even when assuming an optimistic value for the degradation of stone claddings, the estimation errors will not be high enough to compromise the safety of the claddings analysed;
- Regarding the computational methods—artificial neural networks and fuzzy logic—although generally very accurate (with correlation coefficients of 0.839 and 0.882, respectively), they show unappealing results when compared with the other models presented, since these models involve a high level of complexity.

Table 7.4 presents some statistical indicators that assess the estimated service life obtained by each proposed model. The average estimated service life obtained by the different methods varies between 67 and 79 years, which seems a very acceptable and realistic range of values. The exception to this range of values corresponds to the result obtained using the fuzzy logic model, which leads to an average service life of 90 years, a relatively high value, yet a value that does not seem unrealistic due to the durability of the stone claddings. It is important to mention that this value corresponds to the average of the values obtained for the sample test, therefore, if another sample test is chosen, a different result may be achieved. The fuzzy model also leads to a median of the estimated service life for the sample test equal to 71 years, which is in the order of magnitude of the values proposed by the other models. Additionally, the fuzzy logic model presents a very high standard deviation with large dispersion values.

For the comparison between the proposed models, it is relevant to emphasize that the deterministic and computational models are established in order to estimate the severity of degradation of the claddings analysed (predicting the values of S_w), unlike the factor and the graphical methods, which allow obtaining directly the estimated service life of the case studies. The factor and the graphical method lead to average estimated service lives within the range of results obtained by the other models, showing, however, a higher standard deviation. These results were expected since methodologies including, for example, four independent variables lead to estimated service lives resulting from the combination of these four variables, disregarding the other characteristics of the claddings. If a larger number of variables are used, a larger number of combinations will also be possible, leading to a large number of results. Consequently, as the factor method comprises all the characteristics of the stone claddings, combining twelve variables, it necessarily

Table 7.4 Estimated service life according to the different models analysed applied to stone claddings

Model	Average ESL (years)	Maximum ESL (years)	Minimum ESL (years)	Standard deviation (years)	Confidence interval 95 %	Value of the ESL with a probability higher than 95 % to be exceeded*	Value of the ESL with a probability higher than 5 % to be exceeded*
Simple nonlinear regression	68.182	–	–	–	–	–	–
Multiple linear regression	77.213	86.599	62.219	7.650	±1.052	64.630	89.798
Interaction effects	77.222	86.614	62.169	7.646	±1.052	64.646	89.799
Path analysis	74.968	82.495	65.747	4.310	±0.593	67.880	82.057
Polynomial model	69.110	75.609	62.216	2.873	±0.395	64.384	73.836
Gompertz curve	72.775	88.595	57.253	7.489	±1.030	60.456	85.094
Von Bertalanffy curve	73.585	90.464	56.593	8.114	±1.116	60.239	86.931
Richards curve	69.380	75.318	60.014	4.642	±0.639	61.745	77.015
Morgan-Mercer-Flodin curve	76.219	83.124	64.970	5.610	±0.772	66.991	85.447
Weibull curve	68.563	71.390	63.707	1.980	±0.272	65.306	71.821
Brody curve	79.370	92.286	62.185	8.358	±1.150	65.622	93.118
Exponential model	73.193	87.387	60.143	7.422	±1.021	60.985	85.401
Potential model	68.008	70.345	64.026	1.828	±0.252	65.001	71.016
Artificial neural networks	70.256	77.934	59.177	5.802	±1.592	60.713	79.799
Fuzzy systems model	90.275	142.000	62.000	31.230	±8.571	38.395	142.156
Factor method	66.905	97.422	44.815	10.236	±1.408	50.068	83.742
Graphical method	70.545	157.370	31.971	17.097	±2.352	42.331	98.577

*Values obtained assuming a normal distribution

leads to a greater dispersion of the results. Concerning the stochastic models, the aim of these models is to evaluate the probability of each case study reaching the end of its service life. Therefore, assuming that the transition between conditions *C* and *D* corresponds to the instant of the maximum probability of the façades reaching the end of their service life, the following results were obtained: (i) the logistic regression model indicates that after 68 years the probability of a stone cladding reaching the end of its service life is higher than 50 % and at age 80 this probability is higher than 90 %; (ii) for the Markov chain model, the probability of belonging to the condition corresponding to the end of service life reaches its peak between 103 and 104 years.

7.3.1.2 Rendered Façades

Tables 7.5 and 7.6 show the statistical indicators used in the analysis of the predictive ability of the proposed models applied to rendered façades. This study is a research work in continuous progression and, therefore, the models established for rendered façades were defined with different variables and using different samples. For this reason, the comparison between the models proposed must be cautious, leading to the following main conclusions:

- For all the proposed models, a MAPE value lower than 10 % is obtained, thus revealing that the models present a potentially very good predictive capacity (Lewis 1997), and are able to conveniently express the degradation phenomena of rendered façades;
- The simple nonlinear regression model leads to acceptable results, with a determination coefficient of 0.881, revealing a high correlation between the model and the data. Apparently, this model presents a mean percentage error lower than the multiple linear and nonlinear regression models; however, it is important to noticed that this statistical measure is normalized to the maximum value of the severity of degradation observed in the data set: in the first model (simple regression) this value is equal to $S_w = 1.34$; in the multiple (linear and nonlinear) regression analysis, the maximum value of the severity of degradation is $S_w = 0.94$; consequently, if the same sample were used in the definition of both methods, multiple regressions would lead to better results (lower errors);
- Models with the analysis of causal relationships between variables (interaction effects and path analysis) lead to an increase in the complexity of the model but the results obtained do not reveal an improvement in the predictive ability of the model. Nevertheless, these models allow understanding the synergetic effects between variables and their effects on the degradation phenomena of rendered façades;
- Multiple nonlinear regression models allow achieving similar results when the same sample and the explanatory variables are used in the model's definition. The potential model shows the best overall performance (with a lower average percentage error normalized to the maximum); polynomial multiple regression,

Table 7.5 Statistical indicators used in the analysis of the predictive ability of the proposed models applied to rendered façades

Model	R^2	r	MAPE	ε_{max}	$PP_{\varepsilon>5\%}$	$PP_{\varepsilon>10\%}$	$PP_{\varepsilon>20\%}$	$PP_{\varepsilon>30\%}$	MAE	RMSE
Simple nonlinear regression	0.881	0.938	4.286	22.493	26.00	10.00	1.00	0.00	0.057	0.084
Multiple linear regression	0.905	0.951	5.143	23.402	37.37	13.13	2.02	0.00	0.048	0.067
Interaction effects	0.904	0.951	5.164	23.402	38.38	12.12	2.02	0.00	0.049	0.067
Path analysis	0.884	0.940	5.231	30.895	35.35	15.15	5.05	2.02	0.049	0.075
Polynomial model	0.904	0.951	5.197	22.675	38.38	14.14	2.02	0.00	0.049	0.067
Gompertz curve	0.876	0.936	6.326	31.490	52.53	13.13	3.03	1.01	0.060	0.077
Von Bertalanffy curve	0.876	0.936	6.325	31.494	52.53	13.13	3.03	1.01	0.060	0.077
Richards curve	0.877	0.936	6.313	31.398	52.53	13.13	4.04	1.01	0.059	0.077
Morgan-Mercer-Flodin curve	0.902	0.950	5.053	22.968	34.34	14.14	2.02	0.00	0.048	0.067
Weibull curve	0.902	0.950	5.053	22.972	34.34	14.14	2.02	0.00	0.048	0.067
Brody curve	0.905	0.951	5.168	22.668	37.37	14.14	2.02	0.00	0.049	0.067
Exponential model	0.879	0.938	6.296	30.916	52.53	13.13	4.04	1.01	0.059	0.076
Potential model	0.901	0.949	5.282	22.251	36.36	10.10	5.05	0.00	0.050	0.068
Artificial neural networks	0.899	0.948	5.233	28.030	32.00	16.00	4.00	0.00	0.070	0.110
Fuzzy systems model	0.867	0.931	4.050	10.966	23.81	4.76	0.00	0.00	0.054	0.066

Table 7.6 Statistical indicators of the ratio between the values predicted by the proposed models and the observed values applied to rendered façades

Model (%)	Percentage of results $S_{w,\text{predicted}}/S_{w,\text{observed}} > 1$ (%)	Percentage of results $S_{w,\text{predicted}}/S_{w,\text{observed}}$ varying ±25 % (%)	Percentage of results $S_{w,\text{predicted}}/S_{w,\text{observed}} > 1.5$ (%)
Simple nonlinear regression	54.6	39.2	29.9
Multiple linear regression	56.3	41.7	27.1
Interaction effects	56.3	41.7	28.1
Path analysis	54.2	36.5	26.0
Polynomial model	57.3	41.7	27.1
Gompertz curve	59.4	43.8	36.5
Von Bertalanffy curve	59.4	43.8	36.5
Richards curve	59.4	43.8	36.5
Morgan-Mercer-Flodin curve	56.3	40.6	27.1
Weibull curve	56.3	40.6	27.1
Brody curve	58.3	40.6	27.1
Exponential model	59.4	42.7	36.5
Potential model	62.5	39.6	29.2
Artificial neural networks	54.2	41.7	25.0
Fuzzy systems model	28.0	28.0	12.0

Morgan-Mercer-Flodin (MMF), Weibull and Brody curve lead to similar results, with a determination coefficient higher than 0.9, thus revealing a high correlation between these models and the observed values. The Gompertz, Von Bertalanffy, Richards and exponential models present the highest prediction errors, nonetheless with determination coefficients above 0.87;

- For rendered façades, it is found that the multiple linear regression model has a similar performance when compared with the polynomial models, Morgan-Mercer-Flodin (MMF), Weibull and Brody; all these models use the same variables, leading to prediction errors of the same order of magnitude. Therefore, in this case, and as seen in the simple regression analysis (Part 3), the increase of statistical significance conferred by nonlinear models is residual, not justifying the increase of complexity of the model;
- Regarding the computational models, ANNs lead to relatively low errors, with small deviations between the values of severity of degradation predicted by the model and the values observed in the fieldwork;. Fuzzy logic systems lead to the best overall results, even when compared with the deterministic models but the degradation severity values predicted by the model are, in most cases (72 %), lower than the true values; this feature of the model may jeopardize the

users safety due to the optimistic character with which this model evaluates the degradation of rendered façades. Furthermore, computational models are significantly more complex than the other models, despite using only four variables, which may compromise their choice over simpler models.

Table 7.7 presents the estimated service life according to the different models analysed. The mean estimated service life obtained by the proposed models is 18.5 years (with a range of 15.6–22 years). The Gompertz, von Bertalanffy, Richards and exponential models lead to an average estimated service life higher than the other models analysed (around 22 years). Even though these values may seem too optimistic for rendered façades that are not subjected to any maintenance action during their service life, these values are in accordance with the values proposed in the literature: (i) Shohet et al. (1999) suggest an estimated service life for rendered façades of 20 years; (ii) Shohet and Paciuk (2004), for a lower level of requirement, propose an estimated service life for rendered façades of 23 years (with a range of 19–27 years). The factor method allows obtaining directly the estimated service life of each case study, leading to an average estimated service life of 16.2 years. The graphical method, which allows estimating the service life of each case study (as described in Part 6) leads to an average estimated service life of 18.8 years. These two methods only use 85 case studies, and a very high standard deviation value is obtained for these models though. Using the stochastic models, the following results are achieved: (i) the logistic regression model reveals that after 17 years the probability of a rendered façade reaching the end of its service life is higher than 50 % and at age 23 this probability is higher than 95 %; (ii) according to the Markov chains model, the probability of belonging to the condition corresponding to the end of service life of rendered façades reaches its peak at age 19.

7.3.1.3 Painted Surfaces

Regarding painted surfaces, in the definition of the deterministic and stochastic models, the whole sample is used (220 case studies). In the computational models, the test sample (corresponding to the results shown in the further analysis regarding the predictive ability of the models) comprises 32 cases studies. In the factor and graphical method, the final sample consisted of 177 case studies (since some case studies were removed from the original sample, given the model's inability to accurately predict the behaviour of case studies that are in very early degradation states). Tables 7.8 and 7.9 present some indicators applied to assess the predictive ability of the proposed models applied to painted surfaces. The following conclusions can be drawn:

- All the models lead to an MAPE value under 10 %, then it can be said that the models present a potentially very good predictive capacity;
- The nonlinear simple regression model leads to a determination coefficient of 0.874, higher than the values obtained for the multiple linear regression models;

Table 7.7 Estimated service life according to the different models analysed applied to rendered façades

Model	Average ESL (years)	Maximum ESL (years)	Minimum ESL (years)	Standard deviation (years)	Confidence interval 95 %	Value of the ESL with a probability higher than 95 % to be exceeded*	Value of the ESL with a probability higher than 5 % to be exceeded*
Simple nonlinear regression	15.760	–	–	–	–	–	–
Multiple linear regression	15.715	22.225	7.658	3.021	±0.595	10.744	20.685
Interaction effects	15.560	23.199	7.019	3.004	±0.592	10.619	20.501
Path analysis	15.736	19.265	12.181	1.547	±0.305	13.191	18.281
Polynomial model	15.823	22.507	8.032	2.982	±0.587	10.919	20.728
Gompertz curve	21.960	29.000	10.000	4.262	±0.839	14.950	28.969
Von Bertalanffy curve	21.980	29.000	10.000	4.286	±0.844	14.930	29.029
Richards curve	21.970	29.000	10.000	4.294	±0.846	14.907	29.033
Morgan-Mercer-Flodin curve	15.778	21.000	8.000	2.776	±0.547	11.212	20.344
Weibull curve	16.101	21.000	8.000	2.866	±0.565	11.387	20.816
Brody curve	15.960	22.000	8.000	2.972	±0.586	11.071	20.849
Exponential model	21.960	29.000	11.000	4.157	±0.819	15.121	28.798
Potential model	17.495	21.000	6.000	3.348	±0.660	11.988	23.003
Artificial neural networks	17.500	23.000	12.000	2.741	±0.895	12.991	22.009
Fuzzy systems model	18.852	26.700	15.400	2.341	±1.001	15.001	22.704
Factor method	16.244	33.144	6.980	5.869	±1.248	6.591	25.899
Graphical method	18.786	46.256	7.149	9.009	±1.915	3.967	33.605

*Values obtained assuming a normal distribution

Table 7.8 Statistical indicators used in the analysis of the predictive ability of the proposed models applied to painted surfaces

Model	R^2	r	MAPE	ε_{max}	$PP_{\varepsilon>5\%}$	$PP_{\varepsilon>10\%}$	$PP_{\varepsilon>20\%}$	$PP_{\varepsilon>30\%}$	MAE	RMSE
Simple nonlinear regression	0.874	0.935	5.389	38.065	36.27	16.06	4.15	1.55	0.039	0.061
Multiple linear regression	0.833	0.913	6.659	40.353	47.15	22.80	5.18	1.04	0.048	0.069
Interaction effects	0.834	0.913	6.717	40.690	47.15	22.80	5.18	1.04	0.049	0.069
Path analysis	0.813	0.902	7.160	42.009	48.70	25.91	6.22	1.55	0.052	0.073
Polynomial model	0.889	0.943	5.221	39.741	36.27	15.54	3.11	1.04	0.038	0.058
Gompertz curve	0.875	0.935	5.346	38.868	34.20	16.06	4.66	1.04	0.039	0.061
Von Bertalanffy curve	0.890	0.943	5.168	38.681	36.79	13.47	3.63	1.04	0.037	0.057
Richards curve	0.891	0.944	5.185	36.912	35.75	13.47	4.15	1.04	0.038	0.057
Morgan-Mercer-Flodin curve	0.885	0.941	5.242	38.468	36.79	15.03	4.15	1.04	0.038	0.059
Weibull curve	0.885	0.941	5.245	38.483	36.79	15.03	4.15	1.04	0.038	0.059
Brody curve	0.833	0.913	6.670	40.564	48.19	22.28	5.18	1.04	0.048	0.069
Exponential model	0.867	0.931	6.343	45.974	46.63	13.99	3.11	1.04	0.046	0.063
Potential model	0.881	0.939	5.353	38.123	35.23	15.54	4.15	1.04	0.039	0.059
Artificial neural networks	0.959	0.979	4.561	13.746	40.63	12.50	0.00	0.00	0.033	0.044
Fuzzy systems model	0.953	0.976	4.899	22.106	46.88	9.38	3.13	0.00	0.035	0.050

Table 7.9 Statistical indicators of the ratio between the values predicted by the proposed models and the observed values applied to painted surfaces

Model	Percentage of results $S_{w,\text{predicted}}/S_{w,\text{observed}} > 1$ (%)	Percentage of results $S_{w,\text{predicted}}/S_{w,\text{observed}}$ varying ±25 % (%)	Percentage of results $S_{w,\text{predicted}}/S_{w,\text{observed}} > 1.5$ (%)
Simple nonlinear regression	46.1	50.8	17.1
Multiple linear regression	53.9	41.5	22.8
Interaction effects	55.4	43.0	22.3
Path analysis	56.5	36.8	24.9
Polynomial model	45.6	49.2	14.0
Gompertz curve	49.7	51.3	17.6
Von Bertalanffy curve	50.3	53.4	18.1
Richards curve	55.4	47.7	18.7
Morgan-Mercer-Flodin curve	48.2	50.8	15.5
Weibull curve	47.7	50.8	15.5
Brody curve	53.9	41.5	23.8
Exponential model	62.2	42.0	32.1
Potential model	57.5	49.2	24.9
Artificial neural networks	46.9	46.9	18.8
Fuzzy systems model	31.3	53.1	9.4

furthermore, the linear multiple regression models present higher prediction errors;

- The multiple nonlinear regression models lead to better results when compared to simple or multiple linear regression models. All the multiple nonlinear regression models defined for painted surfaces present a high predictive ability, with correlation coefficients higher than 0.9, thus revealing a very strong correlation between the predicted and the observed values. Richards curve is the multiple nonlinear regression model with the best overall performance among all the models analysed (with lower errors and higher correlation and determination coefficients); the exponential model and Gompertz curve are those with lower determination coefficients and higher deviations between the predicted and the observed values;
- In the simple nonlinear regression, polynomial model, Gompertz curve, Morgan-Mercer-Flodin model, Weibull curve, artificial neural networks and fuzzy logic models, the severity of degradation predicted by the model is lower than those observed in the fieldwork, although the deviations obtained are very low (with a mean absolute error lower than 0.04). In the majority of the proposed models, the predicted values are higher than the observed values, thus revealing that the models obtained for the painted surfaces are conservative;

- Computational models lead to better results when compared with the other models, leading to an increase of 10 % in the models' statistical significance, with a reduction of the model estimation errors.

Table 7.10 presents some statistical indicators related with the estimated service life of painted surfaces.

The various models lead to an average estimated service life ranging between 8.5 and 10.4 years. Painted surfaces are less durable than the other types of claddings considered in this Book, thus revealing a lower dispersion of the results, i.e. the proposed models lead to small standard deviations, with a small span of the estimated service life values. The values obtained by the proposed models are in accordance with the data present in the literature. Concerning the stochastic models, the following results are obtained: (i) for the logistic regression model, the probability of a painted surface reaching the end of its service life is higher than 50 % after 10 years and this probability is higher than 90 % after 12 years; (ii) the Markov chain model reveals that painted surfaces reach the maximum probability of belonging to the condition corresponding to the end of the service life between 12 and 13 years.

7.3.1.4 Ceramic Tiling Systems

For ceramic tiling systems, the overall sample (195 case studies) is applied in the definition of the deterministic and stochastic models. In the computational methods, the test sample comprises 29 case studies (which are applied to assess the model's predictive ability) and in the factor models, only 186 case studies are analysed, due to the removal of anomalous cases (with unrealistic estimated service lives, as explained in Part 6 of this Book). In Tables 7.11 and 7.12, the statistical indicators used in the analysis of the predictive ability of the proposed models applied to ceramic tiling systems are shown. The following conclusions can be reached:

- All the proposed models show a potentially very good predictive ability;
- The simple nonlinear regression model leads to a determination coefficient higher than the coefficient obtained using the multiple linear regression models (including the models with the analysis of causal relationships between variables), and necessarily the simple nonlinear regression model presents lower deviations between the predicted and the observed values. This result reveals that nonlinear models may be more appropriate to characterize the ceramic tiles degradation phenomena;
- The multiple nonlinear regression models lead to higher determination coefficients and lower errors when compared with the multiple linear regression models. The Gompertz curve is the multiple nonlinear regression model with the best overall performance among all the multiple nonlinear regression models analysed (presenting the lowest errors and the highest correlation between the predicted and the observed values). The Brody curve is the model that presents higher deviations between the predicted and the observed values;

Table 7.10 Estimated service life according to the different models analysed applied to painted surfaces

Model	Average ESL (years)	Maximum ESL (years)	Minimum ESL (years)	Standard deviation (years)	Confidence interval 95 %	Value of the ESL with a probability higher than 95 % to be exceeded*	Value of the ESL with a probability higher than 5 % to be exceeded*
Simple nonlinear regression	9.664	–	–	–	–	–	–
Multiple linear regression	8.487	9.389	6.009	0.701	±0.093	7.334	9.640
Interaction effects	8.450	9.405	6.028	0.693	±0.092	7.310	9.590
Path analysis	8.503	8.975	8.009	0.341	±0.045	7.942	9.065
Polynomial model	9.724	10.772	8.596	0.502	±0.066	8.899	10.549
Gompertz curve	9.659	10.697	8.481	0.579	±0.076	8.707	10.611
Von Bertalanffy curve	9.620	10.655	8.403	0.574	±0.076	8.676	10.565
Richards curve	9.712	10.761	8.589	0.588	±0.078	8.746	10.679
Morgan-Mercer-Flodin curve	9.558	10.131	8.564	0.399	±0.053	8.903	10.214
Weibull curve	9.553	10.136	8.551	0.404	±0.053	8.889	10.218
Brody curve	8.454	9.381	5.999	0.685	±0.091	7.328	9.582
Exponential model	10.236	11.417	8.765	0.611	±0.081	9.232	11.241
Potential model	9.746	10.032	8.993	0.237	±0.031	9.357	10.136
Artificial neural networks	9.491	10.548	7.977	0.623	±0.213	8.451	10.533
Fuzzy systems model	9.698	10.528	8.349	0.445	±0.152	8.956	10.442
Factor method	9.998	12.583	7.869	1.083	±0.160	8.220	11.781
Graphical method	10.389	25.395	3.936	3.250	±0.479	5.045	15.735

*Values obtained assuming a normal distribution

Table 7.11 Statistical indicators used in the analysis of the predictive ability of the proposed models applied to ceramic tiling systems

Model	R^2	r	MAPE	ε_{max}	$PP_{\varepsilon>5\%}$	$PP_{\varepsilon>10\%}$	$PP_{\varepsilon>20\%}$	$PP_{\varepsilon>30\%}$	MAE	RMSE
Simple nonlinear regression	0.767	0.876	7.922	44.894	55.38	30.77	5.64	1.03	0.049	0.065
Multiple linear regression	0.698	0.836	8.502	44.541	54.87	31.79	9.74	1.03	0.052	0.072
Interaction effects	0.703	0.839	8.533	45.257	56.41	32.31	8.72	2.05	0.052	0.071
Path analysis	0.673	0.820	9.007	44.580	59.49	35.38	11.28	2.05	0.055	0.075
Polynomial model	0.781	0.884	7.625	41.569	54.36	30.77	5.64	0.51	0.047	0.063
Gompertz curve	0.795	0.892	7.438	40.067	56.41	33.33	4.62	0.51	0.046	0.061
Von Bertalanffy curve	0.777	0.882	7.902	39.890	54.36	32.82	4.62	0.51	0.048	0.064
Richards curve	0.789	0.888	7.560	39.359	52.82	34.87	4.62	0.51	0.046	0.062
Morgan-Mercer-Flodin curve	0.731	0.855	8.226	44.572	55.90	31.28	9.23	1.03	0.050	0.069
Weibull curve	0.787	0.887	7.626	41.523	54.36	30.26	6.15	0.51	0.047	0.062
Brody curve	0.707	0.841	8.474	44.925	53.85	33.33	8.72	1.54	0.052	0.071
Exponential model	0.785	0.886	7.965	42.413	56.41	32.31	4.62	1.03	0.049	0.063
Potential model	0.788	0.887	7.645	42.166	54.36	28.21	5.64	0.51	0.047	0.062
Artificial neural networks	0.865	0.930	5.189	17.260	37.93	17.24	0.00	0.00	0.032	0.043
Fuzzy systems model	0.773	0.879	6.940	19.783	48.28	31.03	0.00	0.00	0.043	0.055

Table 7.12 Statistical indicators of the ratio between the values predicted by the proposed models and the observed values applied to ceramic tiling systems

Model	Percentage of results $S_{w,\text{predicted}}/S_{w,\text{observed}} > 1$ (%)	Percentage of results $S_{w,\text{predicted}}/S_{w,\text{observed}}$ varying ±25 % (%)	Percentage of results $S_{w,\text{predicted}}/S_{w,\text{observed}} > 1.5$ (%)
Simple nonlinear regression	55.7	29.1	36.0
Multiple linear regression	48.2	33.3	32.3
Interaction effects	46.7	30.8	32.8
Path analysis	45.1	30.3	31.3
Polynomial model	46.2	31.3	29.2
Gompertz curve	53.8	35.9	30.8
Von Bertalanffy curve	36.9	26.2	22.6
Richards curve	44.1	28.7	23.6
Morgan-Mercer-Flodin curve	45.1	32.3	28.2
Weibull curve	52.8	30.8	29.7
Brody curve	45.1	31.8	29.7
Exponential model	62.1	27.7	41.0
Potential model	52.8	31.8	32.3
Artificial neural networks	34.5	34.5	6.9
Fuzzy systems model	27.6	37.9	13.8

- In the majority of the proposed models (multiple linear regression, interaction effects, path analysis model, polynomial model, von Bertalanffy curve, Richards curve, Morgan-Mercer-Flodin (MMF) and Brody model), the values predicted are lower than the real values, indicating that, in most of the case studies, the model underestimates the ceramic claddings degradation (assuming a severity of degradation for the façades lower than the real one). However, once again this feature of these models does not compromise the analysis of the results, because the deviations between predicted and observed values are small, corresponding to an average absolute error around 0.05;
- The ANNs model is the most effective computational model, with a determination coefficient of 0.865 (higher than the other proposed models) and small estimation errors.

Table 7.13 presents the statistical indicators related with the estimated service life according to the different models analysed applied to ceramic tiling systems. The average estimated service life obtained by the different proposed models varies between 46 and 54 years, a range of values that seems acceptable, with a variation of ±4 years from the average reference service life values considered in the literature (around 50 years). Concerning the stochastic models, the following results are

Table 7.13 Estimated service life according to the different models analysed applied to ceramic tiling systems

Model	Average ESL (years)	Maximum ESL (years)	Minimum ESL (years)	Standard deviation (years)	Confidence interval 95 %	Value of the ESL with a probability higher than 95 % to be exceeded*	Value of the ESL with a probability higher than 5 % to be exceeded*
Simple nonlinear regression	50.144	–	–	–	–	–	–
Multiple linear regression	46.023	50.962	35.209	4.524	±0.635	38.582	53.464
Interaction effects	46.480	55.456	34.615	4.977	±0.699	38.294	54.668
Path analysis	45.977	51.317	41.113	2.642	±0.371	41.631	50.324
Polynomial model	48.497	51.594	42.318	2.540	±0.357	44.319	52.675
Gompertz curve	50.028	57.705	38.255	4.112	±0.577	43.265	56.793
Von Bertalanffy curve	48.043	55.372	36.383	4.033	±0.566	41.409	54.678
Richards curve	48.786	56.132	37.047	3.993	±0.560	42.218	55.354
Morgan-Mercer-Flodin curve	46.358	51.569	35.668	3.516	±0.493	40.576	52.141
Weibull curve	49.721	51.939	45.435	2.158	±0.303	46.171	53.271
Brody curve	46.308	55.879	31.898	5.073	±0.712	37.965	54.652
Exponential model	52.081	60.818	40.565	4.627	±0.649	44.471	59.692
Potential model	49.829	52.225	45.047	2.135	±0.300	46.319	53.341
Artificial neural networks	50.186	53.500	39.800	3.519	±1.281	44.398	55.975
Fuzzy systems model	50.833	67.138	37.809	6.961	±2.534	39.383	62.283
Factor method	49.768	70.806	27.917	8.987	±1.292	34.986	64.551
Graphical method	54.463	118.436	23.077	18.094	±2.600	24.700	84.225

*Values obtained assuming a normal distribution

obtained: (i) the logistic regression model reveals that after 50 years the probability of a ceramic cladding reaching the end of its service life is higher than 50 % and at age 61 this probability is higher than 90 %; (ii) according to the Markov chains model, the maximum probability of belonging to the condition corresponding to the end of the service life of rendered façades occurs between 58 and 59 years.

7.3.2 *Sensitivity, Specificity and Accuracy of the Prediction Models*

As mentioned before, the validation of distinct prediction models can be complex and time consuming. In this study, the receiver operating characteristics (ROC) curves are used to evaluate the accuracy of the service life prediction methods. ROC curves were initially developed during World War II, to quantify the effectiveness of radar operators (originally called receiver operators) (Reiser and Faraggi 1997); when the radar detects some information, the operator must decide whether it is a sign (enemy aircraft) or only noise (for example, a group of birds) (Collinson 1998). Since then, the ROC curves have been used to measure the performance of an operator or a diagnosis system. The models proposed in this study allow predicting the performance of façades over time and according to their characteristics. Ultimately, the most important issue that all models attempt to solve is whether façades reach the end of their service life. Thus, despite being prediction models, they may also be seen as classification or diagnosis models, which can be classified in two mutually exclusive classes: "reached the end of their service life"; and "not reached the end of their service life". Therefore, ROC curves can be used to evaluate the ability of the models to correctly classify each case study in these two classes. In this analysis, it is important to evaluate the cases that are correctly classified by the model: (i) case studies that are identified as having reached the end of their service life and, in reality, did reach it (TP—true positives); (ii) case studies that are identified as not having reached the end of its service life, when in reality they did not reach it (TN—true negatives). The diagnosis or classification models may lead to two types of errors: (i) the "false positives" (FP), when the model assumes that a case study has reached the end of their service life, when in reality it has not; (ii) and the "false negatives" (FN), when the model assumed that a case study has not yet reached the end of its life, when in reality it already did. In some areas of research, e.g. in Medicine, it is desirable that the errors in diagnostic tests do not occur and, when they do, they should preferably be of the type "false alarm" (false positives).

The model's accuracy can be evaluated using two indicators: the sensitivity and the specificity of the model. Sensitivity provides an indicator of the ability of the model to correctly predict the cases that have reached the end of their service life (i.e. the probability of the model predicting that a cladding reached the end of their service life when the cladding actually did). On the other hand, the specificity evaluates the ability of the model to correctly predict the cases that have not reached the end of their service life, i.e. the probability of the model predict that a cladding has not reached the end of their service life, when the coating indeed did not reach it. These two measures can be determined as described in Eqs. (7.1) and (7.2).

$$\text{Sensitivity} = \frac{\text{TP}}{\text{FN} + \text{TP}} \times 100\,(\%) \tag{7.1}$$

$$\text{Especificity} = \frac{\text{TN}}{\text{TN} + \text{FP}} \times 100\,(\%) \tag{7.2}$$

where TP represents the true positives, FN the false negatives, TN the true negatives and FP the false positives.

These two parameters are independent of each other, since they are calculated with different data. The sensitivity is determined using only the case studies that actually reached the end of their service life, while the specificity is determined based on the cases that actually have not reached the end of their life. Models with good performance level present high values of sensitivity and specificity. According to Maroco (2007), a model with good classification capability shows sensitivity and specificity above 80 %; models with sensitivity and specificity between 50 and 80 % present a reasonable capacity; and below 50 % the predictive ability of the model can be considered as poor. Based on these indexes is also possible to determine the accuracy of a given diagnostic model, i.e. the analysis of the overall percentage of cases correctly classified by the model—Eq. (7.3).

$$\text{Accuracy} = \frac{\text{TN} + \text{TP}}{\text{TN} + \text{FN} + \text{TP} + \text{FP}} \times 100\,(\%) \tag{7.3}$$

The ROC curve is a graphical procedure that displays the trade-off between sensitivity (y-axis) and 1-specificity (x-axis) (Altman and Bland 1994). The area under the ROC curve is another measure that allows evaluating the accuracy of the model. The area under the ROC curve varies between 0 and 1; a perfect model is characterized by a curve with the maximum value of sensitivity (1 or 100 %) maintaining low values of 1-specificity, i.e. where the true positive segment is 100 % and the false positive segment is zero. Models with good performance show curves passing close to the upper left corner of the plot, with areas under the curve near 1 (Robertson et al. 1983; Manel et al. 2001).

The practical lower limit of the area under the curve of a diagnostic test is 0.5, representing a model with a ROC curve coincident with the diagonal line, also called as the chance diagonal (Obuchowski 2003). Physically, this means that such a model cannot classify the cases that have reached the end of their service life better than a game of chance, e.g. tossing a coin. Therefore, a model with an area under the curve higher than 0.5 is at least better than relying on pure chance. According to Hosmer and Lemeshow (2000), the area under the ROC curve can be related with the discriminating power of the model; namely: (i) models with an area under the ROC curve of 0.5 show no discriminating power; (ii) models with an area under the ROC curve in the]0.5; 0.7[range have weak discriminating power; (iii) models with an area under the ROC curve in the [0.7; 0.8[range exhibit acceptable discriminating power; (iv) models with an area under the ROC curve in the [0.8; 0.9[range exhibit good discriminating power; and (v) models with an area under the ROC curve greater or equal to 0.9 exhibit an exceptional ability to classify the case studies in the two classes considered.

In the next sections, the sensitivity, specificity and accuracy of the models proposed for the different types of claddings analysed are evaluated.

7.3.2.1 Stone Claddings

Table 7.14 presents the classification ability of the proposed models applied in the service life prediction of stone claddings. Based on the results obtained, the following conclusions can be drawn:

- All the proposed models present an adequate classification capability, with an accuracy higher than 90 % in the majority of the proposed models; however, generally, the proposed models have low sensitivity, revealing that these models are less able to identify the case studies that have actually reached the end of their service life; this can be justified by the optimistic character of the models' predictions, as mentioned above;

Table 7.14 Classification ability of the proposed models for stone claddings

Model	TP (%)	TN (%)	FP (%)	FN (%)	Sensitivity (%)	Specificity (%)	Accuracy (%)
Simple nonlinear regression	4.93	84.24	4.43	6.40	43.48	95.00	89.16
Multiple linear regression	5.91	87.68	0.99	5.42	52.17	98.89	93.60
Interaction effects	5.91	87.68	0.99	5.42	52.17	98.89	93.60
Path analysis	3.94	88.67	0.00	7.39	34.78	100	92.61
Polynomial model	6.40	87.19	1.48	4.93	56.52	98.33	93.60
Gompertz curve	7.88	87.19	1.48	3.45	69.57	98.33	95.07
Von Bertalanffy curve	8.37	87.19	1.48	2.96	73.91	98.33	95.57
Richards curve	7.88	87.19	1.48	3.45	69.57	98.33	95.07
Morgan-Mercer-Flodin curve	4.43	88.67	0.00	6.90	39.13	100	93.10
Weibull curve	5.91	84.24	4.43	5.42	52.17	95.00	90.15
Brody curve	2.46	88.67	0.00	8.87	21.74	100	91.13
Exponential curve	4.93	84.24	4.43	6.40	43.48	95.00	89.16
Potential curve	5.91	84.24	4.43	5.42	52.17	95.00	90.15
Artificial neural networks	8.00	86.00	6.00	2.00	80.00	93.48	94.00
Fuzzy systems model	0.00	80.00	22.00	0.00	–	78.43	80.00

- The multiple nonlinear regression model of von Bertalanffy is the one with the highest classification capacity, with accuracy around 96 %, sensitivity around 74 % and specificity around 98 %; the Gompertz and Richards curves also show high levels of accuracy;
- The fuzzy logic model presents the worst performance, with the lowest level of accuracy, classifying correctly only 80 % of the test sample (50 case studies); it should be stressed though that the test sample is randomly selected and in the model studied none of the case studies contained in the test sample had reached the end of its service life, thus making it impossible to determine the sensitivity of the model.

Table 7.15 shows some statistical indicators related with the ROC curves obtained for the proposed models. As expected, the results obtained confirm the previous analysis, revealing that the von Bertalanffy, Gompertz, Richards and ANN models are those with higher accuracy levels, with areas under the ROC curve near 0.9. The results of the fuzzy logic model do not allow the definition of a ROC curve, since the sensitivity of the model is equal to zero (due to the singularities of the test sample, as mentioned before). The remaining models present an area under

Table 7.15 Area under the ROC curve of the proposed models for stone claddings

Model	Area under the curve	Standard error*	Asymptotic Sig.**	Asymptotic 95 % confidence interval	
				Lower bound	Upper bound
Simple nonlinear regression	0.692	0.069	0.003	0.557	0.827
Multiple linear regression	0.755	0.067	0.000	0.623	0.887
Interaction effects	0.755	0.067	0.000	0.623	0.887
Path analysis	0.674	0.071	0.007	0.535	0.813
Polynomial model	0.774	0.066	0.000	0.645	0.903
Gompertz curve	0.839	0.059	0.000	0.724	0.955
Von Bertalanffy curve	0.861	0.055	0.000	0.752	0.970
Richards curve	0.839	0.059	0.000	0.724	0.955
Morgan-Mercer-Flodin curve	0.696	0.071	0.002	0.557	0.834
Weibull curve	0.736	0.067	0.000	0.605	0.867
Brody curve	0.609	0.071	0.090	0.470	0.747
Exponential curve	0.692	0.069	0.003	0.557	0.827
Potential curve	0.736	0.067	0.000	0.605	0.867
Artificial neural networks	0.867	0.108	0.007	0.656	1.000
Fuzzy systems model	–	–	–	–	–

*Under the nonparametric assumption
**Null hypothesis: true area = 0.5

the ROC curve that varies between 0.6 and 0.8. As mentioned before, part of the proposed models present lower sensitivity, thus leading to relatively small areas under the ROC curve. Nevertheless, the proposed models have a reasonable classification capability (well above 0.5, since below this value the model's predictive ability no better than a result obtained by chance).

7.3.2.2 Rendered Façades

Regarding the rendered façades, Table 7.16 shows the classification ability of the proposed models. From the analysis of the sensitivity, specificity and accuracy of the proposed models, the following conclusions can be achieved:

- The ANNs are the most accurate model, correctly classifying all the case studies within the test sample (25 case studies);
- The simple nonlinear regression model also presents a high classification capacity, with an accuracy of 96 %;
- The multiple nonlinear regression models show a good classifying capacity (with sensitivity and specificity values higher than 90 %); among these models, the Gompertz curve, von Bertalanffy, Richards, Morgan-Mercer-Flodin and Weibull models are those with higher accuracy levels; the Morgan-Mercer-Flodin and Weibull curves have a sensitivity of 100 %, classifying correctly all the cases that have reached the end of their service life;

Table 7.16 Classification ability of the proposed models for rendered façades

Model	TP (%)	TN (%)	FP (%)	FN (%)	Sensitivity (%)	Specificity (%)	Accuracy (%)
Simple nonlinear regression	29.00	67.00	2.00	2.00	93.55	97.10	96.00
Multiple linear regression	29.29	65.66	4.04	1.01	96.67	94.20	94.95
Interaction effects	28.28	64.65	5.05	2.02	93.33	92.75	92.93
Path analysis	28.28	66.67	3.03	2.02	93.33	95.65	94.95
Polynomial model	29.29	64.65	5.05	1.01	96.67	92.75	93.94
Gompertz curve	28.28	68.69	1.01	2.02	93.33	98.55	96.97
Von Bertalanffy curve	28.28	68.69	1.01	2.02	93.33	98.55	96.97
Richards curve	28.28	68.69	1.01	2.02	93.33	98.55	96.97
Morgan-Mercer-Flodin curve	30.30	66.67	3.03	0	100	95.65	96.97
Weibull curve	30.30	66.67	3.03	0	100	95.65	96.97
Brody curve	29.29	65.66	4.04	1.01	96.67	94.20	94.95
Exponential curve	27.27	68.69	1.01	3.03	90.00	98.55	95.96
Potential curve	28.28	67.68	2.02	2.02	93.33	97.10	95.96
Artificial neural networks	40.00	60.00	0	0	100	100	100
Fuzzy systems model	9.52	76.19	0	14.29	40.00	100	85.71

- The multiple linear regression model presents high accuracy levels;
- The model with interaction effects presents the worst performance (with the lowest accuracy among all the proposed models); however, it still correctly classifies around 93 % of the sample;
- Although the fuzzy model is extremely effective in the prediction of the severity of degradation of rendered façades, it presents a sensitivity of 40 % only, i.e. in a sample of 21 case studies, three cases (12 %) are incorrectly classified by the model; furthermore, the error of the model is not conservative, since the model assumes that the three case studies have not reached the end of their service life, when in fact they already did.

In Table 7.17, some statistical indicators related with the area under the ROC curve of the proposed models are shown, for rendered façades.

This table also provides a confidence interval of 95 % for the area under the ROC curve, where narrower intervals represent more reliable models, with higher discriminant capacity. As seen before, the fuzzy logic models present a high

Table 7.17 Area under the ROC curve of the proposed models for rendered façades

Model	Area under the curve	Standard error*	Asymptotic Sig.**	Asymptotic 95 % confidence interval	
				Lower bound	Upper bound
Simple nonlinear regression	0.953	0.028	0.000	0.898	1.000
Multiple linear regression	0.954	0.025	0.000	0.905	1.000
Interaction effects	0.930	0.032	0.000	0.868	0.993
Path analysis	0.945	0.030	0.000	0.886	1.000
Polynomial model	0.947	0.026	0.000	0.895	0.999
Gompertz curve	0.959	0.028	0.000	0.905	1.000
Von Bertalanffy curve	0.959	0.028	0.000	0.905	1.000
Richards curve	0.959	0.028	0.000	0.905	1.000
Morgan-Mercer-Flodin curve	0.978	0.015	0.000	0.950	1.000
Weibull curve	0.978	0.015	0.000	0.950	1.000
Brody curve	0.954	0.025	0.000	0.905	1.000
Exponential curve	0.943	0.033	0.000	0.878	1.000
Potential curve	0.952	0.029	0.000	0.895	1.000
Artificial neural networks	1.000	0.000	0.000	1.000	1.000
Fuzzy systems model	0.700	0.156	0.174	0.395	1.000

*Under the nonparametric assumption
**Null hypothesis: true area = 0.5

number of inaccuracies. However, this model is applied to the test sample only (21 case studies)—which leads to a more penalizing effect of errors when compared, for instance, to the multiple linear and nonlinear regression models that are applied to a sample comprising 99 case studies. The multiple linear and nonlinear regression models lead to areas under the ROC curve higher than 0.9, revealing models with an exceptional discriminating power. Concerning the ANNs, this model correctly classifies all the case studies analysed, thus leading to an area under the curve equal to 1. The fuzzy model is the one with the worst performance, with an area under the curve of 0.7; still, according to the criteria of Hosmer and Lemeshow (2000), it presents an acceptable discriminating power.

7.3.2.3 Painted Surfaces

Regarding painted surfaces, Table 7.18 shows the statistical indicators applied to evaluate the classification ability of the models proposed, leading to the following conclusions:

- All models present high sensitivity (with values between 81 and 98 %), high specificity and also high accuracy levels and the majority of the models correctly classify over 90 % of the cases studied;

Table 7.18 Classification ability of the proposed models for painted surfaces

Model	TP (%)	TN (%)	FP (%)	FN (%)	Sensitivity (%)	Specificity (%)	Accuracy (%)
Simple nonlinear regression	25.39	66.32	6.22	2.07	92.45	91.43	91.71
Multiple linear regression	26.94	59.59	12.95	0.52	98.11	82.14	86.53
Interaction effects	26.94	58.03	14.51	0.52	98.11	80.00	84.97
Path analysis	26.42	56.99	15.54	1.04	96.23	78.57	83.42
Polynomial model	23.83	68.91	3.63	3.63	86.79	95.00	92.75
Gompertz curve	25.39	66.32	6.22	2.07	92.45	91.43	91.71
Von Bertalanffy curve	23.83	65.28	7.25	3.63	86.79	90.00	89.12
Richards curve	23.83	66.32	6.22	3.63	86.79	91.43	90.16
Morgan-Mercer-Flodin curve	24.87	66.84	5.70	2.59	90.57	92.14	91.71
Weibull curve	24.87	66.84	5.70	2.59	90.57	92.14	91.71
Brody curve	26.94	559.59	12.95	0.52	98.11	82.14	86.53
Exponential curve	22.28	67.88	4.66	5.18	81.13	93.57	90.16
Potential curve	24.35	67.36	5.18	3.11	88.68	92.86	91.71
Artificial neural networks	31.25	65.63	0.00	3.13	90.91	100.00	96.88
Fuzzy systems model	31.25	65.63	0.00	3.13	90.91	100.00	96.88

- The path analysis model is the one with the lowest accuracy, correctly classifying 83.4 % of the case studies, and the lowest specificity (78.6 %) among the proposed models;
- In the case of painted surfaces, computational models present the best performance, with an accuracy of 96.9 % and a specificity of 100 %, correctly classifying all the case studies that have not reached the end of their service life (not leading to any false positive); it should be noted that the computational models used the same test sample, and both models correctly classify the same number of case studies.

Table 7.19 shows some statistical indicators to evaluate the ROC curves obtained for the models applied to painted surfaces. The results are consistent with the previous analysis, thus revealing that the computational models are those with the highest discriminating power (with an area under the ROC curve of 0.955). The path analysis model presents the lowest discriminating capacity, with an area under the ROC curve equal to 0.874, nonetheless revealing a good discriminating power [according to the criteria proposed by Hosmer and Lemeshow (2000)].

Table 7.19 Area under the ROC curve of the proposed models for painted surfaces

Model	Area under the curve	Standard error*	Asymptotic Sig.**	Asymptotic 95 % confidence interval	
				Lower bound	Upper bound
Simple nonlinear regression	0.919	0.025	0.000	0.870	0.969
Multiple linear regression	0.901	0.023	0.000	0.856	0.946
Interaction effects	0.891	0.024	0.000	0.844	0.937
Path analysis	0.874	0.026	0.000	0.822	0.926
Polynomial model	0.909	0.029	0.000	0.852	0.966
Gompertz curve	0.919	0.025	0.000	0.870	0.969
Von Bertalanffy curve	0.884	0.031	0.000	0.824	0.944
Richards curve	0.891	0.030	0.000	0.832	0.950
Morgan-Mercer-Flodin curve	0.914	0.027	0.000	0.861	0.966
Weibull curve	0.914	0.027	0.000	0.861	0.966
Brody curve	0.901	0.023	0.000	0.856	0.946
Exponential curve	0.874	0.034	0.000	0.807	0.940
Potential curve	0.908	0.028	0.000	0.852	0.963
Artificial neural networks	0.955	0.052	0.000	0.853	1.000
Fuzzy systems model	0.955	0.052	0.000	0.853	1.000

*Under the nonparametric assumption
**Null hypothesis: true area = 0.5

7.3.2.4 Ceramic Tiling Systems

Table 7.20 presents the classification capability of the proposed models applied to ceramic tiling systems, from which the following conclusions can be drawn:

- All the proposed models present high specificity values (higher than 70 %) and an accuracy higher than 85 %, showing a high classification capability, since more than 85 % of the case studies analysed are correctly classified by the proposed models;
- The multiple linear regression model presents the best overall performance (correctly classifying 89.7 % of the sample), followed by the simple nonlinear regression model, the polynomial model, von Bertalanffy, Richards and Weibull curves, with an accuracy of 89.2 %;
- Computational models, ANNs and fuzzy logic systems, are those with the lowest accuracy, classifying correctly about 86 % of the test sample. ANNs present a low sensitivity, revealing some inaccuracies in the classification of the case studies that actually have reached the end of their service life; in turn, the fuzzy logic model, even though presenting a lower accuracy, has a high sensitivity and specificity (above 80 %).

In Table 7.21, some statistical indicators related with the area under the ROC curve of the proposed models for ceramic tiling systems are presented.

Table 7.20 Classification ability of the proposed models for ceramic tiling systems

Model	TP (%)	TN (%)	FP (%)	FN (%)	Sensitivity (%)	Specificity (%)	Accuracy (%)
Simple nonlinear regression	16.41	72.82	3.59	7.18	69.57	95.30	89.23
Multiple linear regression	20.00	69.74	6.67	3.59	84.78	91.28	89.74
Interaction effects	18.97	68.21	8.21	4.62	80.43	89.26	87.18
Path analysis	20.00	67.69	8.72	3.59	84.78	88.59	87.69
Polynomial model	17.95	71.28	5.13	5.64	76.09	93.29	89.23
Gompertz curve	16.92	71.79	4.62	6.67	71.74	93.96	88.72
Von Bertalanffy curve	18.46	70.77	5.64	5.13	78.26	96.62	89.23
Richards curve	17.95	71.28	5.13	5.64	76.09	93.29	89.23
Morgan-Mercer-Flodin curve	19.49	69.23	7.18	4.10	82.61	90.60	88.72
Weibull curve	17.95	71.28	5.13	5.64	76.09	93.29	89.23
Brody curve	20.00	68.72	7.69	3.59	84.78	89.93	88.72
Exponential curve	16.92	71.79	4.62	6.67	71.74	93.96	88.72
Potential curve	17.44	71.28	5.13	6.15	73.91	93.29	88.72
Artificial neural networks	10.34	75.86	3.45	10.34	50.00	95.65	86.21
Fuzzy systems model	17.24	68.97	10.34	3.45	83.33	86.96	86.21

Table 7.21 Area under the ROC curve of the proposed models for ceramic tiling systems

Model	Area under the curve	Standard error*	Asymptotic Sig.**	Asymptotic 95 % confidence interval	
				Lower bound	Upper bound
Simple nonlinear regression	0.824	0.042	0.000	0.741	0.907
Multiple linear regression	0.880	0.034	0.000	0.814	0.946
Interaction effects	0.848	0.037	0.000	0.776	0.921
Path analysis	0.867	0.034	0.000	0.800	0.924
Polynomial model	0.847	0.039	0.000	0.770	0.924
Gompertz curve	0.828	0.041	0.000	0.747	0.910
Von Bertalanffy curve	0.854	0.038	0.000	0.780	0.929
Richards curve	0.847	0.039	0.000	0.770	0.924
Morgan-Mercer-Flodin curve	0.866	0.035	0.000	0.797	0.935
Weibull curve	0.847	0.039	0.000	0.770	0.924
Brody curve	0.874	0.034	0.000	0.807	0.940
Exponential curve	0.828	0.041	0.000	0.747	0.910
Potential curve	0.836	0.040	0.000	0.757	0.915
Artificial neural networks	0.728	0.136	0.090	0.461	0.996
Fuzzy systems model	0.851	0.098	0.009	0.659	1.000

*Under the nonparametric assumption
**Null hypothesis: true area = 0.5

These results confirm that the multiple linear regression model presents the highest classification capacity, with an area under the ROC curve of 0.88, thus revealing a model with a good discriminant power. The ANNs model presents the lowest classification capability, with an area under the ROC curve of 0.73 (nonetheless, still presenting a reasonable discriminant power).

7.4 Multi-criteria Analysis of Service Life Prediction Models

In this Book, several methodologies for service life prediction were defined, with conceptually different approaches, leading to different types of information. The validity and accuracy of the proposed models have been demonstrated in the previous sections of this chapter, revealing that all the models are, with different levels of accuracy, statistically significant and able to model the deterioration and service life of façade's claddings. This section presents different approaches for service life prediction with practical applications in the construction sector, in which different stakeholders have to decide which model to use based on a set of criteria to support that decision.

Even though the ability to decide is one of the fundamental faculties of the human being, a rational process of decision-making can be complex, vague and of a controversial nature, as it incorporates subjective criteria and different "points of view" (Basel and Brühl 2013). Bana and Costa (1993) state unequivocally that one must accept that subjectivity is ubiquitous in decision-making processes. Belton and Stewart (2002) refer that there is no such thing as the "right answer"; in fact, the authors mentioned that subjectivity is inherent to the decision-making process. As mentioned by Gigerenzer and Selten (2002), in the real world, decisions do not work according to the laws of logic. Damásio (1994) states that it is not possible to make "rational" decisions without the sensitive, affective and "subjective" component of the brain. To make matters more difficult, even when the evaluation criteria of a model are objective, different stakeholders will necessarily have different priorities; for example, a condominium administration company (at a small scale) may prefer simpler models, while a real estate manager, involving high annual maintenance costs, may prefer the most accurate and accurate models (even if these models are more expensive and complex).

In short, in the real world, the decision-making process has a multi-criteria nature, being necessary to evaluate all the factors that directly or indirectly influence the decision to be made. Decision theory—considered by some authors as a branch of operational research (Bana and Costa 1993; Brans 1994)—aims to provide a tool to assist decision-making with well-established criteria, based on the preferences of decision-makers (Roy 1985; Phillips 1986), not judging the way that decision-makers opt for a given alternative. Decision theory is based on a basic principle of preference; according to which alternative A1 is preferable to alternative A2, when the number of criteria in which A1 is better than A2 is greater than the number of criteria in which A2 is better than A1 (Vincke 1992). Yoon and Hwang (1995) refer that compensatory decision models analyse whether the disadvantages of a model according to a given criterion may or may not be compensated by its advantages according to other criteria. The application of a multi-criteria analysis allows determining the best alternative from a set of often conflicting alternatives (Voogd 1983). In the end, stakeholders may choose the best option, based on consistent judgment criteria, thus taking more informed, justified decisions (Goodwin and Wright 1991).

According to Jato-Espino et al. (2014), the definition of a multi-criteria model goes through various stages: (i) characterization of the decision context, typifying the stakeholders profile and the constraints associated with the decision process; (ii) definition of the criteria—identification of relevant parameters to evaluate the various alternatives; (iii) determination of the weighting coefficients, which provide a relative scale of the assessment criteria; (iv) calculation of the total value of each alternative; (v) analysis of the sensitivity and robustness of the results. In the definition of a multi-criteria analysis model of the different service life prediction methods, two main aspects are addressed: (i) the characteristics of each model (decision criteria) and the main motivations and objectives of decision-makers (leading to different weights for these criteria).

7.4.1 Definition of the Evaluation Criteria

To evaluate the applicability of the proposed service life prediction models, some criteria must be analysed. Hovde (2002) suggested that service life prediction methods should fulfil a set of requirements, namely: (i) easy to learn; (ii) easy to use; (iii) accurate; (iv) easy to update; (v) easily adaptable to new data; (vi) supported by data; (vii) related to design methods and existing tools; (viii) free of excessive administrative work; (ix) recognizing the importance of innovation; (x) applicable to various environments; (xi) acceptable to professionals and clients; (xii) reflecting current knowledge; (xiii) having a flexible sophistication level for any detailed planning. Additionally, the applicability and credibility of each model should be carefully analysed, as well as the trade-off between simplicity and accuracy, which is especially conspicuous in the field service life prediction modelling. As mentioned by Miwa and Simon (1993), simple models present the advantage of parsimony, while well-fitting models are more accurate, often losing simplicity in detriment of accuracy. In this study, two points of view are analysed, regarding decision-makers: (i) the programmers' point of view: who defines/creates the model; (ii) and the user's perspective: who applies the model and only wants to get the information related to the service life of façades claddings, without necessarily having any knowledge about the model's design. Regarding the planner's point of view, four criteria are considered: (i) the ease of learning—conceptually, some models are more complicated and more time is needed to acquire the theoretical knowledge necessary to implement the model; (ii) the time required to define the model—while some models require a long period of time to be implemented (e.g. when the creation of an algorithm in specific software is required to apply the model), others can be defined in a few minutes; (iii) the software required—since some models can developed with simple and available online tools, and others require the application of specialized software; and (iv) the number and subjectivity of the variables included in the model. Concerning the user's perspective, only two criteria were considered: (i) interpretability and richness of the information produced by models; (ii) accuracy of the models, which includes the prediction errors and deviations between the predicted and observed values. Generally, the user will be interested above all in the quality of the results, since all the difficulties related with the model's definition must be overtaken by the planner. Figure 7.1 summarizes the six criteria considered as the most relevant to influence decision-making regarding the adoption of a given service life prediction model instead of another one.

As mentioned before, decision-makers usually face various conflicting alternatives; it is therefore necessary to develop methods to establish the preferences' hierarchy, identifying the correct relative weights for the criteria analysed

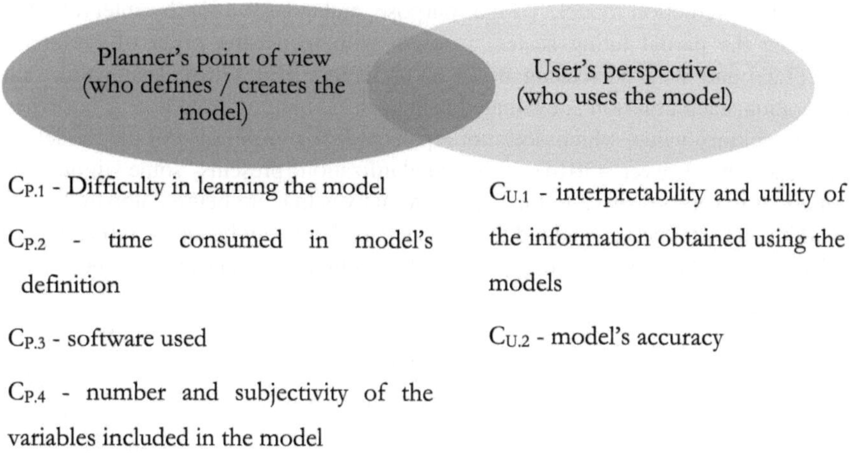

Fig. 7.1 Criteria adopted in multi-criteria analysis of service life prediction models, depending on the decision-maker's point of view

(Bardham et al. 2011). According to Zeleny (1998), decision-making based on a multi-criteria analysis relies on the pursuit of the optimal solution, requiring in some way the assessment of compensation (trade-off) between the criteria analysed. Currently, there are different approaches for multi-criteria decision-making. In this study, an additive aggregation approach with compensatory rationality is adopted, i.e. even if a model has a worse performance/rating in criterion i, its global ranking may be good because of a high classification according to criterion j. Therefore, a score or single rating is determined for each alternative (model), using Eq. (7.4) (Cochran and Chen 2005).

$$X_i = \sum_{j=1}^{m} (\lambda_j \cdot X_{ij}) \text{ with } \sum_{j=1}^{m} \lambda_j = 1 \tag{7.4}$$

where X_{ij} is the classification of model i according to criterion j, λ_j is the criterion weight j and X_i is the measure of the global characterization of the model.

For the application of this approach, it is necessary to ensure that the criteria are mutually independent in terms of the stakeholder's preferences, which implies that the preference for one model in a rating criterion is independent of the model's rating in the remaining criteria. Some of the criteria adopted in this study present a qualitative nature, being measured in relative terms (by comparison between models). A scale needs thus to be developed, as consistent and coherent as possible, based on the knowledge acquired throughout the application of the various methods proposed.

In practice, different hierarchical scales are assigned to the different criteria analysed. Next, it is necessary to standardize the scales adopted for each criterion analysed in order to obtain an overall classification that allows evaluating each service life prediction model. To that purpose, a standardization by interval (standardizing the partial rating scales) is used, with increasing order of preference [Eq. (7.5)] and with decreasing order of preference [Eq. (7.6)]. Using this standardization, each criterion is evaluated using a scale ranging from 0 to 1, but where 0 is an arbitrary value, which does not represent lack of relevance of the model. As mentioned by Carver (2014), this standardization presents some limitations, namely, it is very sensitive to the set of alternatives that are being compared, i.e. if an additional alternative is added to the analysis, there may be an overall change in the standardized results. However, this method is intuitive and can easily be used by any stakeholder.

$$x_{ij} = \frac{X_{ij} - \min_i(X_{ij})}{\max_i(X_{ij}) - \min_i(X_{ij})} \tag{7.5}$$

$$x_{ij} = \frac{\max_i(X_{ij}) - X_{ij}}{\max_i(X_{ij}) - \min_i(X_{ij})} \tag{7.6}$$

Where X_{ij} is the classification of model i according to criterion j, $\max_i(X_{ij})$ the maximum value of criterion j, $\min_i(X_{ij})$ the minimum value of criterion j and x_{ij} the standardized classification of model i according to criterion j.

7.4.1.1 Required Learning Process to Use the Model

Regarding the ease of learning of a model, the average time required to obtain the knowledge to allow its autonomous application is evaluated. The learning process comprises different stages: (i) an extensive analysis of the literature about the statistical tool underlying the prediction model; (ii) an analysis of the potential and applicability of the methodology to be applied as a service life prediction tool; (iii) an evaluation of the results that may be obtained and the best way to achieve those results; (iv) the model experimentation and the statistical software necessary for its implementation; (v) the ability to deduce the results. Table 7.22 shows the hierarchy established based on these parameters, ranging between the simplest to the most complex method.

Simple nonlinear regression is the simplest method, since it is based on fitting a mathematical equation to a sample by minimizing the quadratic deviation between the observed values and the curve representing the equation (Silva et al. 2013). The following methods are increasingly more complex.

The factor method is simple and expedient when applied in a deterministic approach; however, it requires the identification of all the degradation factors that must be included in the analysis; furthermore, the values of the durability factors must be optimized through an iterative process, analysing the physical meaning of the results and ensuring higher levels of accuracy (minimizing prediction errors). The following methods entail increased difficulties; since their application requires

Table 7.22 Ease of learning the service life prediction models

Service life prediction model	Scale	Standardized scale—hierarchy
Simple nonlinear regression	9	1—*simpler*
Factor method (deterministic)	8	0.875 (7/8)
Multiple linear regression[a]	7	0.750 (6/8)
Multiple nonlinear regression[b]	6	0.625 (5/8)
Factor method (stochastic)	5	0.500 (4/8)
Logistic regression	4	0.375 (3/8)
Markov chains	3	0.250 (2/8)
Artificial neural networks (ANNs)	2	0.125 (1/8)
Fuzzy logic model	1	0—*more complex*

[a]Including the model with interaction effects and path analysis
[b]Refers to all types of multiple nonlinear regression

the acquisition of knowledge regarding the theoretical foundations and assumptions inherent to the application of models. The application of multiple linear regression techniques has been widely documented in the literature and it is relatively simple to obtain a step-by-step description of this model application. On the other hand, multiple nonlinear regression may be more complex, since the planner needs to define the mathematical expression that characterizes the phenomena under analysis, implying the identification of the explanatory variables, by an iterative process. The stochastic approach to the factor method, is rather more complex and time consuming: although the theoretical foundation is the same as that of the deterministic approach, various experiments are required to define the best methodology to assign a probabilistic nature of the durability factors; it is also necessary to evaluate the possibility of multiplying probability distributions. The stochastic models, such as Markov chains or logistic regression, represent the next step of complexity. Finally, computational methods are the most complex of all the proposed methods. Nevertheless, artificial neural networks are somewhat simpler than fuzzy logic models. The latter are the most complex tools, with a time-consuming learning process, and whose theoretical assumptions are intricate and difficult to grasp to a user unfamiliar with computer modelling.

7.4.1.2 Time Consumption Associated to the Model

In the evaluation of the time necessary for the construction and processing of the model, the time required to organize the dataset has been ignored. On the contrary—and for simplicity's sake, it is considered that the time to input, organize and process data is spent only once and that such data will be used in all the models. The hierarchical scale related to the time necessary for the creation and processing of the model is shown in Table 7.23.

Marteinsson (2003) refers that, generally, more complex models lead to more accurate results but require more time for their definition and implementation.

Table 7.23 Time consumed in the construction and processing of the service life prediction methods

Service life prediction model	Modelling time	Standardized scale—hierarchy
Simple nonlinear regression	±30 s	1—*faster*
Multiple linear regression[a]	±1 min	0.998
Path analysis	±5 min	0.981
Logistic regression	±15 min	0.939
Multiple nonlinear regression[b]	±30 min	0.877
Factor method (deterministic)	±1 h	0.752
Factor method (stochastic)	±2 h	0.501
Markov chains	±2 h	0.501
Artificial neural networks (ANNs)	±4 h	0
Fuzzy logic model	±4 h	0—*slower*

[a]Including the model with interaction effects and path analysis
[b]Refers to all types of multiple nonlinear regression

In simple nonlinear regression and multiple linear regression (with interaction effects and path analysis), the model's construction and processing is practically instantaneous. Logistic regression is an expedient method, which can be rapidly created using a tool available in SPSS software. The same applies to models of nonlinear multiple regression. However, the latter is more time consuming, since it is necessary to insert the mathematical equation in the software; followed by a sensitivity analysis in which the variables with potential explanatory power will be inserted one by one in the prediction model (it is also necessary to establish the starting values for the constants). The deterministic approach of the factor method is more time consuming than the previous methods; the time necessary to perform the iterations needed to optimize the durability factors is variable, since it depends on the number of durability factors and on the restrictions imposed to the iterative process. The probabilistic approach of the factor method and Markov chain analysis require the creation of an algorithm (in available commercial software, such as Matlab), therefore leading to higher time consumption. In these analysis, only the time required to obtain the mathematical equation that allows the service prediction is analysed, i.e. the time spent formatting charts and tables is not taken into account as well as the time lost when the algorithm created does not produce the result required (time needed to correct the errors). The computational models are the most time-consuming approaches. Fuzzy models need to be defined from scratch by programming in Matlab; however, after the creation of the algorithm, the model produces results very rapidly, i.e. computational time around 10 s, discounting the time spent on the adjustment of parameters. On the other hand, artificial neural networks do not require computational programming since a specific software can be used to implement the model; however, considerable time must be spent testing several architectures and several runs must be performed (in order to achieve the best model).

Table 7.23 presents values measured in real-life situations, corresponding to the application of the model by a qualified researcher who knows exactly what is to be achieved and how to proceed to model each approach. In reality, the definition and development of the models by the authors took much longer than it is shown in Table 7.21, since it represented a learning process, consisting of several stages of trial and error.

7.4.1.3 Software Used in Model's Definition

Concerning the software used for the application of the prediction models analysed in this Book, four programmes were used: (i) Excel (from Microsoft); (ii) SPSS (Statistical Package for Social Sciences, from IBM); (iii) Matlab (from The MathWorks); (iv) NeuroSolutions (from NeuroDimension). In the evaluation of this criterion, two parameters were analysed: (i) availability and prominence of the software; (ii) cost of software's acquisition. Currently, Excel is a tool widely used for calculation, used by all actors in the construction sector (including managers, planners, architects, engineers, and others); moreover, at present, this software can be used online, free of charge. SPSS has been widely used in statistics and social sciences, remaining rarely applied in the engineering field. Matlab is a very common tool in the engineering field. However, unlike SPSS, which already includes a number of functions that allow obtaining results more rapidly, Matlab requires programming skills, based on an interactive environment for numerical computation and algorithm's creation. This software, for a current license, is also more expensive than SPSS. NeuroSolutions is a very specific software, only applied for the definition of neural network models or for the implementation of advanced artificial intelligence and learning algorithms. This software can be used as an Excel add-in (in an easy-to-use interface), being more expensive than Excel but cheaper than SPSS or Matlab. The hierarchical definition of the service life prediction models based on the software used for its construction is presented in Table 7.24. The definition of the standardized scale takes into account the cost of each software applied.

Table 7.24 Software used in the construction and processing of the service life prediction methods

Service life prediction model	Software used	Standardized scale—hierarchy
Simple nonlinear regression	Excel	1—*cheaper*
Factor method (deterministic)	Excel	1
Multiple linear regression[a]	Excel	1
Artificial neural networks (ANNs)	NeuroSolutions	0.893
Logistic regression	SPSS	0.108
Multiple nonlinear regression[b]	SPSS	0.108
Factor method (stochastic)	Matlab	0
Markov chains	Matlab	0
Fuzzy logic model	Matlab	0—*more expensive*

[a]Including the model with interaction effects and path analysis
[b]Refers to all types of multiple nonlinear regression

7.4.1.4 Interpretability and Utility of the Information Obtained Using the Service Life Prediction Models

In the analysis of the interpretability and utility of the information obtained using the proposed models, two parameters are considered: (i) readability of the output and ease of use provided by the mathematical equation models for a user unfamiliar with statistical models; (ii) richness of the information obtained and its usefulness for service life prediction. The proposed service life prediction models lead to different types of results; using these methods, the estimated service life of the façade's claddings can be given by absolute values or probabilistic distributions of values. The output of deterministic and computational methods is an average estimated value. On the other hand, stochastic models allow including the randomness of the degradation process and thus provide information about the service life of the façade's claddings in the form of a probability distribution. With this information, the user not only knows the instant from which is necessary to intervene (given by an interval of probable values) but is also able to quantify the risks associated with the degradation process, i.e. the probability of cladding failure as a function of its age and characteristics. The hierarchical scale adopted for the interpretability of results and usefulness of the information produced by the service life prediction models is presented in Table 7.25.

The factor method (using a deterministic approach), simple nonlinear regression, multiple linear and nonlinear regression models are those whose mathematical expressions are simpler. Logistic regression, Markov chains and the factor method

Table 7.25 Interpretability of the results and usefulness of the information produced by the service life prediction models

Service life prediction model	Interpretability of the results and usefulness of the information produced	Standardized scale—hierarchy
Factor method (deterministic)	1	1—*simpler*
Simple nonlinear regression	1	1
Multiple linear regression[a]	1	1
Multiple nonlinear regression[b]	1	1
Factor method (stochastic)	1	1
Logistic regression	1	1
Markov chains	1	1
Artificial neural networks (ANNs)	2	0.5
Fuzzy logic model	3	0—*more complex*

[a]Including the model with interaction effects and path analysis
[b]Refers to all types of multiple nonlinear regression

using a probabilistic approach come in the next level of the scale; these models provide information in the form of a probability distribution; the post-treatment of the results is more complex, but the information produced is more relevant. Therefore, the trade-off between the interpretability and the richness of the results produced allows assuming that these models present the same value in the hierarchical scale as the deterministic models. ANNs are in the next level of the scale, since they yield a mathematical expression that is difficult to use, providing an average estimated service life associated with a set of dispersion measures. The last level in the scale corresponds to the fuzzy logic models; in this method, the application of the mathematical equation implies obtaining the degree of activation of the fuzzy rules, which requires the use of a Matlab toolbox or graphic reading of the membership functions of the model (more subjective and much less precise). Therefore, the requirement of use Matlab to implement fuzzy models can be a constraint for some users and decisive to exclude this method.

Until now, the essential criteria in the decision-making process relating to the service life prediction models proposed in this Book were analysed. The criteria discussed so far are transversal to all the claddings under analysis, since, for example, the model's processing time is not affected by the dataset, taking into account that the samples used are within the same order of magnitude (between 100 and 250 cases). However, some criteria vary depending on the type of cladding that is being examined; in the following sections, each type of cladding is analysed separately, by performing a partial overall analysis for each type of cladding analysed in this Book.

Regarding the number and subjectivity of the variables included in the service life prediction models, it is noted that, in reality and as mentioned in the Occam's razor principle, the simplest solution is often the most correct (Blumer et al. 1987). Also in life prediction models, solutions that can be easily implemented are preferable to more complex models; thus when two models display similar performance, it is preferable to adopt the model with fewer parameters or variables. Additionally, the inclusion of a large number of variables in the model leads to the need of more comprehensive samples. Besides the number of variables, the subjectivity associated with the quantification of these variables should also be analysed.

Since the analysis of the effectiveness and the accuracy of the proposed models have already been addressed in this chapter, a more detailed analysis is performed for each type of exterior cladding analysed, in the next sections. In particular, the model's accuracy is evaluated according to its ability to correctly predict the estimated service of each case study within the sample, since this is the ultimate goal of a decision-maker when using any of proposed methods. It should be noticed that some of the models (deterministic and computational) have been developed in order to predict the severity of degradation of the façade claddings, and the estimated service life using these models is obtained by determining the instant in time (age) in which the maximum acceptable degradation level is reached (corresponding to a severity of degradation of 20 %). Therefore, it is expected that these models lead to higher deviations between the predicted and the observed values when compared with the factor methods, which were developed in order to directly obtain the estimated service life of the façades claddings.

As mentioned before, in this Book, it is assumed that the graphical method provides the values of estimated service life that represent the observed reality. In this method, a degradation curve (which characterizes graphically the degradation evolution of the claddings over time) is defined for each point in the sample, allowing estimating the service life for each case study. Accepting this assumption, the model's accuracy is analysed based on three parameters: (i) using the ratio between the average service life predicted by the model and the average service life obtained by the graphical method, assessing the deviation of this ratio in relation to the unit; ideally, this ratio would be equal to 1; (ii) through the deviation between the values of the estimated service life with a probability higher than 5 % to be exceeded, the one predicted by the model and the one observed; this index can be very useful to the decision-maker, for it expresses with some certainty the probability threshold from which a satisfactory performance of the façade cannot be guaranteed; (iii) the difference between the probability distribution that characterizes the predicted values and the probability distribution associated with the observed values, assuming a normal distribution for all models. This indicator was obtained through the absolute value of the difference between the integrals of the two distributions, using Wolfram Mathematica software. This index ranges from 0 (for distributions that are coincident) to 2 (when the distribution shows no area under the curve in common, since, for the probability density distribution, the probability of the entire sample space is equal to 1—$\int_{-\infty}^{\infty} f(x)\mathrm{d}x = 1$—which implies that the two distributions do not intersect, i.e. the total area under the curve is equal to 2). These indicators should be carefully analysed and allow some guidance regarding the differences between the predicted and observed values.

In the next sections, the partial evaluations of the proposed models are presented for the different claddings under analysis.

7.4.1.5 Natural Stone Claddings

Table 7.26 shows the hierarchical scale regarding the number and subjectivity of the variables included in the service life prediction models applied to natural stone claddings. Models with the same number of variables may have a distinct rank in the hierarchy of models depending on the subjectivity inherent to the variables included, assuming the following assumptions:

- Age is a parameter that is easily quantifiable (although there may be some doubts due to lack of information regarding the maintenance of the buildings);
- Some variables are objective parameters, easy to quantify, such as the façades' orientation, distance from the sea, height of the building, among others;
- Variables such as exposure to damp and exposure to wind-rain action are more difficult to quantify, especially when there is no possibility to individually assess the actual microclimate exposure conditions of the cases studied.

Table 7.26 Number and subjectivity of the variables included in the service life prediction models (stone claddings)

Service life prediction model	Number of variables	Variables included in the model	Scale	Standardized scale
Simple nonlinear regression	1	Age	1	1
Logistic regression	1	Age	1	1
Markov chains	1	Age	1	1
Artificial neural networks (ANNs)	4	Age; size of the stone plate; damp; type of finishing	2	0.857
Fuzzy logic model, multiple linear regression, Richards curve	4	Age; size of the stone plate; damp; distance from the sea	2	0.857
Morgan-Mercer-Flodin (MMF) curve, Weibull curve, Brody curve, exponential and potential model	5	Age; size of the stone plate; damp; distance from the sea; type of stone	3	0.714
Multiple linear regression with interaction effects	5	Age; size of the stone plate; damp; distance from the sea; wind rain action	4	0.571
Multiple linear regression —path analysis	6	Age; type of stone; type of cladding; orientation; type of finishing; height	5	0.429
Gompertz and von Bertalanffy curves	6	Age; size of the stone plate; damp; distance from the sea; type of stone; orientation	6	0.286
Polynomial model	6	Age; size of the stone plate; damp; distance from the sea; wind-rain action; orientation	7	0.143
Factor method (deterministic and stochastic)	11	Type of stone; colour; type of finishing; size of the stone plates; thickness of stone plates; type of cladding; orientation; distance from the sea; wind-rain action; type of use; height	8	0

Regarding stone claddings, simple nonlinear regression, logistic regression and Markov chains are the simpler methods, contemplating only one variable: the age of the claddings. ANNs, fuzzy logic models, multiple linear regression and Richards curve come in the next level. All these models are placed in the same level in the adopted scale, since they contemplate three variables that are objectively quantifiable and one variable whose quantification is more difficult to quantify and thus prone to some subjectivity (exposure to damp).

The Morgan-Mercer-Flodin (MMF) curve, Weibull curve, Brody curve, exponential and potential model are the models present in the following level, with five explanatory variables, of which two of them are quantified based on subjective criteria (exposure to damp and exposure to wind-action). In the following scale levels, the same assumptions are adopted; the type of stone and the type of cladding are parameters that can be objectively quantified (according to the categories considered and described in Chap. 3), thus giving preference to models that include these variables over models using variables more difficult to quantify. The factor method appears in the last place of the hierarchical scale, because they include all the variables analysed.

Table 7.27 presents the evaluation matrix regarding the accuracy of the service life prediction models applied to natural stone claddings. In this overall analysis, logistic regression is the most accurate model, followed by the factor method and the Gompertz curve. Fuzzy logic systems are the less accurate model, leading to an average estimated service life much higher than the values obtained by the remaining methods analysed in this Book. The values obtained through Fuzzy logic seem too optimistic, but these values are obtained for the test sample, in which the severity of degradation never exceeds 20 % (none of the case studies within the sample has reached the end of its service life) and, furthermore, the results are skewed by case studies particularly durable. Therefore, more than representing one limitation of the model, the results obtained are due to a characteristic of the sample.

The multi-criteria impact profiles evaluated for natural stone claddings are shown in Table 7.28, based on the six criteria analysed. The fuzzy logic model is the one with the worst overall results, assuming the last place in the hierarchical scale. This model is extremely complex and, in the case of stone claddings, has the lowest level of efficiency among the proposed models. The multiple linear regression model presents the best results, offsetting an average degree of complexity with a high efficiency level. Moreover, the inclusion of more than one explanatory variable, namely accounting for other degradation factors besides the age of claddings, leads to a model that allows understanding the influence of these parameters on the claddings' degradation.

7.4.1.6 Rendered Façades

Regarding rendered façades, the number and subjectivity of the variables included in the service life prediction models is presented in Table 7.29.

For rendered façades, the variables eave's protection, level of protection in the balconies, ground floor protection, platbands copings and building's volume are easy to quantify as well; on the other hand, variables such as render type, design level and distance from pollution sources are more difficult to determine. Rather

Table 7.27 Evaluation matrix regarding the accuracy of the service life prediction models applied to natural stone claddings

Model	Average estimated service life (years)	Difference between the $ESL_{predicted}/ESL_{observed}$ ratio and 1	Standardized score	Values of the estimated service life with a probability higher than 5 % to be exceeded	Difference between the ratio between the predicted and observed values and 1	Standardized score	Deviation between probability distributions	Standardized score	Sum of standardized scores	Overall scale
Simple nonlinear regression	68.182	0.033	0.893	121.399	0.232	0.545	2.000	0.000	1.438	0.343
Multiple linear regression	77.213	0.095	0.672	89.798	0.089	0.913	0.817	0.797	2.382	0.878
Interaction effects	77.222	0.095	0.671	89.799	0.089	0.913	0.818	0.797	2.381	0.878
Path analysis	74.968	0.063	0.787	82.057	0.168	0.710	1.181	0.552	2.049	0.689
Polynomial model	69.110	0.020	0.941	73.836	0.251	0.494	1.386	0.414	1.849	0.576
Gompertz curve	72.775	0.032	0.900	85.094	0.137	0.790	0.766	0.832	2.521	0.957
Von Bertalanffy curve	73.585	0.043	0.858	86.931	0.118	0.838	0.708	0.871	2.567	0.983
Richards curve	69.380	0.017	0.955	77.015	0.219	0.578	1.112	0.598	2.131	0.736
Morgan-Mercer-Flodin curve	76.219	0.080	0.723	85.447	0.133	0.799	1.025	0.657	2.179	0.763
Weibull curve	68.563	0.028	0.913	71.821	0.271	0.441	1.547	0.306	1.660	0.469
Brody model	79.370	0.125	0.561	93.118	0.055	1.000	0.808	0.804	2.364	0.868
Exponential model	73.193	0.038	0.879	85.401	0.134	0.798	0.777	0.824	2.500	0.945
Potential model	68.008	0.036	0.884	71.016	0.280	0.420	1.577	0.285	1.590	0.429
Artificial neural networks	70.256	0.004	1.000	79.799	0.190	0.651	0.956	0.704	2.354	0.863
Fuzzy logic model	90.275	0.280	0.000	142.156	0.442	0.000	0.765	0.832	0.832	**0.000**
Factor method	66.905	0.052	0.828	83.742	0.150	0.754	0.516	1.000	2.582	0.991
Logistic regression	67.767	0.039	0.872	83.43	0.154	0.746	0.547	0.979	2.597	**1.000**

Table 7.28 Multi-criteria impact profiles evaluated for natural stone claddings

Model		$C_{P.1}$ Ease of learning	$C_{P.2}$ Time required to build the model	$C_{P.3}$ Necessary software	$C_{P.4}$ Variables included in the model	$C_{U.1}$ Utility of the results	$C_{U.2}$ Model's accuracy	Overall scale	Overall standardized scale
Deterministic models	Simple nonlinear regression	1	1	1	1	1	0.343	5.343	0.969
	Multiple linear regression	0.75	0.998	1	0.857	1	0.878	5.483	**1.000**
	Interaction effects	0.75	0.998	1	0.571	1	0.878	5.197	0.936
	Path analysis	0.75	0.981	1	0.429	1	0.689	4.849	0.859
	Polynomial model	0.625	0.877	0.108	0.143	1	0.576	3.329	0.520
	Gompertz curve	0.625	0.877	0.108	0.286	1	0.957	3.853	0.636
	Von Bertalanffy curve	0.625	0.877	0.108	0.286	1	0.983	3.879	0.642
	Richards curve	0.625	0.877	0.108	0.857	1	0.736	4.203	0.714
	Morgan-Mercer-Flodin curve	0.625	0.877	0.108	0.714	1	0.763	4.087	0.689
	Weibull curve	0.625	0.877	0.108	0.714	1	0.469	3.793	0.623
	Brody model	0.625	0.877	0.108	0.714	1	0.868	4.192	0.712
	Exponential model	0.625	0.877	0.108	0.714	1	0.945	4.269	0.729
	Potential model	0.625	0.877	0.108	0.714	1	0.429	3.753	0.614
Computational methods	Artificial neural networks	0.125	0	0.893	0.857	0.5	0.863	3.381	0.531
	Fuzzy logic model	0	0	0	0.857	0	0.000	1.000	**0.000**

(continued)

Table 7.28 (continued)

Model		$C_{P.1}$	$C_{P.2}$	$C_{P.3}$	$C_{P.4}$	$C_{U.1}$	$C_{U.2}$	Overall scale	Overall standardized scale
		Ease of learning	Time required to build the model	Necessary software	Variables included in the model	Utility of the results	Model's accuracy		
Factor methods	Factor method (deterministic)	0.875	0.752	1	0	1	0.991	5.475	0.998
	Factor method (stochastic)	0.5	0.501	0	0	1	0.991	3.849	0.636
Stochastic models	Logistic regression	0.375	0.939	0.108	1	1	1	4.422	0.763

Table 7.29 Number and subjectivity of the variables included in the service life prediction models (rendered façades)

Service life prediction model	Number of variables	Variables included in the model	Scale	Standardized scale
Simple nonlinear regression	1	Age	1	1
Logistic regression	1	Age	1	1
Markov chains	1	Age	1	1
Fuzzy logic model	4	Age; orientation; render type; height	2	0.833
Artificial neural networks (ANNs)	4	Age; render type; damp; protection level	3	0.667
Multiple linear regression	6	Age; orientation; balconies; render type; damp; protection level	4	0.500
Multiple nonlinear regression—Polynomial, Morgan-Mercer-Flodin (MMF) curve, Weibull curve, Brody curve, exponential model	6	Age; orientation; render type; damp; balconies; protection level	4	0.500
Multiple nonlinear regression—Gompertz curve, Von Bertalanffy (CVB) curve, Richards curve	6	Age; orientation; render type; damp; distance from the sea; protection level	4	0.500
Multiple nonlinear regression—potential model	6	Age; orientation; render type; damp; height; protection level	4	0.500
Multiple linear regression with interaction effects	8	Age; orientation; balconies; render type; damp; protection level; height; distance from the sea	5	0.333
Multiple linear regression—path analysis	9	Age; eave's protection; platbands; design level; height; colour; ground protection; building's volume; distance from the sea	6	0.167
Factor method (deterministic and stochastic)	16	Render type; colour; building's volume; eave's protection; balconies; ground protection; design level; height; orientation; distance from the sea; damp; distance from pollution sources; protection level; type of property	7	0

like it was seen for stone claddings, the simple nonlinear regression, logistic regression and Markov chains are the simpler models in renderings, since they only use age as explanatory variable, which is easy to quantify. The fuzzy model has precedence over neural networks since, although both models present four variables, the former has three variables whose measurement is objective in opposition to the latter that has only one such variable. The multiple linear regression and nonlinear multiple regression models, despite having distinct variables, have each three subjective variables and three variables objectively quantified. The remaining models present an increasing number of variables and therefore a declining rank in the standardized scale.

Table 7.30 shows the evaluation matrix regarding the accuracy of the service life prediction models applied to rendered façades. The different models were defined using different samples and for different purposes; therefore, the results must be analysed carefully. Concerning the rendered façades, the fuzzy logic model presents the best overall results, followed by the factor method, the multiple linear regression and the potential model. The logistic regression and artificial neural networks models also present a very high efficiency level. Simple nonlinear regression is the less accurate model regarding the difference between probability distributions since it provides an average value only, without the indication of the results dispersion. Nonlinear regression models, namely the exponential model, Gompertz curve, Von Bertalanffy curve (CVB) and Richards curve, produce relatively good results, according to the criteria under analysis; however, these models have been created to predict the severity of degradation of rendered façades and not their service life. When analysing the accuracy of the models according to their errors, these models (exponential model, Gompertz curve, Von Bertalanffy curve (CVB) and Richards) are the ones with larger deviations between the predicted and observed values of estimated service lives; furthermore, they lead to an average estimated service life of around 22 years, a relatively high value compared with the remaining models.

Based on the six criteria evaluated it is possible to define the multi-criteria impact profiles for rendered façades—Table 7.31. The simple nonlinear regression model shows the best overall results according to the various criteria analysed. Fuzzy logic model is the one with worse performance, since its high precision does not compensate its complexity. In this analysis, it is concluded that the simplest methods such as simple linear regression gain advantage over the other ones due to the straightforward, time saving approach—either at their creation stage or at the operation stage—despite eventual loss of accuracy.

In fact, although the evaluation criteria are independent of each other, naturally, simpler models consume less time in model's definition; moreover, the inclusion of a greater number of variables leads to more complex models, requiring more skills from the user, even if they can analyse the influence of these variables in the degradation process—which provides a more comprehensive understanding of the factors that influence the service life of rendered façades, well beyond the information provided by models that only consider the influence of age.

Table 7.30 Evaluation matrix regarding the accuracy of the service life prediction models applied to rendered façades

Model	Average estimated service life (years)	Difference between the $ESL_{predicted}/ESL_{observed}$ ratio and 1	Standardized score	Values of the estimated service life with a probability higher than 5 % to be exceeded	Difference between the ratio between the predicted and observed values and 1	Standardized score	Deviation between probability distributions	Standardized score	Sum of standardized scores	Overall scale
Simple nonlinear regression	15.760	0.161	0.063	29.550	0.121	1.000	2.000	0.000	1.063	0.404
Multiple linear regression	15.715	0.163	0.049	20.685	0.384	0.213	0.993	0.652	0.914	0.305
Interaction effects	15.560	0.172	0.000	20.501	0.390	0.197	1.001	0.646	0.843	0.258
Path analysis	15.736	0.162	0.056	18.281	0.456	0.000	1.384	0.399	0.455	**0.000**
Polynomial model	15.823	0.158	0.083	20.728	0.383	0.217	0.998	0.648	0.949	0.328
Gompertz curve	21.960	0.169	0.016	28.969	0.138	0.948	0.740	0.816	1.781	0.880
Von Bertalanffy curve	21.980	0.170	0.010	29.029	0.136	0.954	0.736	0.818	1.782	0.881
Richards curve	21.970	0.169	0.013	29.033	0.136	0.954	0.734	0.819	1.787	0.884
Morgan-Mercer-Flodin curve	15.778	0.160	0.069	20.344	0.395	0.183	1.049	0.616	0.868	0.274
Weibull curve	16.101	0.143	0.171	20.816	0.381	0.225	1.018	0.635	1.031	0.383
Brody model	15.960	0.150	0.127	20.849	0.380	0.228	0.997	0.649	1.004	0.364
Exponential model	21.960	0.169	0.016	28.798	0.143	0.933	0.759	0.803	1.753	0.861
Potential model	17.495	0.069	0.612	23.003	0.315	0.419	0.878	0.726	1.758	0.865
Artificial neural networks	17.500	0.068	0.614	22.009	0.345	0.331	1.023	0.632	1.577	0.745
Fuzzy logic model	18.852	0.004	1.000	22.704	0.324	0.392	1.121	0.569	1.962	**1.000**
Factor method	16.244	0.135	0.216	25.899	0.229	0.676	0.455	1.000	1.892	0.954
Logistic regression	16.540	0.120	0.310	22.960	0.317	0.415	0.664	0.864	1.590	0.753

Table 7.31 Multi-criteria impact profiles evaluated for rendered façades

Model		$C_{P.1}$	$C_{P.2}$	$C_{P.3}$	$C_{P.4}$	$C_{U.1}$	$C_{U.2}$	Overall scale	Overall standardized scale
		Ease of learning	Time required to build the model	Necessary software	Variables included in the model	Utility of the results	Model's accuracy		
Deterministic models	Simple nonlinear regression	1	1	1	1	1	0.404	5.404	**1.000**
	Multiple linear regression	0.75	0.998	1	0.5	1	0.305	4.553	0.762
	Interaction effects	0.75	0.998	1	0.333	1	0.258	4.339	0.702
	Path analysis	0.75	0.981	1	0.167	1	0	3.898	0.578
	Polynomial model	0.625	0.877	0.108	0.5	1	0.328	3.438	0.449
	Gompertz curve	0.625	0.877	0.108	0.5	1	0.880	3.990	0.604
	Von Bertalanffy curve	0.625	0.877	0.108	0.5	1	0.881	3.991	0.604
	Richards curve	0.625	0.877	0.108	0.5	1	0.884	3.994	0.605
	Morgan-Mercer-Flodin curve	0.625	0.877	0.108	0.5	1	0.274	3.384	0.434
	Weibull curve	0.625	0.877	0.108	0.5	1	0.383	3.493	0.465
	Brody model	0.625	0.877	0.108	0.5	1	0.364	3.474	0.460
	Exponential model	0.625	0.877	0.108	0.5	1	0.861	3.971	0.599
	Potential model	0.625	0.877	0.108	0.5	1	0.865	3.975	0.600
Computational methods	Artificial neural networks	0.125	0	0.893	0.667	0.5	0.745	2.930	0.307
	Fuzzy logic model	0	0	0	0.833	0	1	1.833	**0.000**
Factor methods	Factor method (deterministic)	0.875	0.752	1	0	1	0.954	4.581	0.770
	Factor method (stochastic)	0.5	0.501	0	0	1	0.954	2.955	0.314
Stochastic models	Logistic regression	0.375	0.939	0.108	1	1	0.753	4.175	0.656

7.4.1.7 Painted Surfaces

Table 7.32 shows the hierarchy regarding the number and subjectivity of the variables included in the service life prediction models applied to painted surfaces. Models that include only one explanatory variable—simple nonlinear regression, logistic regression and Markov chains—appear in the first position of the adopted scale. The computational methods (ANNs and fuzzy logic), which include three explanatory variables—age, orientation and the distance from the sea—come in second place in the scale.

Table 7.32 Number and subjectivity of the variables included in the service life prediction models (painted surfaces)

Service life prediction model	Number of variables	Variables included in the model	Scale	Standardized scale
Simple nonlinear regression	1	Age	1	1
Logistic regression	1	Age	1	1
Markov chains	1	Age	1	1
Fuzzy logic model and Artificial neural networks (ANNs)	3	Age; orientation; distance from the sea	2	0.833
Multiple linear regression	4	Age; orientation; distance from the sea; colour	3	0.667
Richards curve, Gompertz curve, Von Bertalanffy (CVB) curve, Morgan-Mercer-Flodin (MMF) curve, Weibull curve, Brody curve, exponential and potential model	5	Age; orientation; distance from the sea; colour; wind-rain action	4	0.500
Multiple linear regression with interaction effects	5	Age; orientation; distance from the sea; colour; type of paint	5	0.333
Multiple linear regression —path analysis	5	Age; wind-rain action; height; damp; building's volume	5	0.333
Polynomial model	7	Age; orientation; colour; distance from the sea; texture; wind-rain action	6	0.167
Factor method (deterministic and stochastic)	11	Type of paint; colour; texture; building's volume; orientation; wind-rain action; distance from the sea; damp; distance from pollutants; type of use; height	7	0

In the next place, appears the multiple linear regression model, with four explanatory variables, followed by the multiple nonlinear regression models, which include as explanatory variables age, colour, distance from the sea and exposure to wind-rain action. In these models, only the variable exposure to wind-rain action displays some subjectivity in its quantification.

The multiple linear regression models with interaction effects and path analysis are in the next levels (both models include two variables with a subjective measurement, the first includes exposure to damp and type of paint, the second exposure to damp and exposure to wind-rain action). In the end of the hierarchical scale come the multiple polynomial regression model, with six explanatory variables, and the factor models, with eleven explanatory variables.

Table 7.33 shows the evaluation matrix regarding the accuracy of the service life prediction models applied to painted surfaces. The logistic regression model is the most effective according to the criteria examined, followed by the factor method and multiple exponential regression method. For painted surfaces, the path analysis model presents the worst overall performance. The multi-criteria impact profiles evaluated for painted surfaces are shown in Table 7.34. Once again, the fuzzy logic model shows the worse overall performance, according to the six criteria analysed (since it is a very complex model, with high costs). For painted surfaces, the simple nonlinear regression model presents the best overall performance; it is also the simpler model with very acceptable levels of accuracy. Logistic regression models and multiple linear regression analysis contemplating interaction effects also show a good performance in the description of the degradation phenomena of paintings.

7.4.1.8 Ceramic Tiling Systems

Table 7.35 shows the hierarchical scale defined for the number and subjectivity of the variables included in the service life prediction models applied to ceramic tiling systems. The simple nonlinear regression model, the logistic regression and the Markov chains model are in the first position in this scale, since they only include age as explanatory variable. In the next place come the artificial neural networks and fuzzy logic models, which include three independent variables: age; distance from the sea and; tiles size (whose quantification is objective). In third place, there are the multiple linear regression model and the Weibull curve, which include as explanatory variables age, tiles size and exposure to damp (the last variable presents some uncertainty in its measurement). In the following levels, come the multiple nonlinear models—Gompertz, von Bertalanffy, Morgan-Mercer-Flodin (MMF), Richards, Brody, exponential and potential curves—and the multiple linear regression model that includes the interaction effects, which consider five explanatory variables. The polynomial model also includes five explanatory variables, but is penalized in the hierarchical scale due to the inclusion of the variable wind-rain action, more subjective than the ones included in the previous models. In the last two positions, come the path analysis model, with seven explanatory variables and the factor method, with eleven explanatory variables.

Table 7.33 Evaluation matrix regarding the accuracy of the service life prediction models applied to painted surfaces

Model	Average estimated service life (years)	Difference between the $ESL_{predicted}/ESL_{observed}$ ratio and 1	Standardized score	Values of the estimated service life with a probability higher than 5 % to be exceeded	Difference between the ratio between the predicted and observed values and 1	Standardized score	Deviation between probability distributions	Standardized score	Sum of standardized scores	Overall scale
Simple nonlinear regression	9.664	0.070	0.680	21.303	0.354	0.247	2.000	0.000	0.927	0.244
Multiple linear regression	8.487	0.183	0.021	9.640	0.387	0.129	1.350	0.432	0.582	0.116
Interaction effects	8.450	0.187	0.000	9.590	0.391	0.118	1.358	0.427	0.544	0.102
Path analysis	8.503	0.182	0.030	9.065	0.424	0.000	1.636	0.242	0.272	**0.000**
Polynomial model	9.724	0.064	0.713	10.549	0.330	0.333	1.434	0.376	1.422	0.429
Gompertz curve	9.659	0.070	0.677	10.611	0.326	0.346	1.368	0.420	1.443	0.437
Von Bertalanffy curve	9.620	0.074	0.655	10.565	0.329	0.336	1.374	0.416	1.407	0.424
Richards curve	9.712	0.065	0.707	10.679	0.321	0.362	1.359	0.426	1.495	0.456
Morgan-Mercer-Flodin curve	9.558	0.080	0.621	10.214	0.351	0.257	1.535	0.309	1.187	0.342
Weibull curve	9.553	0.080	0.618	10.218	0.351	0.258	1.530	0.312	1.188	0.342
Brody model	8.454	0.186	0.002	9.582	0.391	0.116	1.363	0.423	0.541	0.101
Exponential model	10.236	0.015	1.000	11.241	0.286	0.488	1.328	0.446	1.934	0.620
Potential model	9.746	0.062	0.726	10.136	0.356	0.240	1.696	0.202	1.168	0.334
Artificial neural networks	9.491	0.086	0.583	10.533	0.331	0.329	1.338	0.440	1.352	0.403
Fuzzy logic model	9.698	0.067	0.699	10.442	0.336	0.309	1.487	0.341	1.348	0.402
Factor method	9.998	0.038	0.867	11.781	0.251	0.609	0.975	0.681	2.157	0.704
Logistic regression	10.148	0.023	0.951	13.527	0.140	1.000	0.495	1.000	2.951	**1.000**

Table 7.34 Multi-criteria impact profiles evaluated for painted surfaces

Model		$C_{P.1}$	$C_{P.2}$	$C_{P.3}$	$C_{P.4}$	$C_{U.1}$	$C_{U.2}$	Overall scale	Overall standardized scale
		Ease of learning	Time required to build the model	Necessary software	Variables included in the model	Utility of the results	Model's accuracy		
Deterministic models	Simple nonlinear regression	1	1	1	1	1	0.244	5.244	**1.000**
	Multiple linear regression	0.75	0.998	1	0.857	1	0.116	4.721	0.869
	Interaction effects	0.75	0.998	1	0.571	1	0.102	4.421	0.795
	Path analysis	0.75	0.981	1	0.429	1	0.000	4.160	0.730
	Polynomial model	0.625	0.877	0.108	0.143	1	0.429	3.182	0.486
	Gompertz curve	0.625	0.877	0.108	0.286	1	0.437	3.333	0.523
	Von Bertalanffy curve	0.625	0.877	0.108	0.286	1	0.424	3.320	0.520
	Richards curve	0.625	0.877	0.108	0.857	1	0.456	3.923	0.670
	Morgan-Mercer-Flodin curve	0.625	0.877	0.108	0.714	1	0.342	3.666	0.606
	Weibull curve	0.625	0.877	0.108	0.714	1	0.342	3.666	0.606
	Brody model	0.625	0.877	0.108	0.714	1	0.101	3.425	0.546
	Exponential model	0.625	0.877	0.108	0.714	1	0.620	3.944	0.676
	Potential model	0.625	0.877	0.108	0.714	1	0.334	3.658	0.604
Computational methods	Artificial neural networks	0.125	0	0.893	0.833	0.5	0.403	2.755	0.379
	Fuzzy logic model	0	0	0	0.833	0	0.402	1.235	**0.000**
Factor methods	Factor method (deterministic)	0.875	0.752	1	0	1	0.704	4.331	0.772
	Factor method (stochastic)	0.5	0.501	0	0	1	0.704	2.705	0.366
Stochastic models	Logistic regression	0.375	0.939	0.108	1	1	1.000	4.422	0.795

Table 7.35 Number and subjectivity of the variables included in the service life prediction models (ceramic tiling systems)

Service life prediction model	Number of variables	Variables included in the model	Scale	Standardized scale
Simple nonlinear regression	1	Age	1	1
Logistic regression	1	Age	1	1
Markov chains	1	Age	1	1
Fuzzy logic model and Artificial neural networks (ANNs)	3	Age; tiles dimension; distance from the sea	2	0.833
Multiple linear regression and Weibull curve	3	Age; tiles dimension; damp	3	0.667
Richards curve, Gompertz curve, Von Bertalanffy (CVB) curve, Morgan-Mercer-Flodin (MMF) curve, Brody curve, exponential and potential model	5	Age; tiles dimension; damp; distance from the sea; orientation	4	0.500
Multiple linear regression with interaction effects	5	Age; tiles dimension; damp; distance from the sea; orientation; substrate	5	0.333
Polynomial model	5	Age; tiles dimension; distance from the sea; damp; wind-rain action	5	0.333
Multiple linear regression —path analysis	7	Age; wind-rain action; distance from the sea; height; peripheral protection; type of surface; substrate	6	0.167
Factor method (deterministic and stochastic)	11	Type of surface; colour; tiles dimension; substrate; peripheral joints; peripheral protection; orientation; distance from the sea; wind-rain action; damp; height	7	0

Table 7.36 presents the evaluation matrix regarding the accuracy of the service life prediction models applied to ceramic tiling systems. In an overall analysis, the logistic regression model is the most accurate one, followed by the factor method and the fuzzy logic model. The path analysis model shows the worse overall performance, based on the parameters evaluated, with higher deviations between the average estimated service and the values obtained by the graphical method (closer to the observed reality). Table 7.37 shows the multi-criteria impact profiles evaluated for ceramic tiling systems. Once again, the fuzzy logic model presents the

Table 7.36 Evaluation matrix regarding the accuracy of the service life prediction models applied to ceramic tiling systems

Model	Average estimated service life (years)	Difference between the $ESL_{predicted}/ESL_{observed}$ ratio and 1	Standardized score	Values of the estimated service life with a probability higher than 5 % to be exceeded	Difference between the ratio between the predicted and observed values and 1	Standardized score	Deviation between probability distributions	Standardized score	Sum of standardized scores	Overall scale
Simple nonlinear regression	50.144	0.079	0.683	66.398	0.212	1.000	2.000	0.000	1.683	0.592
Multiple linear regression	46.023	0.155	0.007	53.464	0.365	0.195	1.233	0.577	0.780	0.182
Interaction effects	46.480	0.147	0.082	54.668	0.351	0.270	1.169	0.625	0.978	0.272
Path analysis	45.977	0.156	0.000	50.324	0.403	0.000	1.497	0.378	0.378	**0.000**
Polynomial model	48.497	0.110	0.413	52.675	0.375	0.146	1.490	0.384	0.943	0.256
Gompertz curve	50.028	0.081	0.664	56.793	0.326	0.402	1.240	0.572	1.638	0.572
Von Bertalanffy curve	48.043	0.118	0.338	54.678	0.351	0.271	1.271	0.549	1.158	0.354
Richards curve	48.786	0.104	0.460	55.354	0.343	0.313	1.268	0.550	1.323	0.429
Morgan-Mercer-Flodin curve	46.358	0.149	0.062	52.141	0.381	0.113	1.364	0.479	0.654	0.125
Weibull curve	49.721	0.087	0.613	53.271	0.368	0.183	1.546	0.341	1.138	0.345
Brody model	46.308	0.150	0.054	54.652	0.351	0.269	1.160	0.632	0.955	0.262
Exponential model	52.081	0.044	1.000	59.692	0.291	0.583	1.155	0.636	2.218	0.836
Potential model	49.829	0.085	0.631	53.341	0.367	0.188	1.550	0.339	1.158	0.354
Artificial neural networks	50.186	0.079	0.690	55.975	0.335	0.352	1.325	0.508	1.549	0.532
Fuzzy logic model	50.833	0.067	0.796	62.283	0.261	0.744	0.879	0.843	2.383	0.910
Factor method	49.768	0.086	0.621	64.551	0.234	0.885	0.689	0.986	2.492	0.960
Logistic regression	50.127	0.080	0.680	64.800	0.231	0.901	0.670	1.000	2.580	**1.000**

Table 7.37 Multi-criteria impact profiles evaluated for ceramic tiling systems

Model		$C_{P.1}$	$C_{P.2}$	$C_{P.3}$	$C_{P.4}$	$C_{U.1}$	$C_{U.2}$	Overall scale	Overall standardized scale
		Ease of learning	Time required to build the model	Necessary software	Variables included in the model	Utility of the results	Model's accuracy		
Deterministic models	Simple nonlinear regression	1	1	1	1	1	0.592	5.592	**1.000**
	Multiple linear regression	0.75	0.998	1	0.667	1	0.182	4.597	0.741
	Interaction effects	0.75	0.998	1	0.5	1	0.272	4.520	0.721
	Path analysis	0.75	0.981	1	0.167	1	0.000	3.898	0.560
	Polynomial model	0.625	0.877	0.108	0.333	1	0.256	3.200	0.378
	Gompertz curve	0.625	0.877	0.108	0.5	1	0.572	3.682	0.504
	Von Bertalanffy curve	0.625	0.877	0.108	0.5	1	0.354	3.464	0.447
	Richards curve	0.625	0.877	0.108	0.5	1	0.429	3.539	0.467
	Morgan-Mercer-Flodin curve	0.625	0.877	0.108	0.5	1	0.125	3.235	0.388
	Weibull curve	0.625	0.877	0.108	0.667	1	0.345	3.622	0.488
	Brody model	0.625	0.877	0.108	0.5	1	0.262	3.372	0.423
	Exponential model	0.625	0.877	0.108	0.5	1	0.836	3.946	0.572
	Potential model	0.625	0.877	0.108	0.5	1	0.354	3.464	0.447
Computational methods	Artificial neural networks	0.125	0	0.893	0.833	0.5	0.532	2.883	0.296
	Fuzzy logic model	0	0	0	0.833	0	0.910	1.744	**0.000**
Factor methods	Factor method (deterministic)	0.875	0.752	1	0	1	0.960	4.587	0.739
	Factor method (stochastic)	0.5	0.501	0	0	1	0.960	2.961	0.316
Stochastic models	Logistic regression	0.375	0.939	0.108	1	1	1.000	4.422	0.696

worse overall performance; this model although relatively accurate (with predicted values closer to the observed values and therefore with small deviations and a high coefficient of determination) is penalized due to its high complexity, both in the model's definition and in the interpretability of the results obtained. Regarding the adherent ceramic tiling systems, the simple nonlinear regression model shows the best overall performance, according to the criteria analysed. Thus, in this case, a simple model that only contemplates the age of the ceramic tiles can conveniently describe the degradation of this type of cladding. The multiple linear regression and the factor method using a deterministic approach are the best placed models in the hierarchy scale, immediately after the simple nonlinear regression model.

7.4.2 Relative Weighting of Decision Criteria

Until now, the multi-criteria analysis performed considers that all the criteria analysed have the same weight in the decision-making process for all stakeholders. However, in general, in the decision-making process, some criteria are more relevant than others, sometimes even decisive, as is, for example, the case of the construction costs of tight budgetary restrictions. Therefore, the assessment of the relative importance of the criteria analysed and the importance of each criterion for stakeholders should be evaluated. The ranking of criteria (criteria weight) as well as a hierarchy of models can be defined using expert opinion. There are different methods in the literature addressing this issue; one of them is the "swing weights" approach that is based on the opinion of a team of experts that assist the decision-making process.

In this study, three hypothetical scenarios for the preferences of decision-makers are analysed:

- Scenario 1—the decision-maker prefers the ease of the model's application in detriment of other criteria; thus a higher weight is assigned to the criteria related with the ease of learning, the time required to build the model and the number and subjectivity of the variables included in the model;
- Scenario 2—the decision-maker is fundamentally interested in the model's accuracy and the utility of the results, thus higher weights are assigned to these criteria;
- Scenario 3—the decision-maker wants the best model (most accurate), in the shortest possible time (he/she wants a model that is easily learned and easy to apply).

There are several possibilities and combinations to establish the scenarios to describe the preferences of decision-makers. In fact, as mentioned before, the decision-making process is subjective and depends, in some situations, on irrational criteria. Therefore, a sensitivity analysis is performed in which the relative weights of the criteria vary according to the preference of the decision-maker (evaluating the different scenarios), in order to understand how criteria weights influence the ranking of the models. The overall evaluation of each model is obtained through an additive aggregation approach in accordance with Eq. (7.4), which lists the weights of the criteria for their classification. In the next sections, the influence of the relative weighting of the decision criteria in the overall assessment of the service life prediction methods is analysed for the different types of claddings analysed.

7.4.2.1 Natural Stone Claddings

Table 7.38 presents the sensitivity analysis of the weights assigned to the first scenario for natural stone claddings. As would be expected, when the decision-maker prefers the ease of the model's application, the simplest models gain prevalence. In this analysis, the simple nonlinear regression model, multiple linear regression model and factor method (deterministic approach) always show the best overall performance. Regardless of the magnitude of the weights assigned in this scenario, the fuzzy logic model always has the worst performance, since, as already referred, it is the most complex one, requiring more time to acquire the knowledge necessary to apply and understand it.

The sensitivity analysis performed for the second scenario is shown in Table 7.39. When the decision-maker is only interested in the model's accuracy and in the richness of the results produced, the factor method (deterministic approach), the multiple linear regression and the model with interaction effects are those with better overall performance. In fact, multiple linear regressions show a very good performance in the description of the degradation phenomena of natural stone claddings, combining simplicity with efficiency, thereby constituting a very interesting tool. As in the previous scenario, the fuzzy logic model shows the worst overall performance.

Table 7.38 Sensitivity analysis of weights assigned to scenario 1 for natural stone claddings

Criteria's weight	Overall assessment of the models
$\lambda_{CP.1} = 2$; $\lambda_{CP.2} = 2$ $\lambda_{CP.3} = 1$; $\lambda_{CP.4} = 2$ $\lambda_{CU.1} = 1$; $\lambda_{CU.2} = 1$	1st simple nonlinear regression 2nd multiple linear regression 3rd multiple linear regression (with interaction effects) *Worst performance—fuzzy logic model*
$\lambda_{CP.1} = 6$; $\lambda_{CP.2} = 2$ $\lambda_{CP.3} = 1$; $\lambda_{CP.4} = 2$ $\lambda_{CU.1} = 1$; $\lambda_{CU.2} = 1$	1st simple nonlinear regression 2nd factor methods (deterministic) 3rd multiple linear regression *Worst performance—fuzzy logic model*
$\lambda_{CP.1} = 8^*$; $\lambda_{CP.2} = 2$ $\lambda_{CP.3} = 1$; $\lambda_{CP.4} = 2$ $\lambda_{CU.1} = 1$; $\lambda_{CU.2} = 1$	1st simple nonlinear regression 2nd factor method (deterministic) 3rd multiple linear regression *Worst performance—fuzzy logic model*

*Above this value, even increasing significantly the weight value $\lambda_{CP.1}$ and maintaining the remaining weights unchanged, the hierarchical scale of models remains the same

Table 7.39 Sensitivity analysis of weights assigned to scenario 2 for natural stone claddings

Criteria's weight	Overall assessment of the models
$\lambda_{CP.1} = 1$; $\lambda_{CP.2} = 1$ $\lambda_{CP.3} = 1$; $\lambda_{CP.4} = 1$ $\lambda_{CU.1} = 2$; $\lambda_{CU.2} = 2$	1st factor method (deterministic) 2nd multiple linear regression 3rd multiple linear regression (with interaction effects) *Worst performance—fuzzy logic model*
$\lambda_{CP.1} = 1$; $\lambda_{CP.2} = 1$ $\lambda_{CP.3} = 1$; $\lambda_{CP.4} = 1$ $\lambda_{CU.1} = 5$; $\lambda_{CU.2} = 15$	1st factor method (deterministic) 2nd logistic regression 3rd multiple linear regression *Worst performance—fuzzy logic model*
$\lambda_{CP.1} = 1$; $\lambda_{CP.2} = 1$ $\lambda_{CP.3} = 1$; $\lambda_{CP.4} = 1$ $\lambda_{CU.1} = 5$; $\lambda_{CU.2} = 20^{*}$	1st factor method (deterministic) 2nd multiple linear regression 3rd multiple linear regression (with interaction effects) *Worst performance—fuzzy logic model*
$\lambda_{CP.1} = 1$; $\lambda_{CP.2} = 1$ $\lambda_{CP.3} = 1$; $\lambda_{CP.4} = 1$ $\lambda_{CU.1} = 5$; $\lambda_{CU.2} = 2$	1st factor method (deterministic) 2nd multiple linear regression 3rd multiple linear regression (with interaction effects) *Worst performance—fuzzy logic model*
$\lambda_{CP.1} = 1$; $\lambda_{CP.2} = 1$ $\lambda_{CP.3} = 1$; $\lambda_{CP.4} = 1$ $\lambda_{CU.1} = 20^{**}$; $\lambda_{CU.2} = 5$	1st factor method (deterministic) 2nd multiple linear regression 3rd multiple linear regression (with interaction effects) *Worst performance—fuzzy logic model*

*Above this value, even increasing significantly the weight value $\lambda_{CU.1}$ and maintaining the remaining weights unchanged, the hierarchical scale of models remains the same

**Above this value, even increasing significantly the weight value $\lambda_{CU.2}$ and maintaining the remaining weights unchanged, the hierarchical scale of models remains the same

Table 7.40 shows the sensitivity analysis performed for the third scenario. When a decision-maker aims to balance simplicity with accuracy of the model, in the case of natural stone claddings, the multiple linear regression models and factor method (deterministic approach) lead to the best overall results. Once again, the fuzzy logic model presents the worst overall results.

7.4.2.2 Rendered Façades

Table 7.41 shows the sensitivity analysis performed for the first scenario, revealing that the simple nonlinear regression model prevails over all others due to its higher simplicity. The three models with better performance when preference is given to simplicity are simple the nonlinear regression, multiple linear regression (with and without interaction effects) and factor method with a deterministic approach. In this scenario, whatever the variation of the weights, the fuzzy logic model always has the worst performance.

When the decision-maker's profile is represented by scenario 2 (Table 7.42), it appears that the simple nonlinear regression may be preferred, followed by the multiple linear regression and factor model (using a deterministic approach). Also in this case, the fuzzy logic model presents the worst overall results.

Table 7.40 Sensitivity analysis of weights assigned to scenario 3 for natural stone claddings

Criteria's weight	Overall assessment of the models
$\lambda_{CP.1} = 2$; $\lambda_{CP.2} = 2$ $\lambda_{CP.3} = 1$; $\lambda_{CP.4} = 1$ $\lambda_{CU.1} = 1$; $\lambda_{CU.2} = 2$	1st multiple linear regression 2nd factor method (deterministic) 3rd multiple linear regression (with interaction effects) *Worst performance—fuzzy logic model*
$\lambda_{CP.1} = 3$; $\lambda_{CP.2} = 3$ $\lambda_{CP.3} = 1$; $\lambda_{CP.4} = 1$ $\lambda_{CU.1} = 1$; $\lambda_{CU.2} = 2$	1st multiple linear regression 2nd factor method (deterministic) 3rd simple nonlinear regression *Worst performance—fuzzy logic model*
$\lambda_{CP.1} = 5^*$; $\lambda_{CP.2} = 5^*$ $\lambda_{CP.3} = 1$; $\lambda_{CP.4} = 1$ $\lambda_{CU.1} = 1$; $\lambda_{CU.2} = 3$	1st simple nonlinear regression 2nd Multiple linear regression 3rd multiple linear regression (with interaction effects) *Worst performance—fuzzy logic model*
$\lambda_{CP.1} = 2$; $\lambda_{CP.2} = 2$ $\lambda_{CP.3} = 1$; $\lambda_{CP.4} = 1$ $\lambda_{CU.1} = 1$; $\lambda_{CU.2} = 3^{**}$	1st factor method (deterministic) 2nd multiple linear regression 3rd multiple linear regression (with interaction effects) *Worst performance—fuzzy logic model*

*Above this value, even increasing significantly the weights values $\lambda_{CP.1}$ and $\lambda_{CP.2}$ and maintaining the remaining weights unchanged, the hierarchical scale of models remains the same
**Above this value, even increasing significantly the weight value $\lambda_{CU.2}$ and maintaining the remaining weights unchanged, the hierarchical scale of models remains the same

Table 7.41 Sensitivity analysis of weights assigned to scenario 1 for rendered façades

Criteria's weight	Overall assessment of the models
$\lambda_{CP.1} = 2$; $\lambda_{CP.2} = 2$ $\lambda_{CP.3} = 1$; $\lambda_{CP.4} = 2$ $\lambda_{CU.1} = 1$; $\lambda_{CU.2} = 1$	1st simple nonlinear regression 2nd multiple linear regression 3rd multiple linear regression (with interaction effects) *Worst performance—fuzzy logic model*
$\lambda_{CP.1} = 6$; $\lambda_{CP.2} = 2$ $\lambda_{CP.3} = 1$; $\lambda_{CP.4} = 2$ $\lambda_{CU.1} = 1$; $\lambda_{CU.2} = 1$	1st simple nonlinear regression 2nd multiple linear regression 3rd multiple linear regression (with interaction effects) *Worst performance—fuzzy logic model*
$\lambda_{CP.1} = 9^*$; $\lambda_{CP.2} = 2$ $\lambda_{CP.3} = 1$; $\lambda_{CP.4} = 2$ $\lambda_{CU.1} = 1$; $\lambda_{CU.2} = 1$	1st simple nonlinear regression 2nd multiple linear regression 3rd factor method (deterministic) *Worst performance—fuzzy logic model*

*Above this value, even increasing significantly the weight value $\lambda_{CP.1}$ the maintaining the remaining weights unchanged, the hierarchical scale of models remains the same

Table 7.43 shows the sensitivity analysis performed for the third scenario. When the stakeholder gives prevalence to both simplicity and accuracy of the model, trying to balance these two criteria, it is concluded that the simple nonlinear regression, factor method with a deterministic approach and multiple linear regression are preferred. The fuzzy logic model is penalized due of its higher complexity.

Table 7.42 Sensitivity analysis of weights assigned to scenario 2 for rendered façades

Criteria's weight	Overall assessment of the models
$\lambda_{CP.1} = 1$; $\lambda_{CP.2} = 1$ $\lambda_{CP.3} = 1$; $\lambda_{CP.4} = 1$ $\lambda_{CU.1} = 2$; $\lambda_{CU.2} = 2$	1st simple nonlinear regression 2nd multiple linear regression 3rd multiple linear regression (with interaction effects) *Worst performance—fuzzy logic model*
$\lambda_{CP.1} = 1$; $\lambda_{CP.2} = 1$ $\lambda_{CP.3} = 1$; $\lambda_{CP.4} = 1$ $\lambda_{CU.1} = 5$; $\lambda_{CU.2} = 20^{*}$	1st simple nonlinear regression 2nd multiple linear regression 3rd factor method (deterministic) *Worst performance—fuzzy logic model*
$\lambda_{CP.1} = 1$; $\lambda_{CP.2} = 1$ $\lambda_{CP.3} = 1$; $\lambda_{CP.4} = 1$ $\lambda_{CU.1} = 5$; $\lambda_{CU.2} = 2$	1st simple nonlinear regression 2nd multiple linear regression 3rd multiple linear regression (with interaction effects) *Worst performance—fuzzy logic model*
$\lambda_{CP.1} = 1$; $\lambda_{CP.2} = 1$ $\lambda_{CP.3} = 1$; $\lambda_{CP.4} = 1$ $\lambda_{CU.1} = 20^{**}$; $\lambda_{CU.2} = 8$	1st simple nonlinear regression 2nd multiple linear regression 3rd factor method (deterministic) *Worst performance—fuzzy logic model*

*Above this value, even increasing significantly the weight value $\lambda_{CU.1}$ and maintaining the remaining weights unchanged, the hierarchical scale of models remains the same
**Above this value, even increasing significantly the weight value $\lambda_{CU.2}$ and maintaining the remaining weights unchanged, the hierarchical scale of models remains the same

Table 7.43 Sensitivity analysis of weights assigned to scenario 3 for rendered façades

Criteria's weight	Overall assessment of the models
$\lambda_{CP.1} = 2$; $\lambda_{CP.2} = 2$ $\lambda_{CP.3} = 1$; $\lambda_{CP.4} = 1$ $\lambda_{CU.1} = 1$; $\lambda_{CU.2} = 2$	1st simple nonlinear regression 2nd multiple linear regression 3rd multiple linear regression (with interaction effects) *Worst performance—fuzzy logic model*
$\lambda_{CP.1} = 3^{*}$; $\lambda_{CP.2} = 3^{*}$ $\lambda_{CP.3} = 1$; $\lambda_{CP.4} = 1$ $\lambda_{CU.1} = 1$; $\lambda_{CU.2} = 2$	1st simple nonlinear regression 2nd multiple linear regression 3rd multiple linear regression (with interaction effects) *Worst performance—fuzzy logic model*
$\lambda_{CP.1} = 2$; $\lambda_{CP.2} = 2$ $\lambda_{CP.3} = 1$; $\lambda_{CP.4} = 1$ $\lambda_{CU.1} = 1$; $\lambda_{CU.2} = 3$	1st simple nonlinear regression 2nd multiple linear regression 3rd multiple linear regression (with interaction effects) *Worst performance—fuzzy logic model*
$\lambda_{CP.1} = 2$; $\lambda_{CP.2} = 2$ $\lambda_{CP.3} = 1$; $\lambda_{CP.4} = 1$ $\lambda_{CU.1} = 1$; $\lambda_{CU.2} = 8^{**}$	1st simple nonlinear regression 2nd multiple linear regression 3rd factor method (deterministic) *Worst performance—fuzzy logic model*

*Above this value, even increasing significantly the weights values $\lambda_{CP.1}$ and $\lambda_{CP.2}$ and maintaining the remaining weights unchanged, the hierarchical scale of models remains the same
**Above this value, even increasing significantly the weight value $\lambda_{CU.2}$ and maintaining the remaining weights unchanged, the hierarchical scale of models remains the same

7.4.2.3 Painted Surfaces

The weights of the sensitivity analysis assigned to the first scenario (privileging the ease of application of the model) are presented in Table 7.44. For this scenario, the models with better overall performance are the simple nonlinear regression, multiple linear regression (with and without interaction effects), factor method with a deterministic approach and logistic regression. The fuzzy logic model has the worst overall performance for scenario 1.

Table 7.45 shows the sensitivity analysis of the weights assigned to the second scenario. In this scenario, the decision-maker aims to combine the effectiveness and quality of the results achieved by each one of the proposed models. According to the results, the models with better performance are the simple nonlinear regression, logistic regression and factor method (with a deterministic and a stochastic approach). The fuzzy logic model presents the worst overall performance.

Finally, the sensitivity analysis performed for the third scenario is shown in Table 7.46. The simple nonlinear regression, logistic regression, factor method (with a deterministic approach) and multiple linear regression are the models with better overall performance. Once again, the fuzzy logic model is the one with the worst overall performance according to the criteria evaluated.

7.4.2.4 Ceramic Tiling Systems

Table 7.47 shows the sensitivity analysis of the weights assigned to the first scenario. In this scenario, the simple nonlinear regression, multiple linear regression (with and without interaction effects) and factor method (with a deterministic approach) are the models with better overall performance, i.e. they are models whose implementation is in fact very easy. The fuzzy logic model always shows the worst performance, given the difficulties inherent to creating and applying this model.

Table 7.44 Sensitivity analysis of weights assigned to scenario 1 for painted surfaces

Criteria's weight	Overall assessment of the models
$\lambda_{C_{P.1}} = 2$; $\lambda_{C_{P.2}} = 2$ $\lambda_{C_{P.3}} = 1$; $\lambda_{C_{P.4}} = 2$ $\lambda_{C_{U.1}} = 1$; $\lambda_{C_{U.2}} = 1$	1st simple nonlinear regression 2nd multiple linear regression 3rd multiple linear regression (with interaction effects) and logistic regression *Worst performance—fuzzy logic model*
$\lambda_{C_{P.1}} = 9^*$; $\lambda_{C_{P.2}} = 2$ $\lambda_{C_{P.3}} = 1$; $\lambda_{C_{P.4}} = 2$ $\lambda_{C_{U.1}} = 1$; $\lambda_{C_{U.2}} = 1$	1st simple nonlinear regression 2nd multiple linear regression 3rd factor method (deterministic) *Worst performance—fuzzy logic model*

*Above this value, even increasing significantly the weight value $\lambda_{C_{P.1}}$ and maintaining the remaining weights unchanged, the hierarchical scale of models remains the same

Table 7.45 Sensitivity analysis of weights assigned to scenario 2 for painted surfaces

Criteria's weight	Overall assessment of the models
$\lambda_{C_{P.1}} = 1$; $\lambda_{C_{P.2}} = 1$ $\lambda_{C_{P.3}} = 1$; $\lambda_{C_{P.4}} = 1$ $\lambda_{C_{U.1}} = 2$; $\lambda_{C_{U.2}} = 2$	1st simple nonlinear regression 2nd logistic regression 3rd factor method (deterministic) *Worst performance—fuzzy logic model*
$\lambda_{C_{P.1}} = 1$; $\lambda_{C_{P.2}} = 1$ $\lambda_{C_{P.3}} = 1$; $\lambda_{C_{P.4}} = 1$ $\lambda_{C_{U.1}} = 5$; $\lambda_{C_{U.2}} = 15^{*}$	1st logistic regression 2nd factor method (deterministic) 3rd factor method (stochastic) *Worst performance—fuzzy logic model*
$\lambda_{C_{P.1}} = 1$; $\lambda_{C_{P.2}} = 1$ $\lambda_{C_{P.3}} = 1$; $\lambda_{C_{P.4}} = 1$ $\lambda_{C_{U.1}} = 5$; $\lambda_{C_{U.2}} = 2$	1st simple nonlinear regression 2nd logistic regression 3rd factor method (deterministic) *Worst performance—fuzzy logic model*
$\lambda_{C_{P.1}} = 1$; $\lambda_{C_{P.2}} = 1$ $\lambda_{C_{P.3}} = 1$; $\lambda_{C_{P.4}} = 1$ $\lambda_{C_{U.1}} = 20^{**}$; $\lambda_{C_{U.2}} = 5$	1st logistic regression 2nd factor method (deterministic) 3rd simple nonlinear regression *Worst performance—fuzzy logic model*

*Above this value, even increasing significantly the weight value $\lambda_{C_{U.1}}$ and maintaining the remaining weights unchanged, the hierarchical scale of models remains the same

**Above this value, even increasing significantly the weight value $\lambda_{C_{U.2}}$ and maintaining the remaining weights unchanged, the hierarchical scale of models remains the same

Table 7.46 Sensitivity analysis of weights assigned to scenario 3 for painted surfaces

Criteria's weight	Overall assessment of the models
$\lambda_{C_{P.1}} = 2$; $\lambda_{C_{P.2}} = 2$ $\lambda_{C_{P.3}} = 1$; $\lambda_{C_{P.4}} = 1$ $\lambda_{C_{U.1}} = 1$; $\lambda_{C_{U.2}} = 2$	1st simple nonlinear regression 2nd logistic regression 3rd factor method (deterministic) *Worst performance—fuzzy logic model*
$\lambda_{C_{P.1}} = 5^{*}$; $\lambda_{C_{P.2}} = 5^{*}$ $\lambda_{C_{P.3}} = 1$; $\lambda_{C_{P.4}} = 1$ $\lambda_{C_{U.1}} = 1$; $\lambda_{C_{U.2}} = 2$	1st simple nonlinear regression 2nd multiple linear regression 3rd multiple linear regression (with interaction effects) *Worst performance—fuzzy logic model*
$\lambda_{C_{P.1}} = 2$; $\lambda_{C_{P.2}} = 2$ $\lambda_{C_{P.3}} = 1$; $\lambda_{C_{P.4}} = 1$ $\lambda_{C_{U.1}} = 1$; $\lambda_{C_{U.2}} = 3$	1st logistic regression and simple nonlinear regression 2nd factor method (deterministic) 3rd multiple linear regression *Worst performance—fuzzy logic model*
$\lambda_{C_{P.1}} = 2$; $\lambda_{C_{P.2}} = 2$ $\lambda_{C_{P.3}} = 1$; $\lambda_{C_{P.4}} = 1$ $\lambda_{C_{U.1}} = 1$; $\lambda_{C_{U.2}} = 5^{**}$	1st logistic regression 2nd factor method (deterministic) 3rd simple nonlinear regression *Worst performance—fuzzy logic model*

*Above this value, even increasing significantly the weights values $\lambda_{C_{P.1}}$ and $\lambda_{C_{P.2}}$ and maintaining the remaining weights unchanged, the hierarchical scale of models remains the same

**Above this value, even increasing significantly the weight value $\lambda_{C_{U.2}}$ and maintaining the remaining weights unchanged, the hierarchical scale of models remains the same

Table 7.47 Sensitivity analysis of weights assigned to scenario 1 for ceramic tiling systems

Criteria's weight	Overall assessment of the models
$\lambda_{CP.1} = 2$; $\lambda_{CP.2} = 2$ $\lambda_{CP.3} = 1$; $\lambda_{CP.4} = 2$ $\lambda_{CU.1} = 1$; $\lambda_{CU.2} = 1$	1st simple nonlinear regression 2nd multiple linear regression 3rd multiple linear regression (with interaction effects) *Worst performance—fuzzy logic model*
$\lambda_{CP.1} = 6$; $\lambda_{CP.2} = 2$ $\lambda_{CP.3} = 1$; $\lambda_{CP.4} = 2$ $\lambda_{CU.1} = 1$; $\lambda_{CU.2} = 1$	1st Simple nonlinear regression 2nd Multiple linear regression 3rd multiple linear regression (with interaction effects) *Worst performance—fuzzy logic model*
$\lambda_{CP.1} = 8^{*}$; $\lambda_{CP.2} = 2$ $\lambda_{CP.3} = 1$; $\lambda_{CP.4} = 2$ $\lambda_{CU.1} = 1$; $\lambda_{CU.2} = 1$	1st simple nonlinear regression 2nd multiple linear regression 3rd factor method (deterministic) *Worst performance—fuzzy logic model*

*Above this value, even increasing significantly the weight value $\lambda_{CP.1}$ and maintaining the remaining weights unchanged, the hierarchical scale of models remains the same

The sensitivity analysis of the weights assigned to scenario 2 for ceramic tiling systems is shown in Table 7.48. Decision-makers belonging to this profile may favour the logistic regression and factor methods, according to the analysis carried out. In this scenario, the fuzzy logic model presents the worst overall performance.

Table 7.49 shows the sensitivity analysis of the weights assigned to scenario 3 for ceramic tiling systems. When the decision-makers' profile is represented by scenario 3, they should choose models such as the simple nonlinear regression,

Table 7.48 Sensitivity analysis of weights assigned to scenario 2 for ceramic tiling systems

Criteria's weight	Overall assessment of the models
$\lambda_{CP.1} = 1$; $\lambda_{CP.2} = 1$ $\lambda_{CP.3} = 1$; $\lambda_{CP.4} = 1$ $\lambda_{CU.1} = 2$; $\lambda_{CU.2} = 2$	1st simple nonlinear regression 2nd factor method (deterministic) 3rd logistic regression *Worst performance—fuzzy logic model*
$\lambda_{CP.1} = 1$; $\lambda_{CP.2} = 1$ $\lambda_{CP.3} = 1$; $\lambda_{CP.4} = 1$ $\lambda_{CU.1} = 5$; $\lambda_{CU.2} = 15^{*}$	1st logistic regression 2nd factor method (deterministic) 3rd factor method (stochastic) *Worst performance—fuzzy logic model*
$\lambda_{CP.1} = 1$; $\lambda_{CP.2} = 1$ $\lambda_{CP.3} = 1$; $\lambda_{CP.4} = 1$ $\lambda_{CU.1} = 5$; $\lambda_{CU.2} = 2$	1st simple nonlinear regression 2nd factor method (deterministic) 3rd logistic regression *Worst performance—fuzzy logic model*
$\lambda_{CP.1} = 1$; $\lambda_{CP.2} = 1$ $\lambda_{CP.3} = 1$; $\lambda_{CP.4} = 1$ $\lambda_{CU.1} = 20^{**}$; $\lambda_{CU.2} = 5$	1st factor method (deterministic) and logistic regression 2nd simple nonlinear regression 3rd multiple nonlinear regression—exponential model *Worst performance—fuzzy logic model*

*Above this value, even increasing significantly the weight value $\lambda_{CU.1}$ and maintaining the remaining weights unchanged, the hierarchical scale of models remains the same

**Above this value, even increasing significantly the weight value $\lambda_{CU.2}$ and maintaining the remaining weights unchanged, the hierarchical scale of models remains the same

Table 7.49 Sensitivity analysis of weights assigned to scenario 3 for ceramic tiling systems

Criteria's weight	Overall assessment of the models
$\lambda_{C_{P.1}} = 2$; $\lambda_{C_{P.2}} = 2$ $\lambda_{C_{P.3}} = 1$; $\lambda_{C_{P.4}} = 1$ $\lambda_{C_{U.1}} = 1$; $\lambda_{C_{U.2}} = 2$	1st simple nonlinear regression 2nd factor method (deterministic) 3rd logistic regression *Worst performance—fuzzy logic model*
$\lambda_{C_{P.1}} = 5^*$; $\lambda_{C_{P.2}} = 5^*$ $\lambda_{C_{P.3}} = 1$; $\lambda_{C_{P.4}} = 1$ $\lambda_{C_{U.1}} = 1$; $\lambda_{C_{U.2}} = 2$	1st simple nonlinear regression 2nd factor method (deterministic) 3rd multiple linear regression (with interaction effects) *Worst performance—fuzzy logic model*
$\lambda_{C_{P.1}} = 2$; $\lambda_{C_{P.2}} = 2$ $\lambda_{C_{P.3}} = 1$; $\lambda_{C_{P.4}} = 1$ $\lambda_{C_{U.1}} = 1$; $\lambda_{C_{U.2}} = 3$	1st simple nonlinear regression 2nd factor method (deterministic) 3rd logistic regression *Worst performance—fuzzy logic model*
$\lambda_{C_{P.1}} = 2$; $\lambda_{C_{P.2}} = 2$ $\lambda_{C_{P.3}} = 1$; $\lambda_{C_{P.4}} = 1$ $\lambda_{C_{U.1}} = 1$; $\lambda_{C_{U.2}} = 5^{**}$	1st factor method (deterministic) 2nd logistic regression 3rd simple nonlinear regression *Worst performance—fuzzy logic model*

*Above this value, even increasing significantly the weight value $\lambda_{C_{P.1}}$ and maintaining the remaining weights unchanged, the hierarchical scale of models remains the same

**Above this value, even increasing significantly the weight value $\lambda_{C_{U.2}}$ and maintaining the remaining weights unchanged, the hierarchical scale of models remains the same

factor method (with a deterministic approach) and logistic regression, which combine a relatively simple application to achieve results close to the reality observed for the ceramic tiles analysed in the fieldwork. Once again, the fuzzy logic model is the one with the worst performance in this analysis.

7.5 Discussion of Results and Recommendations

The multi-criteria analysis performed in this chapter intends to provide some recommendations that allow a more rational and informed selection of a service life prediction model, based on the purpose of the model, the planner's profile and the user's perspective. In the construction sector, there are different actors, managers, economists, architects, engineers and others, with different backgrounds and capacities. The criteria that influence the preferences of a given decision-maker are subjective and difficult to model; there are infinite possible scenarios, thus the contemplation of all of them is virtually impossible and is beyond the scope of this study. Nevertheless, this chapter provides some information regarding the advantages and limitations of the models proposed, assisting the decision-making process of the stakeholders in the construction industry.

First, it is important to realize that the proposed models apply significantly different concepts, leading naturally to different results. However, the analysis performed in Sect. 7.3 of this chapter reveals that all of them can be successfully applied in the service life prediction of façade's claddings (and their application can be extended to other construction elements). The comparative analysis performed in this chapter intends to analyse a set of criteria that influence the selection process of a particular service life prediction model. Hierarchical scales are set assuming that the user does not have any previous knowledge about predictive models or the statistical tools used. Naturally, if this assumption is not fulfilled, different hierarchical scales would be obtained. In reality, the proficiency and the professional experience of the planner can influence the decision-making process; for example, a planner who is familiar with specific software can opt for a model that uses this software even if this model is apparently more complex. In fact, users of this type of models tend to choose the model that best fits their knowledge and expertise, even if the selected model is not the best choice (according to the hierarchical scales defined).

Regardless of the complexity and time required to obtain the results, the richness of the results produced and the accuracy of the model are crucial parameters for the users of service life prediction models, and as such should be analysed. Therefore, a critical analysis of the different proposed models should be performed. Table 7.50 presents a comparative analysis of the proposed models, considering only the accuracy of the model and based on a global analysis of the six criteria evaluated (difficulty in learning the model, time consumed in the model's definition, software used; number and subjectivity of the variables included in the model, interpretability and utility of the information obtained using the models and model's accuracy). The model's accuracy is evaluated according to three indicators, described in Sect. 7.4.1: (i) ratio between the average estimated service life predicted by the model and the average estimated service life obtained by the graphical method; (ii) deviation between the values of the estimated service life with a probability higher than 5 % of being exceeded predicted by the model and

Table 7.50 Comparative analysis of the proposed models and their applicability to the façade claddings analysed

Type of cladding	Most accurate model	Method with higher deviations between the observed and the predicted values	Overall analysis (based on the six criteria adopted)	
			Best	Worst
Renderings	Factor method	Path analysis	SNLR[a]	Fuzzy logic
Natural stone	Logistic regression	Fuzzy logic	MLR[b]	
Ceramic claddings	Logistic regression	Path analysis	SNLR[a]	
Painted surfaces	Logistic regression	Path analysis	SNLR[a]	

[a] *SNLR* means simple nonlinear regression
[b] *MLR* means multiple linear regression

observed; (iii) difference between the probability distribution that characterizes the predicted values and those associated with the observed values.

In an overall analysis, it is possible to conclude that all the proposed models are effective and accurate to model the service life of the façade claddings under analysis. In fact, all the service life prediction models proposed in this study lead to mean absolute percentage error values (MAPE) lower than 10 %, thus revealing a potentially very good predictive capacity (Lewis 1997). In addition, the coefficients of determination (R^2) and correlation (r) are always higher than 0.6 and 0.8, respectively, indicating that there is a strong correlation between the observed data in fieldwork and the values predicted by the models. Regarding the classification capability of the proposed models, these models perform well in identifying the case studies that have reached the end of their service life (with an accuracy higher than 80 % for all the models established in this Book).

From the analysis of the results obtained, it is possible to conclude that the fuzzy logic model presents the worst results for the four types of cladding analysed. This model is still extremely complex (at all stages of its implementation, from the acquisition of knowledge to the modelling itself), intending to model subjective concepts using mathematical equations. The use of this model involves programming and the application of specific algorithms, which can be unintelligible for some users. Furthermore, the output of the model results in a mathematical equation that cannot be directly applied by the user, since it requires the use of a toolbox to obtain the degrees of activation of the membership functions. Based on the indicators considered, the following main conclusions can be drawn:

- The factor method is the most accurate in the service life prediction of rendered façades. On the other hand, the path analysis model is the less accurate, including nine independent variables, whose numerical translation is subjected to some subjectivity. Based on the six criteria analysed, simple nonlinear regression leads to the best results for renderings, combining a high simplicity with a very acceptable numerical accuracy, leading to a coefficient of determination of 0.881, a MAPE of 4.3 % and an accuracy of 96 %;
- Regarding natural stone claddings, logistic regression is the most accurate method, in all the three criteria adopted, with an average estimated service life of 67.8 years, near the average estimated service life obtained by the graphical method—70.5 years, with a small deviation of the probability distribution that characterizes the results obtained by this method and the values observed in field work. The fuzzy logic model leads to the worst results, with the highest deviation between the average estimated service life of the model (90 years) and the average estimated service life obtained by the graphical method (70.5 years). From the analysis of the six criteria, it is possible to conclude that the multiple linear regression model leads to the best overall results, with a determination coefficient of 0.711, a MAPE of 7.1 % and an accuracy of 93.6 %;
- Concerning painted surfaces, logistic regression shows the higher accuracy level, leading to small deviations between the statistical distributions that characterizes the predicted and the estimated service life of painted façades, with

an average estimated service life of 10 years (similar to the result achieved by the graphical method). The path analysis model is the less accurate, with an average estimated service life of 8.5 years and a significant deviation from the value obtained by the graphical method; nonetheless, this model leads to a coefficient of determination of 0.813, showing an accuracy of 83 %;

- Logistic regression is also the most accurate model in the service life prediction of ceramic tiling systems, with the smallest deviation between the probability distributions that characterize the results obtained by this method and the observed values. Contrarily, the path analysis method is the less effective with an accuracy of 87.7 % and a coefficient of determination of 0.673. In an overall analysis, contemplating the six criteria, the simple nonlinear regression analysis presents the best overall performance, since is simpler and easier to apply, leading to a coefficient of determination of 0.767 (revealing a good correlation between the values predicted by the model and the observed values) and an accuracy of 89 %.

Table 7.51 presents the comparative analysis of the proposed models according to the scenarios analysed.

In terms of simplicity, the simple regression models have clear advantages in comparison with other models; however, these models only allow obtaining an absolute value, i.e. an estimated service life for the whole sample. When using a simple regression model, the value of the estimated service life will always be the same in spite of the exposure conditions. Furthermore, this model does not provide indications of how the claddings' characteristics influence their degradation process or their estimated service lives.

For decision-makers whose model's simplicity is the most relevant parameter to take into consideration (decision-makers with the profile of scenario 1), it is suggested that use an intermediate model, for example multiple linear regression, whose modelling is quite simple and which allows encompassing the significant variables in the degradation process—unlike simple regression, which only

Table 7.51 Comparative analysis of the proposed models according to the scenarios analysed

Type of cladding	Scenario 1		Scenario 2		Scenario 3	
	Best	Worst	Best	Worst	Best	Worst
Renderings	SNLR and MLR	Fuzzy logic	SNLR and MLR	Fuzzy logic	SNLR and MLR	Fuzzy logic
Natural stone	SNLR, MLR and FM		FM, MLR, LR		MLR, SNLR and FM	
Ceramic claddings	SNLR and MLR		SNLR, FM and LR		SNLR, FM and LR	
Painted surfaces	SNLR and MLR		SNLR, FM and LR		SNLR, LR and MLR	

where *SNLR* represents the simple nonlinear regression, *MLR* the multiple linear regression, *FM* the factor method and *LR* the logistic regression encompasses age and the factor method that includes more variables, in spite of their relevance to the phenomenon under study—though it may not be the most accurate model. Although more complex, models that include more than one variable can analyse the degradation phenomenon in more than one dimension, including for example exposure to damp, distance from the sea or exposure to wind-rain action.

In some situations—for stakeholders that use service life prediction models in insurance companies or in the definition of maintenance policies—it is important to obtain information related with the façade's degradation in the form of a probability distribution, thereby evaluating the risk associated to the façade's condition. In this case, the decision-maker may choose a stochastic model. These methods are essential when the decision-maker intends to obtain the service life of an element in the form of a characteristic value, evaluating with a well-known precision the instant from which the element under analysis is no longer able to meet conveniently the requirements for which it was designed. From the analysis of the stochastic models proposed, logistic regression seems to be the right choice, since not only it is available in statistical software (unlike the Markov chains model which requires the definition of a specific algorithm for the construction element in question), but also because it allows analysing the probability of a façade reaching the end of its service life over time and according to its characteristics.

The computational methods are definitely the most complex models. However, they provide more accurate results that are obtained through a validation process that ensures the generalization ability of the models. For users unfamiliar with this type of model, the model's accuracy may not be sufficient to prefer this type of methodology, since accuracy does not compensate the complexity of the model. On the other hand, for situations in which security and performance may be of paramount importance, such as with infrastructural asset management or for the management of a very large stock of buildings for which every decision may require large sums of money, accuracy may turn out to be the preferred characteristic of the model, regardless of the complexity and cost of the latter.

7.6 Conclusions

Throughout the present study, the applicability of different methods to model the degradation and service life of the facades claddings under analysis has been presented and discussed. In this chapter, a comparative analysis of the different proposed models is performed, providing useful information to stakeholders to assist the decision-making process in the selection of a given service life prediction model for a given application. Therefore, the criteria considered relevant to the decision-making process are analysed, aggregating preferences depending on the decision-maker's profile.

One of the main conclusions that can be drawn is that all the prediction methods are able to conveniently model the phenomena under analysis; each model has advantages and limitations, and none of them is unquestionably better. Generally speaking, simpler models require less time to be developed and can be equally accurate. However, these models analyse the problem only in one dimension, i.e. they only determine the loss of performance of the buildings as a function of their age. More complex models take longer to be implemented and acquire the data necessary for their application; furthermore they are more difficult to learn and require specific software to be implemented (that may not be available to all the stakeholders). Computational methods are the most complex and time-consuming approaches as well as the more precise and accurate. Different methods analyse distinct parameters that influence the degradation of façades; they may also provide interesting insight regarding the choice of a particular type of coating over others, depending on site specific environmental exposure conditions. Likewise, stochastic models produce different information and results, since in these models the estimated service life is given by a probabilistic distribution, providing information related to the end of its service life and the risk associated with this event.

The multi-criteria analysis performed in this chapter allows a multidimensional analysis of the choice of a specific model to predict the service life of façade claddings. The criteria evaluated in this chapter are used to benchmark the various models, defining a ranking of preferences.

Regardless of the criteria adopted, some subjectivity associated with the decision-making process will always remain, since this process is conditioned by individual perceptions, context and actual preferences of the decision-makers. In practice, a stakeholder may not be able to define precisely the level of importance of each criterion (criteria weights) and, even when experts' opinion is used, different values to the weights of the criteria may be assigned. Finally, knowing the characteristics of the models as presented in this study, the decision-maker may be able to choose the best model according to the results that he/she wants to achieve.

References

Altman DG, Bland JM (1994) Diagnostic tests 2: predictive values. Br Med J 309(6947):102

Bana, Costa CA (1993) Les problématiques dans le cadre de l'activité d'aide à la décision. Document du LAMSADE no. 74, Université Paris Dauphine

Bardhan T, Ngeru J, Pitts R Jr (2011) A Delphi-multi-criteria decision making approach in the selection of an enterprise-wide integration strategy. In: Second international conference on information management and evaluation, Toronto, Canada

Basel JS, Brühl R (2013) Rationality and dual process models for reasoning in managerial cognition and decision making. Eur Manage J 31(6):745–754

Belton V, Stewart TJ (2002) Multiple criteria decision analysis: an integrated approach. Kluwer Academic Publishers, London, UK

Blumer A, Ehrenfeucht A, Haussler D, Warmuth MK (1987) Occam's razor. Inf Process Lett 24:377–380

Bordass W, Cohen R, Staneven M, Leaman A (2000) Assessing building performance in use 2: technical performance of the Probe buildings. Build Res Inf 29(2):103–113

Brans J-P (1994) The PROMCALC & GAIA decision support system for multicriteria decision aid. Decis Support Syst 12:297–310
Carver SJ (2014) Integrating multi-criteria evaluation with geographical information systems. Int J Geogr Inf Syst 5(3):321–339
Cochran JK, Chen H-N (2005) Fuzzy multi-criteria selection of object-oriented simulation software for production system analysis. Computers & Operations Research 32:153–168
Collinson P (1998) Of bombers, radiologists, and cardiologists: time to ROC. Heart 80(3):215–217
Damásio A (1994) Descartes' error: emotion, reason, and the human brain, 1st edn. Grosset/Putnam, New York, USA
Dieterle F (2003) Multianalyte quantifications by means of integration of artificial neural networks, genetic algorithms and Chemometrics for time-resolved analytical data. Ph.D Thesis in Chemical and Pharmacy, Eberhard Karls Tübingen University, Tübingen
Einstein A (1921) Geometry and Experience. Prussian Academy of Science, Berlin, Germany
Frangopol DM, Lin K-Y, Estes AC (1997) Reliability of reinforced concrete girders under corrosion attack. J Struct Eng 123(3):286–297
Gigerenzer G, Selten R (eds) (2002) Bounded rationality: the adaptive toolbox. MIT Press, Cambridge, pp 1–12
Goodwin P, Wright G (1991) Decision analysis for management judgement. Wiley, Chichester, England
Halpern S (1978) The assurance sciences: an introduction to quality control and reliability, 1st edn. Prentice Hall, Englewood Cliffs, USA
Hosmer DW, Lemeshow S (2000) Applied logistic regression, 2nd edn. Wiley, New York, USA
Hovde PJ (2000) Factor methods for service life prediction: a state-of-the-art. Draft Report, Norwegian University of Science and Technology, Trondheim, Norway
Hovde PJ (2002) The factor method for service life prediction from theoretical evaluation to practical implementation. In: 9th International conference on durability of buildings materials and components, Brisbane, Australia, Paper 232.
ISO 15686-1 (2011) Buildings and constructed assets—service life planning—part 1: general principles and framework. International Organization for Standardization, Switzerland
Jato-Espino D, Rodriguez-Hernandez J, Andrés-Valeri VC, Ballester-Muñoz F (2014) A fuzzy stochastic multi-criteria model for the selection of urban pervious pavements. Expert Syst Appl 41(15):6807–6817
Jiang Y, Sinha KC (1989) Bridge service life prediction model using the Markov chain. Transportation Research Record No. 1223, Transportation Research Board. National Research Council, Washington D.C., USA, pp 24–30
Lacasse MA, Sjöström C (2004) Recent advances in methods for service life prediction of buildings materials and components—an overview. In: CIB World Building Congress, Canada, pp 1–10
Lewis CD (1997) Demand forecasting and inventory control. A computer aided learning approach. 1st Edn, Woodhead Publishing Ltd., In association with the Institute of Operations Management, Cambridge, England.
Limbourg P, Aponte DES (2005) An optimization algorithm for imprecise multi-objective problem functions. IEEE Congress on Evolutionary Computation (CEC 2005), Edinburgh, pp 459–466
Lounis Z, Mirza MS (2001) Reliability-based service life prediction of deteriorating concrete structures. In: 3rd international conference on concrete under severe conditions, vol 1. Vancouver, Canada, pp 965–972
Lourenço PB (2002) Computations of historical masonry constructions. Prog Struct Mat Eng 4(3):301–309
Lourenço PB, Krakowiak KJ, Fernandes FM, Ramos LF (2007) Failure analysis of Monastery of Jerónimos, Lisbon: how to learn from sophisticated numerical models. Eng Fail Anal 14(2):280–300
Madanat S, Mishalani R, Ibrahim WHW (1995) Estimation of infrastructure transition probabilities from condition rating data. J Infrastruct Syst 1(2):120–125

Manel S, Williams HC, Ormerod SJ (2001) Evaluating presence-absence models in ecology: the need to account for prevalence. J Appl Ecol 38:921–931

Marteinsson B (2003) Assessment of service lives in the design of buildings—development of the factor method. Licentiate Thesis, KTH's Research School, University of Gävle, Gävle

Maroco J (2007) Statistical analysis using SPSS. (in Portuguese), 3rd edn. Sílabo Editions, Lisbon.

Miwa K, Simon HA (1993) Production system modeling to represent individual differences: tradeoff between simplicity and accuracy in simulation behavior. Prospects for artificial intelligence. Frontiers in artificial intelligence and applications. In: Sloman A, Hogg D, Humphreys G, Ramsay A, Partridge D (eds) Proceedings of AISB93, the ninth biennial conference of the society of the study of artificial intelligence and the simulation of behaviour, IOS Press, 1993, pp 158–167

Morcous G, Lounis Z, Mirza MS (2002) Service-life prediction of concrete bridge decks using case-based reasoning. In: 6th international conference on short and medium span bridges. Vancouver, Canada, pp 769–776

Moser K (2004) Engineering design methods for service life prediction. In: CIB W080/RILEM 175 SLM: service life methodologies prediction of service life for buildings and components, task group: performance based methods of service life prediction. Trondheim, Norway, pp 52–95

Moser K, Edvardsen C (2002) Engineering design method for service life prediction. In: 9th international conference on the durability of building materials and components 2002. Brisbane, Australia, paper 222

Oberkampf WL, Helton JC, Joslyn CA, Wojtkiewicz SF, Ferson S (2004) Challenge problems: uncertainty in system response given uncertain parameters. Reliab Eng Syst Saf 85(1–3):11–19

Obuchowski NA (2003) Receiver operating characteristic curves and their use in radiology. Radiology 229:3–8

Phillips LD (1986) Decision analysis and its application in industry, 1st edn. Elsevier Science Publishers B.V, Netherlands

Ratkowsky DA (1983) Nonlinear regression modeling: a unified practical approach, 1st edn. Marcel Dekker Editors, New York, USA

Reiser B, Faraggi D (1997) Confidence intervals for the generalized ROC criterion. Biometrics 53(2):644–652

Robertson EA, Zweig MH, Van Steirteghem MD (1983) Evaluating the clinical accuracy of laboratory tests. Am J Clin Pathol 79:78–86

Ross SM (1996) Stochastic processes, 2nd edn. Wiley, New York, USA

Roy B (1985) Méthodologie multicritère d'aide à la décision, 1st edn. Economica, Paris

Rudbeck C (1999) Assessing the service life of building envelope construction. In: 8th international conference on durability of building materials and components. Vancouver, Canada, pp 1051–1061

Sanders DH, Zhang YJ (1994) Bridge deterioration models for states with small bridge inventories. Transportation Research Record No. 1442, Transportation Research Board. National Research Council, Washington D.C., USA, pp 101–109

Setunge S, Hasan MS (2011) Concrete bridge deterioration prediction using Markov chain approach. In: International conference on structural engineering, construction and management (ICSECM), Kandy, Sri Lanka, Paper 95

Shohet IM, Paciuk M (2004) Service life prediction of exterior cladding components under standard conditions. Constr Manage Econ 22(10):1081–1090

Shohet I, Rosenfeld Y, Puterman M, Gilboa E (1999) Deterioration patterns for maintenance management—a methodological approach. 8th international conference on durability of building materials and components. Vancouver, Canada, pp 1666–1678

Silva A, de Brito J, Gaspar PL (2012) Application of the factor Method to maintenance decision support for stone cladding. Autom Constr 22(3):165–174

Silva A, Dias JLR, Gaspar PL, de Brito J (2013) Statistical models applied to service life prediction of rendered façades. Autom Constr 30:151–160

Vergouwe Y (2003) Validation of clinical prediction models. Theory and applications in testicular germ cell cancer. Ph.D Thesis in Biomedicine, Erasmus University Rotterdam, Rotterdam, The Netherlands

Vincke P (1992) Multicriteria decision-aid, 1st edn. Wiley, New York, USA

Voogd H (1983) Multicriteria evaluation for urban and regional planning. Pion, London

Yoon K, Hwang C-L (1995) Multiple attribute decision making an introduction, 1st edn. Sage Publications, Thousand Oaks, USA

Zeleny M (1998) Multiple criteria decision making: eight concepts of optimality. Human Syst Manage 17:97–107

Zimmermann H-J (2000) An application-oriented view of modeling uncertainty. Eur J Oper Res 122(2):190–198

Chapter 8
Conclusions

8.1 Introduction

In the near future, it is expected that the construction sector starts requiring detailed information concerning the environmental and economic impact of buildings, to be implemented during the design stage (Marteinsson 2003). Therefore, it is increasingly important that different constructive alternatives are compared regarding their total cost and their environmental impact during the different phases of buildings' life cycle. Currently, the available information on durability and service life of buildings and their components is still incipient. In most cases, the data on service life of materials are based on acquired experience (relative to the performance of materials in different environments) and on the behaviour of materials tested in laboratory and subjected to a specific set of degradation agents. However, the information from laboratory testing has to be carefully analysed, since the degradation mechanisms are analysed individually, when real life degradation phenomena are associated with the simultaneous occurrence of various agents, thus making it difficult to adjust the results obtained in laboratory to real in-use situations. Furthermore, most of the information available concerning the service life of one building element cannot be directly applied to other buildings, since all constructions are very complex prototypes with unique responses to unrepeatable conditions.

In addition to the information regarding the environmental and economic performance of buildings over time, the knowledge concerning their durability and service life can be extremely useful to practitioners in the construction sector. In fact, it can be applied by manufacturers who want to set warranty periods for materials, by designers who wish to select the most suitable material for a given application or buy facilities managers that need to establish regular periods of

A. Silva et al., *Methodologies for Service Life Prediction of Buildings*,
Green Energy and Technology, DOI 10.1007/978-3-319-33290-1_8

inspection, maintenance and repair (Cole and Corrigan 2009). For this purpose, it is crucial to define models to establish viable relationships between the service life of the building components and their specific conditions. Based on this information, it is possible to optimize the definition of maintenance strategies since the design stage or for already existing buildings. It is also possible to estimate their remaining service life of the building stock and its components, which is a fundamental dimension to managers, insurers, owners and users. Finally, various authors refer the need to develop simple and easy-to-use tools for the service life prediction of the construction elements, thus enabling a more rational management of the building stock (Bower 1999; Tolman and Tolman 2003).

This book intends to answer this demand, providing a set of mathematical and statistical methodologies which can be applied to the service life of nonstructural elements. The proposed models are applied to façade claddings, whose relevance is referred in the book. In fact, claddings act as the "skin" of the building, with a key role in its overall performance, thus improving the durability of the structure, protecting it from environmental agents. Furthermore, claddings have a huge influence on the perception of the urban space, with direct consequences on the quality and image of cities. This book is based on the performance of construction elements in real in-use conditions, whose degradation condition is evaluated based on visual inspections. The service life is estimated based on the knowledge of the characteristics of the construction element and its degradation condition. This information is then translated into a numerical index, called severity of degradation (described in Part 2). The severity of degradation describes the overall degradation level of a façade, based on the defects detected during the fieldwork—their extent and severity and their location on the façades—thus acting as an indicator of level of performance of the façade. In this work, this indicator is also used to validate and define the different models presented and discussed.

The methodologies proposed in this study allow a better understanding of the influence of the specific conditions of each case study on its degradation process. The existence of such tools is an essential element for the maintenance of building stock, with numerous practical applications for the construction sector. In this chapter, the main findings and conclusions from the study are discussed, as well as some perspectives for the practical application of this study.

8.2 General Conclusions

This section summarizes some lessons learned after the completion of this research work. These findings are related with the methodology adopted, i.e. the research assumptions, which are common to all the proposed models and all types of cladding.

Regarding the state of knowledge within the service life prediction methods, in recent decades, it is possible to witness a great development in this area. In fact, the number of studies addressing the durability and service life of buildings is increasing and already are there enforced standards on this subject. However, there are no widely accepted methodologies. The existing methods correspond to general methodologies, which do not consider the singularities of each material subjected to a set of degradation agents. The factor method (which is already recommended in ISO 15686: 2011) embodies these issues, but has been criticized in the literature for being too simplistic (not reflecting the complexity of the degradation phenomena) and does not include all the factors that contribute to the end of the service life of building materials. Furthermore, in practice, service life prediction tools are not yet widely used, although most, if not all, decision-makers recognize the importance of these tools. The current trend is towards the definition of simple methodologies that can be applied by all the actors in the construction sector, allowing the design of buildings with sustainable criteria and promoting proper maintenance of the built heritage, thus justifying research-based works as this book.

All the models proposed in this book are able to translate the physical reality observed during fieldwork. These methodologies allow: (i) describing the progressive loss of performance of claddings over time; (ii) estimating the service life of the façades claddings; (iii) identifying the relevant degradation factors to the degradation process of façades. However, when applying these models it is necessary to take into account their constraints and limitations, such as:

- **Service life prediction is not an exact science**—first of all, it must be taken into account that all the proposed models function as an approximation of reality —as of today, it is not possible to predict the future! In fact, reality is extremely complex and difficult to be modelled; what is more, each building is unique, showing an unrepeatable reaction to the specific set of conditions to which it is subjected. Therefore, as much as defining accurate models that can translate the observed reality, it is extremely important to know the error associated with the estimates produced by the models;
- **The maximum acceptable degradation level is a subjective concept, subjected to uncertainty in its own definition**—this theoretical limit is a subjective concept, with high variability, changing over time and according to the building context, either social, economic, environmental, or other (as mentioned in Chap. 2);
- **All the models react to the data**—regardless of the sample size, there is always uncertainty associated with the estimation of the model's parameters, since these are estimated from a finite sample. In fact, all models are defined based on a dataset and can present different results if applied to a new universe of cases; however, these models are also easily complemented and adaptable to new case studies. Furthermore, when a large and representative sample is used, the introduction of new case studies leads to small deviations in the results obtained;
- **The model's accuracy depends on the data obtained during the fieldwork**—regardless of the objectivity adopted during the fieldwork, different factors can

influence the results obtained: (i) in the present approach, the fieldwork is based on visual inspections only in which it is impossible to identify/characterize certain properties of the claddings (e.g. the bonding material applied in ceramic claddings); (ii) the difficulty to identify defects arising from design and execution errors, leading often to early failure and affecting the natural loss of performance of claddings over time; (iii) the difficulty to measure some claddings' properties and identify defects in elevated parts of the façade; (iv) the difficulty to identify defects in darker claddings; (v) the difficulty to obtain information regarding maintenance and rehabilitation actions; (vi) the subjectivity inherent to the surveyor (depending significantly on the proficiency of this agent)—to overcome this limitation, each type of cladding is analysed by the same inspector and applying well-established parameters;

- **The subjectivity associated to the quantification of the variables included in the model**—some characteristics of the claddings are not easily quantifiable; first, because some variables are qualitative and are associated with different categories, e.g. the case of façades protection or wind-rain action, whose adopted categories depend on criteria that should be well explained and defined clearly, in this case they depend on the height of the building and the surrounding conditions as vegetation and adjacent buildings that may promote the façade's protection against aggressive agents. The qualitative variables need to be translated into numerical values for the model's application; for this translation, in this study the knowledge acquired regarding the positive or negative influence of each category in the claddings' service life is used.

In spite of these limitations, the models proposed in this book present a high predictive and classification capability, as it has been discussed throughout this book. Naturally, some models have higher levels of accuracy, thus leading to small deviations between the predicted and the observed values. Therefore, it can be said with some confidence that the proposed models can be successfully used to model the service life of façade claddings and that it is also possible to apply these methodologies to other construction elements.

Apart from the description of the different models proposed, this book also intends to provide some indications and recommendations concerning the selection of the best model for a given application. As mentioned in previous chapters, the singularities of each model, as well as the advantages, limitations and the richness of the results produced influence the selection of the more adequate response to each particular context. Finally, it is important to recognize that the various models proposed lead to different service lives. However, service life prediction models must be flexible, and as important as getting an estimated service life it is paramount to know the scatter of the results and risks associated with the estimates. Thus, the results of each model should not be analysed only as a mean value, which varies from model to model, but as a range of estimated service life values, whose standard deviation and interval of confidence are known. As discussed in the following sections, the results obtained by the various methods lead to a credible range of variation, in line with similar results in the literature.

8.3 Partial Conclusions

In this book, different approaches are proposed to model the service life of façade claddings. Therefore, in the following paragraphs some considerations particular to each approach are provided.

Concerning the deterministic models, described in Chap. 3, the following conclusions can be drawn:

- **Simple regression models** are expedient techniques, easy to learn and apply. These models are based on the analysis of the evolution of degradation of claddings over time, whose service life can be estimated through a graphical procedure, based on the intersection between the degradation curves and the maximum acceptable degradation level. This method has been used by a number authors (Shohet et al. 1999; Shohet and Paciuk 2004; Gaspar 2009), demonstrating good results in the description of the degradation of construction elements and leading to consistent and realistic results. However, these models only relate deterioration with the age of the element under analysis—and, in reality, the age of the claddings is unable to explain 100 % of the variability of the degradation—and individually with other claddings' characteristics; thus it can be said that they analyse the degradation problem in two dimensions only. In other words, this simple regression analysis does not allow including more than one variable in the mathematical description of the façades degradation, thus not providing any indication regarding the identification of the most significant variables in the description of the degradation phenomena;
- The **multiple linear regression models** are an extension of the simple linear regression analysis, based on the same assumptions, but more inclusive. The multiple linear regression models are more flexible since they allow incorporating more than one variable in the regression equation, evaluating, out of the claddings' characteristics, those that best contribute to the explanation of the variability of façades' degradation. Besides, these models allow analysing the causal relationships between the claddings characteristics, thus providing relevant information regarding the sequence of effects that lead to the end of the service life of façades claddings;
- The **multiple nonlinear regression models** allow encompassing all the relevant variables for the degradation of claddings using nonlinear mathematical equations (e.g. polynomial models). However, as described in Sect. 8.3, it is not always possible to identify with certainty which model best describes the relationship between the independent variables and the target variable. Moreover, the determination of the variables to be included in the model and the estimation of the model's parameters need to be implemented based on an iterative process. These models are more complex than the previous ones, but generally lead to better results, since the real phenomena, such as the degradation of façade claddings, are in most cases best represented by nonlinear models.

Chapter 4 addresses the application of stochastic models, including **logistic regression** and **Markov chains** to the estimation of the service life of claddings. These models are more complex than deterministic models and require specific software to be implemented. However, stochastic models produce more rich information when compared with the former, since they allow analysing the service life of claddings using probability distribution functions; therefore, these methods allow dealing with the uncertainty associated to the degradation phenomena; additionally, they allow estimating: (i) the probability of the façades claddings belonging to a given degradation condition according to their age and their characteristics; (ii) the probability of failure at a given age; (iii) the mean time of permanence in a given degradation condition; and (iv) the probability of claddings to reach the end of their service life after a period of time and depending on their characteristics.

Regarding computational models, described in Chap. 5, the following conclusions are obtained:

- The **artificial neural networks** (ANNs) models function as an emulation of the human brain, being extremely effective in the definition of nonlinear functions to describe the degradation of façade claddings. Furthermore, ANNs allow encompassing all the relevant variables in the claddings degradation; the determination of the variables to be included in the model can be a lengthy procedure when there is no previous knowledge concerning the degradation of the claddings analysed though. As a further disadvantage, this model is usually associated with the concept of "black box", since the explanation of how the model achieves the solutions is not always obvious or clear to unfamiliar users;
- The **fuzzy logic systems** are very complex techniques that require specific software to be implemented and applied. These models are extremely useful when: (i) the problem to model is subject to high levels of uncertainty, since in reality it is not always possible to describe accurately all the characteristics of the claddings analysed or understand the synergy between all factors and degradation mechanisms; (ii) the phenomena to be modelled are characterized by inaccurate or naturally vague data, difficult to describe. These models are capable of modelling extremely complex nonlinear functions, applying for that effect an approach to human reasoning.

The application of factor models, described in Chap. 6, leads to the following conclusions:

- The **classical approach** (deterministic) is a relatively simple method for the estimation of the service life of building components based on their characteristics, widely used by the international scientific community. This model provides an average estimated service life for the claddings analysed, given by an absolute value. However, this approach is incapable of capturing the variability associated with the claddings ageing process, neglecting the degradation condition at the time in which the service life estimation is performed.
- The **stochastic approach** combines the use of a simple method with a stochastic approach to the degradation phenomenon. This model allows obtaining the

estimated service life for each case study, given by a probability distribution. Despite the relative conceptual simplicity of this method, its practical application can be complex, requiring the use of a specific algorithm and software.

In Chap. 7 of this book, a comparative analysis between the various methodologies proposed is performed. That chapter intends to provide some information regarding the different models, assisting decision-makers and stakeholders in the selection of the best approach for a given application. To perform this comparative analysis, the relevant criteria that influence the selection of a service life prediction model are analysed, such as: (i) difficulty in learning the model; (ii) the time consumed in the model's definition; (iii) the software applied; (iv) the number and subjectivity of the variables included in the model; (v) the richness and interpretability of the information obtained using the models; and (vi) the model's accuracy. Through this analysis it can be concluded that none of the methodologies proposed is unequivocally better than the others, since all of them present advantages and limitations. Every actor must therefore decide the required accuracy level to describe the phenomena to be modelled—e.g. if the stakeholder needs to get information in a probabilistic way, he/she should apply a stochastic model; in turn, if the stakeholder just wants to obtain an average value for the estimated service life of the cladding, with an acceptable precision, he/she should use simpler models (such as the deterministic models); on the other hand, when high levels of accuracy are required, the stakeholder may choose more complex models.

Generally, simpler models tend to be easy to implement in real situations and can be adopted even by inexperienced users. In an overall analysis, simpler models may be more interesting in most situations, since all the proposed models present high levels of precision and sometimes more complex models lead to a residual increment only in the model's accuracy. Often it seems more reasonable to sacrifice the accuracy of the model in exchange for greater applicability and simplicity. On the other end of the spectrum, more complex models allow encompassing the uncertainty associated with the degradation phenomena, including the parameters that influence the degradation of the façades claddings and may be a useful tool in some specific, critical situations that demand a high level of accuracy.

8.4 Conclusions Regarding Each Type of Cladding Studied

In this section, the results obtained are discussed, based on the various models proposed and for the four types of claddings analysed. So far, the estimated service life has been presented in the form of a mathematical result obtained through the different formulations. In the following analyses, the estimated service life values are rounded to unit, transforming it on an operational service life value, which can be easily comprehended and applied in the definition of maintenance strategies.

It is important to remember as well that the different approaches proposed in this book necessarily lead to different results. This dispersion of results is higher in the cases of natural stone claddings, and ceramic tiling systems, due to a greater complexity of the cladding system when compared to renderings or painted surfaces. Moreover, natural stone claddings, unlike the other types of claddings, are made of natural materials, whose characteristics are unique and unrepeatable.

8.4.1 Natural Stone Claddings

Regarding natural stone claddings, the proposed models lead to the results presented in Table 8.1.

From the analysis of the results obtained, the following conclusions can be drawn:

- The proposed models present a good predictive capacity (with coefficients of determination and correlation higher than 0.6 and 0.8, respectively), showing a strong correlation between the degradation observed during the fieldwork and the values predicted by the models. Moreover, the models have a classification capacity potentially very good, with an accuracy higher than 80 %;
- The fuzzy logic model presents the highest estimated service life; this value seems optimistic when compared to other results (even though stone is an extremely durable material, especially when subjected to a favourable set of conditions). However, it must be taken into account that this model is defined using a training sample and are thus validated with a test sample; the average estimated service life is obtained only for the test sample; therefore, the results obtained are in some way conditioned by the particularities of the test sample;
- For the remaining models, the average estimated service life varies between 67 and 79 years. In the literature there are several studies that attest to the durability of this type of cladding (Brand 1997), even though there is not a maximum value for the durability and service life of natural stone. Various studies (BSI 7543 1992; Bell 1993) reveal that stone claddings are capable of meeting the minimum performance requirements for at least 60 years; on the other hand, Shohet and Paciuk (2004) estimate a service life of 64 years for stone claddings (with a variation range between 59 and 70 years);
- Moreover, the proposed models allow identifying the more relevant characteristics for the degradation process of stone claddings. Among the variables analysed in this study, age is the most important factor to explain the degradation phenomena of natural stone claddings. Additionally, factors such as the size of the stone plates, the distance from the sea, the exposure to damp and the type of natural stone are present in the majority of the proposed models. From the analysis of the relationship between the estimated service life and the claddings' characteristics, the following conclusions can be drawn:

Table 8.1 Values of the estimated service life obtained by the proposed models applied to natural stone claddings

Methods		Average ESL (years)	Standard deviation (years)	Confidence interval 95 %	Value of the ESL with a probability higher than 95 % of being exceeded[a]	Value of the ESL with a probability higher than 5 % of being exceeded[a]
Part 3—Deterministic models	Simple nonlinear regression	68	–	–	–	–
	Multiple linear regression	77	8	±1.1	65	90
	Interaction effects	77	8	±1.1	65	90
	Path analysis	75	4	±0.6	68	82
	Polynomial model	69	3	±0.4	64	74
	Gompertz curve	73	8	±1.0	60	85
	Von Bertalanffy curve	74	8	±1.1	60	87
	Richards curve	69	5	±0.6	62	77
	Morgan–Mercer–Flodin curve	76	6	±0.8	67	85
	Weibull curve	69	2	±0.3	65	72
	Brody curve	79	8	±1.2	66	93
	Exponential model	73	7	±1.0	61	85
	Potential model	68	2	±0.3	65	71
Part 4—Stochastic models	Logistic regression	After 68 years the probability of a stone cladding reaching the end of its service life is higher than 50 % and at age 80 this probability is higher than 95 %				
	Markov chains	The probability of belonging to the condition corresponding to the end of service life reaches its peak between 103 and 104 years				
Part 5—Computational models	Artificial neural networks	70	6	±1.6	61	80
	Fuzzy logic model	90	31	±8.6	38	142

(continued)

Table 8.1 (continued)

Methods		Average ESL (years)	Standard deviation (years)	Confidence interval 95 %	Value of the ESL with a probability higher than 95 % of being exceeded[a]	Value of the ESL with a probability higher than 5 % of being exceeded[a]
Part 6—Factorial models	Classic approach	67	10	±1.4	50	84
	Stochastic approach	After 67 years the probability of a stone cladding reaching the end of its service life is higher than 50 %				

[a]Values obtained assuming a normal distribution

- Claddings with medium-sized plates have estimated service lives higher than claddings with large-sized plates. As mentioned before, claddings with large-sized plates present a small area occupied by the joints, thus being subject to higher stress concentration. Furthermore, larger stones usually correspond to heavier plates, also requiring more thickness to prevent their fracture during transportation and placement into use. Thus, claddings with large stone plates are more susceptible to cracking and detachment defects. Finally, the bonding of smaller elements can be easily ensured.
- Stone claddings located in coastal areas (less than 5 km from the sea) present smaller estimated service lives when compared with claddings than 5 km away from the sea, since claddings on the oceanfront are subjected to salt spray, which is extremely harmful to natural stone, promoting the presence of defects such as loss of integrity;
- Claddings with high exposure to damp present lower estimated service lives;
- Claddings of granite are the most durable, with higher estimated service lives, followed by limestone and marble.

8.4.2 Rendered Façades

Table 8.2 presents the values of the estimated service life obtained by the proposed models applied to rendered façades.

From the analysis of the results obtained by the different methodologies, the following conclusions can be drawn:

- All the proposed models have a high predictive capacity, with correlation coefficient higher than 0.9, thus indicating a very strong correlation between the data and the model. Moreover, the models present a high classification ability (correctly classifying the cases that have either reached or not the end of their

Table 8.2 Values of the estimated service life obtained by the proposed models applied to rendered façades

Methods		Average ESL (years)	Standard deviation (years)	Confidence interval 95 %	Value of the ESL with a probability higher than 95 % of being exceeded[a]	Value of the ESL with a probability higher than 5 % of being exceeded[a]
Part 3—Deterministic models	Simple nonlinear regression	16	–	–	–	–
	Multiple linear regression	16	3	±0.6	11	21
	Interaction effects	16	3	±0.6	11	21
	Path analysis	16	2	±0.30	13	18
	Polynomial model	16	3	±0.6	11	21
	Gompertz curve	22	4	±0.8	15	29
	Von Bertalanffy curve	22	4	±0.8	15	29
	Richards curve	22	4	±0.8	15	29
	Morgan–Mercer–Flodin curve	16	3	±0.5	11	20
	Weibull curve	16	3	±0.6	11	21
	Brody curve	16	3	±0.6	11	21
	Exponential model	22	4	±0.8	15	29
	Potential model	17	3	±0.7	12	23
Part 4—Stochastic models	Logistic regression	After 17 years the probability of a rendered façade reaching the end of its service life is higher than 50 % and at age 23 this probability is higher than 95 %				
	Markov chains	The probability of belonging to the condition corresponding to the end of service life reaches its peak at 19 years				
Part 5—Computational models	Artificial neural networks	18	3	±0.9	13	22
	Fuzzy logic model	19	2	±1.0	15	23

(continued)

Table 8.2 (continued)

Methods		Average ESL (years)	Standard deviation (years)	Confidence interval 95 %	Value of the ESL with a probability higher than 95 % of being exceeded[a]	Value of the ESL with a probability higher than 5 % of being exceeded[a]
Part 6—Factorialmodels	Classic approach	16	6	±1.3	7	26
	Stochastic approach	After 16 years the probability of a rendered façade reaching the end of its service life is higher than 50 %				

[a]Values obtained assuming a normal distribution

estimated service life). All the proposed models present an accuracy higher than 92 %;

- The models of Gompertz, von Bertalanffy, Richards and the exponential model lead to an average estimated service life higher than the values obtained by the other models. These values seem too optimistic for rendered façades that are not subjected to any maintenance action during their life cycle;
- Not considering the four above mentioned models, the estimated service life obtained by the remaining models, varies between 15 and 19 years. These values are similar to those obtained by Gaspar (2009), who obtained through simple regression an average estimated service life of 15 years. The study performed by Shohet and Paciuk (2004) led to an estimated service life of 15 years (with a range between 12 and 19 years);
- Service life prediction models should also provide information regarding the influence of the environmental exposure conditions and the building's context in the degradation of rendered façades. The majority of the proposed models (with the exception of simple regression) identify the significant variables in the degradation process. The age and type of mortar are present in all the proposed models, thereby confirming the high importance of these two variables in the degradation of renderings. Exposure to damp and façades orientation is also relevant variable, present in the majority of the models. From the analysis of these variables, the following results were obtained:

  - Concerning the type of mortar, renderings with crushed marble present the higher estimated service lives (however, it should be noted that in the sample analysed, these façades are located in buildings with superior design levels, located in a noble area of the city). Single-layer renderings (which are industrialized mortars, with a well-established composition) present higher estimated service lives than current cement renderings. Lime–cement renderings present the lowest estimated service lives (but once again context may determine these results for the lime–cement renders studied were applied in poorly maintained low income neighbourhoods);

- Regarding the orientation of the façades, by analysing the sample used in Chap. 6 (without outliers), renderings facing south have the highest estimated service lives; renderings facing North present the lowest estimated service lives, as would be expected, taking into account the climatic particularities of Portugal, where the North (more humid and cold) and West (with the presence of prevailing winds) quadrants are the most aggressive;
- Concerning exposure to damp, the façades with favourable exposure are those with the best performance, with higher estimated service lives.

8.4.3 Painted Surfaces

The results obtained using the different models applied to painted surfaces are presented in Table 8.3.

Based on the analysis of these results, the following conclusions can be drawn:

- The analysis of the accuracy of the proposed models reveals a strong correlation between real degradation (as registered during the fieldwork) and the values predicted by the models, with a Pearson correlation coefficient and a determination coefficient of 0.9 and 0.8, respectively. All the proposed models have an accuracy higher than 83 %, showing a potentially good classification capability in the description of the case studies that have either reached or not the end of their service life;
- The estimated service life obtained using the various models varies between 8 and 10 years. The predicted values are credible and coherent with the observed reality. Furthermore, the obtained results' range is narrow and is in accordance with the values present in the literature (Adalberth 1997; Keoleian et al. 2001; Hed 1999; Fay et al. 2000; AIJ 1993; Mithraratne and Vale 2004);
- Like the other types of claddings, age is the most relevant parameter in the explanation of the degradation of painted surfaces. Besides age, the orientation of the façade and distance from the sea are two variables that should be considered on the service life prediction of painted surfaces. From the analysis of these two variables, the following conclusions can be drawn:

  - Painted surfaces facing South are those with lower estimated service lives, since these coatings are more exposed to UV radiation, which accelerates their degradation process. Painted surfaces facing South and West present a higher incidence of defects of colour change, chalking, cracking and loss of adhesion. In turn, claddings facing North and East are those with the highest estimated service lives;
  - Regarding distance from the sea, coatings more than 5 km away from the sea have higher estimated service lives than claddings located in coastal areas.

Table 8.3 Values of the estimated service life obtained by the proposed models applied to painted surfaces

Methods		Average ESL (years)	Standard deviation (years)	Confidence interval 95 %	Value of the ESL with a probability higher than 95 % of being exceeded[a]	Value of the ESL with a probability higher than 5 % of being exceeded[a]
Part 3—Deterministic models	Simple nonlinear regression	10	–	–	–	–
	Multiple linear regression	8	1	±0.1	7	10
	Interaction effects	8	1	±0.1	7	10
	Path analysis	9	0	±0.0	8	9
	Polynomial model	10	1	±0.1	9	11
	Gompertz curve	10	1	±0.1	9	11
	Von Bertalanffy curve	10	1	±0.1	9	11
	Richards curve	10	1	±0.1	9	11
	Morgan-Mercer-Flodin curve	10	0	±0.1	9	10
	Weibull curve	10	0	±0.1	9	10
	Brody curve	8	1	±0.1	7	10
	Exponential model	10	1	±0.1	9	11
	Potential model	10	0	±0.0	9	10
Part 4—Stochastic models	Logistic regression	After 10 years the probability of a painted surface reaching the end of its service life is higher than 50 % and at age 12 this probability is higher than 95 %				
	Markov chains	The probability of belonging to the condition corresponding to the end of service life reaches its peak between 12 and 13 years				

(continued)

Table 8.3 (continued)

Methods		Average ESL (years)	Standard deviation (years)	Confidence interval 95 %	Value of the ESL with a probability higher than 95 % of being exceeded[a]	Value of the ESL with a probability higher than 5 % of being exceeded[a]
Part 5—Computational models	Artificial neural networks	9	1	±0.2	8	11
	Fuzzy logic model	10	0	±0.2	9	10
Part 6—Factorial models	Classic approach	10	1	±0.2	8	12
	Stochastic approach	After 10 years the probability of a stone cladding reaching the end of its service life is higher than 50 %				

[a]Values obtained assuming a normal distribution

8.4.4 Ceramic Tiling Systems

In Table 8.4 the values of the estimated service life obtained by the proposed models applied to ceramic tiling systems are shown. Based on the results obtained, the following conclusions can be drawn:

- The proposed models reveal a strong correlation between the sample analysed during the fieldwork and the values predicted by the models, with determination and correlation coefficients of 0.6 and 0.8, respectively. The classification capability of the models is always higher than 86 %, thus indicating high accuracy levels in the characterization of the case studies that have either reached or not the end of their service life;
- Concerning the estimated service life by the different proposed models, the scatter of results is relatively small, ranging between 46 and 52 years. These results are coherent with the values found in the literature, which suggest an estimated average service life of 50 years for the adherent ceramic tiling systems (Tam et al. 1993; BCIS 2001; Galbusera et al. 2014);
- Regarding adherent ceramic claddings, the variables with higher explanatory power in the description of their degradation process are age (the most relevant factor), the size of the ceramic tiles (included in all the proposed models) an distance from the sea. Based on an overall analysis of the results obtained according to these three variables, the following conclusions are achieved:
  - As in the case of stone claddings, for ceramic claddings, façades with large-sized tiles ($L \leq 20$ cm) are more prone to suffer from detachment, thus presenting lower estimated service lives;
  - Concerning distance from the sea, as expected, claddings in coastal areas present lower estimated service lives.

8.5 Principal Contributions of This Study

The main contribution of this book is the definition of a set of methodologies to estimate the service life of construction elements. Even though this book applies these methods to the specific case of four façade claddings, all the information required for implementing the different methods is documented throughout the book, which allows the adaptation of these tools to other constructive elements.

These models intend to provide a better understanding of the degradation phenomenon of the elements under analysis, allowing relating the characteristics of claddings, their exposure, use and maintenance conditions with their performance over time. The proposed models lead to various types of information, complementing each other. Essentially, the proposed models allow to: (i) identify degradation patterns according to the cladding's age and different degradation factors;

Table 8.4 Values of the estimated service life obtained by the proposed models applied to ceramic tiling systems

Methods		Average ESL (years)	Standard deviation (years)	Confidence interval 95 %	Value of the ESL with a probability higher than 95 % of being exceeded[a]	Value of the ESL with a probability higher than 5 % of being exceeded[a]
Chapter 3—Deterministic models	Simple nonlinear regression	50	–	–	–	–
	Multiple linear regression	46	5	±0.6	39	53
	Interaction effects	46	5	±0.7	38	55
	Path analysis	46	3	±0.4	42	50
	Polynomial model	48	3	±0.4	44	53
	Gompertz curve	50	4	±0.6	43	57
	Von Bertalanffy curve	48	4	±0.6	41	55
	Richards curve	49	4	±0.6	42	55
	Morgan-Mercer-Flodin curve	46	4	±0.5	41	52
	Weibull curve	50	2	±0.3	46	53
	Brody curve	46	5	±0.7	38	55
	Exponential model	52	5	±0.6	44	60
	Potential model	50	2	±0.3	46	53
Chapter 4–Stochastic models	Logistic regression	After 50 years the probability of a ceramic cladding reaching the end of its service life is higher than 50 % and at age 61 this probability is higher than 95 %				
	Markov chains	The probability of belonging to the condition corresponding to the end of service life reaches its peak between 58 and 59 years				
Chapter 5—Computational models	Artificial neural networks	50	4	±1.3	44	56
	Fuzzy logic model	51	7	±2.5	39	62
Chapter 6—Factorial models	Classic approach	50	9	±1.3	35	65
	Stochastic approach	After 50 years the probability of a ceramic cladding reaching the end of its service life is higher than 50 %				

[a]Values obtained assuming a normal distribution

(ii) obtain an estimated service life based on the cladding's characteristics; (iii) analyse the probability of transition between degradation conditions and the mean time of permanence in each degradation condition, expressing the progressive loss of performance through probability functions; (iv) obtain the estimated service life characterized by a probability distribution, for each case study. These models have many practical applications for the construction sector, due to the wealth of information produced, namely:

- **The direct application of models in the definition of maintenance policies**—the knowledge of the estimated service life of building elements (associated with a set of measures of dispersion) can be applied to define in a rational and technically informed way, a set of maintenance strategies over the building's life cycle. Additionally, if the stakeholder has information regarding the durability of the various building components, it is possible to establish joint maintenance strategies, in which the intervention in the building brings together various measures of improvement, thus optimizing the resources' consumption;
- **The application of the information obtained using stochastic models (Chap.** 4) **to define insurance policies**—this type of analysis allows encompassing the notion of risk associated with the failure of claddings, thus enabling the definition of warranty periods. According to a research carried on in England by the Building Research Establishment (BRE), the owners and occupants of the buildings accept they have a legal duty to ensure the safety of buildings. If buildings or their components inflict any damage on third parties, the insurers must cover that damage, but the insurance companies are not legally responsible for the financial loss caused by the failure of construction elements, usually discarding any liability if it is proved that the owners were negligent with respect to the buildings maintenance needs (in the cases of lack of proper maintenance, the insurance premium may increase significantly and the renewal of the insurance may even be denied). The stochastic modelling of the service life of claddings may therefore be crucial in the context of insurance policies, since it allows evaluating, for example, the most probable time to failure of claddings according to their characteristics, as well as the age until which it is extremely unlikely that these elements may have reached the end of their service life;
- **The application of the data related to the service life of buildings and components to life cycle costing analysis (LCC)**—to accurately analyse the life cycle cost of building components, the knowledge regarding the durability and service life of the elements under analysis is paramount. The LCC method is used to evaluate the economic performance of a building or its parts throughout their life cycle (acquisition, development, operation, maintenance, repair and deconstruction/demolition), allowing comparing different investment scenarios (Woodward 1997). A study performed by Chai et al. (2014) applies a LCC method to compare the costs of various maintenance strategies applied to painted surfaces, concluding that the knowledge of the estimated service life of these coatings opens a path to the development of an informed planning of

maintenance actions (using more rationally the resources and costs involved in maintenance actions), showing that the maintenance strategies based solely on visual perception and empirical criteria entail higher costs;

- **The application of the data related to the service life of buildings and components to life cycle assessment (LCA)**—the LCA method considers the environmental impacts of a product throughout its life cycle, based on the identification and quantification of their environmental emissions and the consumption of energy and materials. Construction elements are products with long life cycles and it is therefore essential to establish relationships between the durability and LCA studies of the construction elements. Standards ISO 15686-6 (2004) and FprEN 15804 (CEN 2011) describe some recommendations to consider the service life of buildings in LCA studies. Strand and Hovde (1999) report that reliable data on the service life of the construction elements improve the results of LCA studies (mainly concerning the building's use phase). In fact, to establish realistic scenarios in the definition of LCA models, the data concerning the service life of the construction elements must be included in these models. Silvestre et al. (2015) present an interdisciplinary study of LCA analysis and service life prediction models, encompassing the uncertainty associated with the determination of the reference service life used in life cycle assessment; it also highlights the dependence between the durability of buildings and their environmental impacts throughout their life cycle, allowing comparing different construction options based on their service life or environmental performance.

References

Adalberth K (1997) Energy use during the life cycle of buildings: a method. Build Environ 32 (4):317–320

AIJ (1993) The English edition of the principal guide for service life planning of buildings. Architectural Institute of Japan, Tokyo

BCIS (2001) Life expectancy of buildings components: surveyor's experiences of building in use —a practical guide. Building Cost Information Service, London

Bell FG (1993) Durability of carbonate rocks as building stone with comments on its preservation. Environ Geol 21:187–200

Bower B (1999) Simple minds, smart choices. Sci News 155(22):348–350

Brand S (1997) How buildings learn: what happens after they're built?, 1st edn. Phoenix Illustrated, London

BSI 7543 (1992) Guide to durability of buildings and building elements, products and components. British Standards Institution, London

CEN (2011) Sustainability of construction works—environmental product declarations—core rules for the product category of construction products. FprEN 15804, Comité Européen de Normalisation, Brussels, Belgium

Chai C, de Brito J, Gaspar P, Silva A (2014) Predicting the service life of exterior wall painting: techno-economic analysis of alternative maintenance strategies. J Constr Eng Manage 140(3):04013057

Cole IS, Corrigan PA (2009) Development of a range of methods for estimating the service life of buildings and engineered structures. In: 18th World IMACS/MODSIM Congress, Cairns, Australia, pp. 310–316

Fay R, Treloar G, Iyer-Raniga U (2000) Life-cycle energy analysis of buildings: a case study. Build Res Inf 28(1):31–41

Galbusera MM, de Brito J, Silva A (2014) Application of the factor method to the prediction of the service life of ceramic external wall claddings. J Build Perform Constr Facil 19–29. doi:10.1016/j.conbuildmat.2014.05.045

Gaspar P (2009) Service life of constructions: development of a method to estimate the durability of construction elements. Application to renderings of current buildings (in Portuguese). Doctor Thesis in Sciences of Engineering, Instituto Superior Técnico, Technical University of Lisbon, Portugal

Hed G (1999) Service life planning of building components. In: Proceedings of the 8th international conference on durability of building materials and components, Vancouver, Canada, pp 1543–1551

ISO 15686-6 (2004) Buildings and constructed assets—service life planning—Part 6: procedures for considering environmental impacts. International Organization for Standardization

Keoleian GA, Blanchard S, Reppe P (2001) Life-cycle energy, costs, and strategies for improving a single-family house. J Ind Ecol 4(2):135–156

Marteinsson B (2003) Assessment of service lives in the design of buildings—development of the factor method. Licentiate Thesis, KTH's Research School—HiG, Centre of Built Environment, University of Gävle, Sweden

Mithraratne N, Vale B (2004) Life cycle analysis model for New Zealand houses. Build Environ 39(4):483–492

Shohet IM, Paciuk M (2004) Service life prediction of exterior cladding components under standard conditions. Constr Manage Econ 22(10):1081–1090

Shohet I, Rosenfeld Y, Puterman M, Gilboa E (1999) Deterioration patterns for maintenance management—a methodological approach. In: Proceedings of the 8th international conference on durability of buildings materials and components, Vancouver, Canada, pp. 1666–1678

Silvestre JD, Silva A, de Brito J (2015) Uncertainty modelling of service life and environmental performance to reduce risk in building design decisions. J Civ Eng Manage 21(3):308–322

Strand SM, Hovde PJ (1999) Use of service life data in LCA of building materials. In: Proceedings of the 8th international conference on durability of building materials and components, Ottawa, Canada, pp 1948–1958

Tam C, Loo Y, Quek S, Saw W (1993) Simulated thermal fatigue testing of wall-to-tile bond. In Nagataky S, Nireki T, Tomosawa F (eds) Durability of building materials and components, E & FN Spon

Tolman F, Tolman A (2003) Optimization and decision-making in sustainable construction and operation. 2ILCDESS, pp 123–129

Woodward D (1997) Life cycle costing–theory, information acquisition and application. Int J Proj Manage 15(6):335–344

FSC
www.fsc.org
MIX
Papier aus verantwortungsvollen Quellen
Paper from responsible sources
FSC® C105338

If you have any concerns about our products,
you can contact us on
ProductSafety@springernature.com

In case Publisher is established outside the EU,
the EU authorized representative is:
Springer Nature Customer Service Center GmbH
Europaplatz 3, 69115 Heidelberg, Germany

Printed by Libri Plureos GmbH
in Hamburg, Germany